北京理工大学"双一流"建设精品出版工程

新形态教材
扫描书内二维码

Tactical Missile Guidance and Control System Design
(Second Edition)

战术导弹制导控制系统设计
（第2版）

林德福　何绍溟　李虹言　祁载康　著

北京理工大学出版社
BEIJING INSTITUTE OF TECHNOLOGY PRESS

版权专有　侵权必究

图书在版编目（CIP）数据

战术导弹制导控制系统设计 = Tactical Missile Guidance and Control System Design：英文 / 林德福等著. -- 2 版. -- 北京：北京理工大学出版社，2024.9.

ISBN 978 - 7 - 5763 - 4464 - 6

Ⅰ. TJ765

中国国家版本馆 CIP 数据核字第 20242YH392 号

责任编辑：李颖颖　　**文案编辑**：宋　肖
责任校对：刘亚男　　**责任印制**：李志强

出版发行 /	北京理工大学出版社有限责任公司
社　　址 /	北京市丰台区四合庄路 6 号
邮　　编 /	100070
电　　话 /	（010）68944439（学术售后服务热线）
网　　址 /	http：//www.bitpress.com.cn

版 印 次 /	2024 年 9 月第 2 版第 1 次印刷
印　　刷 /	廊坊市印艺阁数字科技有限公司
开　　本 /	787 mm × 1092 mm　1/16
印　　张 /	18.5
字　　数 /	434 千字
定　　价 /	62.00 元

图书出现印装质量问题，请拨打售后服务热线，负责调换

Preface

The 20th National Congress of the Communist Party of China is a meeting of great importance. The theme of this Congress is holding high the great banner of socialism with Chinese characteristics, fully implementing the Thought on Socialism with Chinese Characteristics for a New Era, carrying forward the great founding spirit of the Party, staying confident and building strength, upholding fundamental principles and breaking new ground, forging ahead with enterprise and fortitude, and striving in unity to build a modern socialist country in all respects and advance the great rejuvenation of the Chinese nation on all fronts. Guided by the spirit of the 20th National Congress of the Communist Party of China, we have published this book.

This book is pitched as an introductory text for tactical missiles, covering the missile systems and design processes that must be considered. The main contents of this book are the summary and extension of the authors' 20 years' research works in the field of missile guidance and control.

We start from Chapter 1 with the basics of missile guidance by exposing the reader to critical concepts such as lateral acceleration as the preferred guidance command, and the interplay between guidance, navigation, and control within the context of missile systems. Chapter 2 discusses the missile dynamics with detailed aerodynamic models. Chapter 3 presents a brief overview of several pertinent missile sub-systems and sensors, supposedly covering actuator dynamics, gyroscopes, accelerometers, and integrated navigation systems. Chapter 4 presents a discussion of the radar guidance system and attempts to extract this system's performance characteristics with its greatest impact on guidance loop design and overall system performance. Chapter 5 describes commonly used seekers, the parasitic loop, and the real seeker model. Details of how to design a proper missile autopilot are presented in Chapter 6, including the most well-known two-loop and three-loop autopilot; Chapter 7 introduces the line-of-sight guidance. The optimality of proportional navigation and its variants are discussed in Chapter 8. In the final chapter, we introduce some modern optimal guidance laws, including optimal trajectory shaping guidance and gravity-turn-assisted guidance.

The book is a textbook for undergraduate students majoring in aerospace engineering. It can also benefit researchers, engineers, and graduate students in the field of Guidance Navigation and Control. It is our hope that this book will serve as a useful step toward further advances in the field of missile guidance and control. An online virtual simulation system is attached to the textbook to facilitate the understanding of the working process of guided projectiles. Please scan the QR codes at the bottom for the virtual simulation system.

The authors have carefully reviewed the content of this book before the printing stage. However, it does not mean that this book is completely free from any possible errors. Consequently, the authors would be grateful to readers who will call attention to errors they might discover. If you find any errors in this book, please contact us via email at hongyan_ae@126.com.

The first three authors would like to express their sincere gratitude to Prof. Zaikang Qi for his invaluable inspiration, support, and patience throughout the research, as well as for his mentorship in their overall professional development. The innovative content of this book is developed under the guidance of Prof. Qi and is closely linked to his involvement in domestic engineering practice.

Finally, the authors would like to express sincere thanks to colleagues from Institute of Autonomous UAV Control, Beijing Institute of Technology for providing valuable and constructive comments. Without their support, the book would not have been a success.

Please scan this QR code for the
online virtual simulation system.

Contents

1 Basics of Missile Guidance Control · 001

§ 1.1 Overview · 001
§ 1.2 Missile Control Methods · 002

2 Missile Mathematical Model · 007

§ 2.1 Symbols and Definitions · 007
§ 2.2 Euler Equations of the Missile Rigid Body Motion · 009
§ 2.3 Configuration of the Control Surfaces · 013
§ 2.4 Aerodynamic Derivatives and the Missile Control Dynamic Coefficients · 014
§ 2.5 The Transfer Function of a Missile as the Controlled Object · 019

3 Basic Missile Control Component Mathematical Models · 028

§ 3.1 Seeker · 028
§ 3.2 Actuator · 028
§ 3.3 Gyroscope · 029
§ 3.4 Accelerometer · 031
§ 3.5 Inertial Navigation Components and Integrated Inertial Navigation Module · 034

4 Guidance Radar · 035

§ 4.1 Introduction · 035
§ 4.2 Motion Characteristic of the Target LOS · 035
§ 4.3 Control Loop of the Guidance Radar · 039
§ 4.4 Effect of Receiver Thermal Noise on Guidance Radar Performance · 046
§ 4.5 Effect of Target Glint on Guidance Radar Performance · 049
§ 4.6 Effect of Other Disturbances on Guidance Radar Performance · 050

4.6.1	Effect of Disturbance Moment on Tracking Radar Performance	050
4.6.2	Effect of Target Maneuvers	052

5 Seekers ... 054

§ 5.1 Overview ... 054
§ 5.2 Electromechanical Structure of Commonly Used Seekers ... 055
    5.2.1 Dynamic Gyro Seeker ... 055
    5.2.2 Stabilized Platform-based Seeker ... 060
    5.2.3 Detector Strap-down Stabilized Optic Seeker ... 064
    5.2.4 Semi-strap-down Platform Seeker ... 065
    5.2.5 Strap-down Seeker ... 065
    5.2.6 Roll-pitch Seeker ... 068
§ 5.3 Mechanism Analysis of the Anti-disturbance Moment of the Seeker's Stabilization Loop and Tracking Loop ... 070
§ 5.4 Transfer Function of Body Motion Coupling and the Parasitic Loop ... 073
    5.4.1 Transfer Function of Body Motion Coupling ... 073
    5.4.2 Seeker-missile Coupling Introduced Guidance Parasitic Loop ... 076
§ 5.5 A Real Seeker Model and Testing Methods ... 079
    5.5.1 A Real Seeker Model ... 079
    5.5.2 Testing Methods for Modeling the Real Seeker ... 082
§ 5.6 Other Parasitic Loop Models ... 083
    5.6.1 Parasitic Loop Model for a Phase Array Strap-down Seeker ... 083
    5.6.2 Parasitic Loop Due to Radome Slope Error ... 084
    5.6.3 Beam Control Gain Error ΔK_B of the Phased Array Seeker and the Radome Slope Error R_{dom} Effect on the Seeker's Performance ... 085
    5.6.4 A Novel Online Estimation and Compensation Method for Strap-down Phased Array Seeker Disturbance Rejection Effect Using Extended State Kalman Filter ... 088
§ 5.7 Stabilization Loop and Tracking Loop Design of the Platform-based Seeker ... 095
    5.7.1 Stabilization Loop Design ... 095
    5.7.2 Tracking Loop Design ... 096

6 Autopilot Design ... 099

§ 6.1 Acceleration Autopilot ... 099
    6.1.1 Two-loop Acceleration Autopilot ... 099
    6.1.2 Two-loop Autopilot with PI Compensation ... 103
    6.1.3 Three-loop Autopilot with Pseudo Angle of Attack Feedback ... 106
    6.1.4 Classic Three-loop Autopilot ... 112

6.1.5	Discussion of Variable Acceleration Autopilot Structures	116
6.1.6	Hinge Moment Autopilot	117
6.1.7	Several Considerations in Acceleration Autopilot Design	120
6.1.8	Acceleration Autopilot Design Methods	125
§6.2	Pitch/Yaw Attitude Autopilot	136
§6.3	Flight Path Angle Autopilot	139
§6.4	Roll Attitude Autopilot	140
§6.5	BTT Autopilot	145
§6.6	Thrust Vector Control and Thruster Control	154
§6.7	Spinning Missiles Control	160
6.7.1	Aerodynamic Coupling	161
6.7.2	Control Coupling	163

7 LOS Guidance ... 166

§7.1 LOS Guidance System ... 166
§7.2 Analysis of the Required Acceleration for the Missile with LOS Guidance ... 167
§7.3 Analysis of the LOS Guidance Loop ... 172
§7.4 Lead Angle Method ... 179

8 Proportional Navigation and Extended Proportional Navigation Guidance Laws ... 182

§8.1 Proportional Navigation Guidance Law ... 182
  8.1.1 Proportional Navigation Guidance Law (PN Guidance Law) ... 182
  8.1.2 Analysis of PN Guidance Law with No Guidance System Lag ... 188
  8.1.3 PNG Characteristics with the Missile Guidance Dynamics Included ... 193
  8.1.4 Adjoint Method ... 205
§8.2 Extended Proportional Navigation (Optimal Proportional Navigation, OPN) Guidance Laws ... 209
  8.2.1 Optimal Proportional Navigation Guidance Law (OPN1): Consideration of Missile Guidance Dynamics ... 209
  8.2.2 Optimal Proportional Navigation Guidance Law (OPN2): Consideration of the Constant Target Maneuver ... 214
  8.2.3 Optimal Proportional Navigation Guidance Law (OPN3): Consideration of Both Constant Target Maneuvers and Missile Guidance Dynamics ... 217
  8.2.4 Estimation of Target Maneuver Acceleration ... 221
  8.2.5 Estimation of t_{go} ... 222
  8.2.6 Proportional Navigation Guidance Law with Impact Angle Constraint ... 222
§8.3 Other Types of Proportional Navigation Laws ... 224

8.3.1　Gravity Over-Compensated Proportional Navigation Law 224
8.3.2　Lead Angle Proportional Navigation Guidance Law 228
§ 8.4　Target Maneuver Acceleration Estimation 230
§ 8.5　Optimum Trajectory Control Design 242

9　Optimal Guidance for Trajectory Shaping 248

§ 9.1　Optimality of Error Dynamics in Missile Guidance 248
9.1.1　Optimal Error Dynamics 249
9.1.2　Analysis of Optimal Error Dynamics 251
§ 9.2　Optimal Predictor-Corrector Guidance 253
9.2.1　General Approach for Guidance Law Design 253
9.2.2　Impact Angle Control 253
9.2.3　Impact Time Control 256
§ 9.3　Graity-Turn-Assisted Optimal Guidance Law 259
9.3.1　Zero-Control-Effort Trajectory Considering Gravity 260
9.3.2　Optimal Guidance Law Design and Analysis 263
9.3.3　Characteristics Analysis by Simulations 267
§ 9.4　3D Optimal Impact Time Guidance for Anti-ship Missiles 271
9.4.1　Problem Formulation 272
9.4.2　3D Optimal Impact Time Guidance Law Design 274
9.4.3　Analysis of Proposed Guidance Law 277
9.4.4　Numerical Simulations 280

References 286

1

Basics of Missile Guidance Control

§ 1.1 Overview

Missile control aims to ensure the missile reaches the target by the end of its flight. To achieve this, the missile must continuously gather data on both its own motion and that of the target. It then applies a guidance law to determine how to adjust its velocity direction based on the current relative motion between the missile and the target. This approach enables the missile to strike the target accurately. The relationship between the angular velocity $\dot{\theta}$ of the missile's velocity vector and its normal acceleration a is described as follows:

$$\dot{\theta} = \frac{a}{V} (V \text{ is the missile's velocity}) \qquad (1.1-1)$$

Therefore, the command of a guidance law that is generated to change the missile velocity vector direction is usually the normal acceleration a_c of the missile. The missile-target interception control loop differs significantly from conventional tracking control loops. While the latter typically involves time-invariant control systems that can be analyzed using standard linear time-domain and frequency-domain methods, the missile-target interception control loop is a time-varying system. As a result, its analysis requires different approaches. To distinguish it from other control systems, this specific outer loop in missile control is often referred to as the "guidance loop".

With the help of autopilots, the missile's output acceleration a will follow the above guidance acceleration command a_c. Under the assumptions of small perturbations, linearization, and constant system parameters, the autopilot loop behaves as a linear time-invariant (LTI) system. This allows for the application of various traditional control theory design methods. Consequently, the autopilot loop, which functions as the inner loop within the guidance system, is still commonly referred to as the "control loop".

The missile's position and velocity information required for guidance are obtained through an inertial navigation system or an integrated inertial navigation system. This process of determining the missile's position and orientation is referred to as navigation. It is important to note that, in this context, "navigation" does not pertain to the traditional definition of steering the course of a ship or aircraft. Fig. 1.1-1 shows the relationships between the terms navigation, guidance, and control in missile control loops.

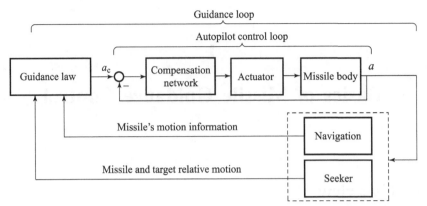

Fig. 1.1 – 1   Block diagram of the missile guidance and control loops

§ 1.2 Missile Control Methods

As previously discussed, the missile control system's objective is to adjust the missile's velocity direction using normal acceleration in accordance with the guidance law command. For tactical missiles flying within the atmosphere, this normal acceleration is provided by aerodynamic forces. Specifically, when the missile has an angle of attack relative to its velocity vector, the resulting lift generates normal acceleration. To maintain a stable angle of attack, the aerodynamic moment caused by this angle must be counterbalanced by the control moment generated by the deflection of the control surfaces.

When the missile's center of gravity is positioned ahead of the center of pressure, the aerodynamic moment induced by the angle of attack will reduce the existing angle of attack. Simultaneously, the missile's x-axis will align with the velocity axis. This aerodynamic layout is referred to as a statically stable aerodynamic configuration (Fig. 1.2 – 1). However, when the center of pressure of the missile is in front of its center of gravity, the existing angle of attack will continuously increase under the action of its corresponding destabilizing aerodynamic moment. Therefore, the missile is in a divergent state. This aerodynamic layout is called a statically unstable aerodynamic configuration (Fig. 1.2 – 2).

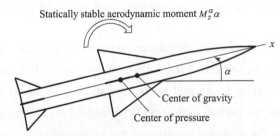

Fig. 1.2 – 1   Missile in a statically stable aerodynamic configuration

1 Basics of Missile Guidance Control

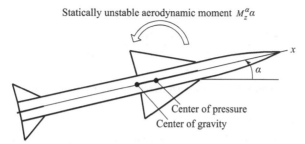

Fig. 1.2 - 2 Missile in a statically unstable aerodynamic configuration

In general, there are three types of aerodynamic configurations for the generation of a missile control moment.

1) Normal aerodynamic configuration

In this aerodynamic configuration, the missile actuator is arranged at the tail of the missile (Fig. 1.2 - 3). The advantage of this configuration is that when the control moment is balanced by the aerodynamic moment caused by the angle of attack, the effective angle of incidence of the control surface becomes the difference between the control surface deflection angle and the angle of attack. This setup optimizes the use of the control deflection angle, allowing for larger control surface deflections and higher angles of attack, which enhances maneuverability. However, a drawback is that the actuator's position coincides with the rear end of the motor, which limits the size of the actuator. Additionally, during maneuvering, the control surface force acts in opposition to the normal force produced by the angle of attack, leading to a reduction in the total normal force. Despite these disadvantages, this aerodynamic configuration remains the most commonly used for tactical missiles due to its overall effectiveness.

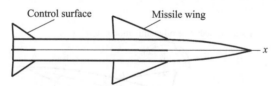

Fig. 1.2 - 3 Normal aerodynamic configuration

2) Canard aerodynamic configuration

In this aerodynamic configuration, the actuator is positioned at the head of the missile (Fig. 1.2 - 4). The advantage of this arrangement is that it allows the missile motor to be positioned independently, freeing up space for other subsystems. Additionally, during maneuvers, the control surface force aligns with the normal force produced by the angle of attack, leading to more efficient utilization of maneuvering force. However, in this configuration, the control surface incident angle is the sum of the actuator deflection angle and the missile's angle of attack. Due to the limitation on the maximum allowed control surface incident angle, large-angle maneuvers are not feasible. As a result, this aerodynamic configuration is less commonly used in missile applications today.

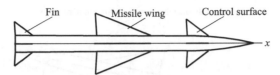

Fig. 1.2-4 Canard aerodynamic configuration

3) Moving-wing aerodynamic configuration

In this aerodynamic configuration, the missile wing can serve as a control surface (Fig. 1.2-5), with the center of pressure positioned in front of the center of gravity, similar to a canard configuration but with a shorter control arm. The wing's large lifting surface primarily provides the necessary lift for missile maneuvering, allowing for a smaller angle of attack. This setup is particularly advantageous when the missile's cruise engine cannot operate at large angles of attack. However, the high power demands of the wing actuator limit its operating frequency bandwidth and, consequently, the response speed of the autopilot. As a result, this aerodynamic configuration is rarely used in current engineering practice.

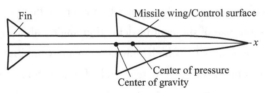

Fig. 1.2-5 Moving-wing aerodynamic configuration

Fig. 1.2-6 and Fig. 1.2-7 illustrate the equilibrium conditions where the control moment and aerodynamic moment balance each other, corresponding to a steady-state angle of attack for statically stable and statically unstable missiles, respectively.

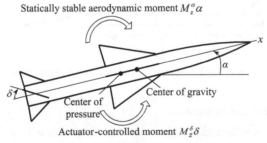

Fig. 1.2-6 Moment equilibrium of a statically stable missile

It is noteworthy that for a statically stable missile, the control moment generated by the actuator deflection angle δ will make the missile rotate in the required direction to produce an angle of attack. When the aerodynamic stabilizing moment, which increases with the angle of attack, reaches the same level as the control moment, the missile achieves an equilibrium state at that angle of attack. Consequently, missiles with sufficient static stability can be designed without an autopilot. However, this aerodynamic feedback approach offers less precise control over the missile's normal acceleration compared to an acceleration autopilot. For statically unstable missiles, a steady-state angle of attack can only be maintained through closed-loop autopilot control, which ensures the necessary balance between the control moment and the aerodynamic moment produced by the angle of attack.

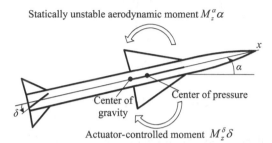

Fig. 1.2 –7 **Moment equilibrium of a statically unstable missile**

As mentioned above, a steady-state angle of attack α is achieved when the control moment and the aerodynamic moment are in equilibrium, which can be expressed as

$$\frac{M_z^\delta \delta}{(\text{Control moment})} = \frac{M_z^\alpha \alpha}{(\text{Aerodynamic moment})}.$$

The transfer function, which takes the actuator deflection angle δ as the input and the angle of attack α as the output, represents the controlled object for the autopilot (Fig. 1.2 –8).

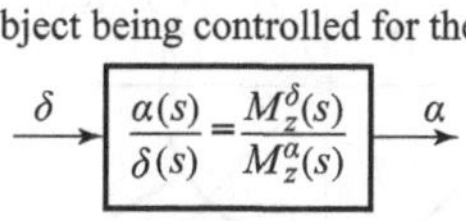

The object being controlled for the autopilot

$$\delta \longrightarrow \boxed{\frac{\alpha(s)}{\delta(s)} = \frac{M_z^\delta(s)}{M_z^\alpha(s)}} \longrightarrow \alpha$$

Fig. 1.2 –8 **The object being controlled for the autopilot**

The missile's static stability is directly proportional to the distance between its center of gravity and its center of pressure. For missiles with low static stability, this distance is relatively small. Therefore, when the center of gravity or center of pressure of the missile with low static stability deviates from its designed value, the value of M_z^δ and the gain of the transfer function $\frac{M_z^\alpha}{M_z^\delta}$ from the actuator δ to the angle of attack α will change greatly from its designed value, which means that the open loop gain of the autopilot loop will also change greatly. This is unacceptable for a normally designed control loop. Therefore, to reduce the autopilot open loop gain change, the missile's static stability is often taken at around 4% ~ 8%. For missiles that must have low static stability aerodynamic configurations for other consideration, the gain from δ to α could be stabilized by designing a pseudo angle of attack feedback loop. For a detailed discussion of this option, see the autopilot design section.

At present, a skid-to-turn (STT) control scheme is adopted for most tactical missiles. That is, in the Cartesian coordinate system, a missile pitch turn is achieved by the generation of the angle of attack α, and a yaw turn is achieved by the generation of the sideslip angle β, as shown in Fig. 1.2 –9.

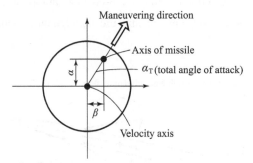

Fig. 1.2 – 9　STT polar diagram

This control scheme offers a very fast response but requires roll stabilization. It is particularly well-suited for aerodynamically symmetrical missiles.

Another scheme, bank-to-turn (BTT), is typically used for surface-symmetrical missiles, especially when there is a significant difference between the pitch and yaw lift surface areas. In this scheme, the missile must turn the main lift surface by an angle φ with the help of a roll control autopilot to have the missile's angle of attack in the required maneuvering direction (Fig. 1.2 – 10).

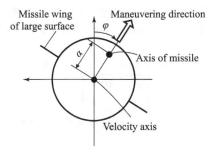

Fig. 1.2 – 10　BTT polar diagram

For missiles using the BTT control scheme, changing the maneuvering direction may require the missile to roll through a large angle to reach the new direction. This results in a slower maneuvering response. Consequently, BTT control is more suitable for the missile's midcourse guidance phase.

2

Missile Mathematical Model

§ 2.1 Symbols and Definitions

The origin of the missile body coordinate system $Ox_b y_b z_b$ is defined as the center of gravity of the missile. The axes are oriented as follows (assuming the missile is either an axisymmetric or plane-symmetric rigid body; Fig. 2.1 – 1).

The roll axis Ox_b, lies in the symmetry plane. Pointing forward is positive.

The yaw axis Oy_b, locates in the symmetry plane of the missile body, with the upward direction defined as positive.

The pitch axis Oz_b, forms the right-handed rectangular coordinate system together with axes Ox_b and Oy_b.

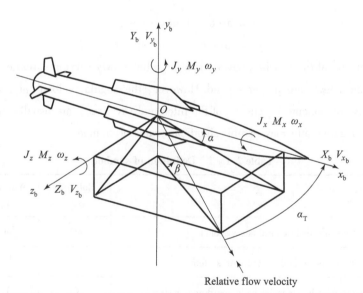

Fig. 2.1 – 1 Definitions of the missile's aerodynamic force, moment, and other related concepts
NOTE: O is the center of gravity of the missile.

Table 2.1 – 1 defines the symbols for aerodynamic forces, moments acting on the missile, linear velocities, and angular velocities, as well as moments of inertia (Fig. 2.1 – 1). The moment of inertia around each axis is defined as

$$J_x = \sum m_i(y_i^2 + z_i^2), \qquad (2.1-1)$$

$$J_y = \sum m_i(z_i^2 + x_i^2), \qquad (2.1-2)$$

$$J_z = \sum m_i(x_i^2 + y_i^2). \qquad (2.1-3)$$

The product of inertia around each axis is defined as

$$J_{yz} = \sum m_i y_i z_i, \qquad (2.1-4)$$

$$J_{zx} = \sum m_i z_i x_i, \qquad (2.1-5)$$

$$J_{xy} = \sum m_i x_i y_i. \qquad (2.1-6)$$

The plane $Ox_b y_b$ is the pitch plane, and the plane $Ox_b z_b$ is the yaw plane. The relevant angles are defined as follows:

α——angle of attack in the pitch plane;
β——angle of attack in the yaw plane (angle of sideslip);
α_T——total angle of attack;
λ——angle of attack plane angle.

Therefore

$$\tan\alpha = \tan\alpha_T \cdot \cos\lambda, \qquad (2.1-7)$$

$$\tan\beta = \tan\alpha_T \cdot \sin\lambda. \qquad (2.1-8)$$

That is

$$\alpha = \arctan(\tan\alpha_T \cdot \cos\lambda), \qquad (2.1-9)$$

$$\beta = \arctan(\tan\alpha_T \cdot \sin\lambda). \qquad (2.1-10)$$

The axial velocity of the missile body V_{x_b} is a large but slowly varying variable, and its variation is usually less than a few percent per second. However, the angular velocity ω_x, ω_y, and ω_z, as well as the velocity components V_{y_b} and V_{z_b} of the pitch and yaw axes, are usually small. They can be either positive or negative and may exhibit significant rates of change.

Table 2.1-1 Definition of symbols

Items	Roll axis x_b	Yaw axis y_b	Pitch axis z_b
Angular velocity (missile body coordinate system)	ω_x	ω_y	ω_z
Velocity component (missile body coordinate system)	V_{x_b}	V_{y_b}	V_{z_b}
Forces acting on the missile (missile body coordinate system)	X_b	Y_b	Z_b
Moments acting on the missile (missile body coordinate system)	M_x	M_y	M_z
Moments of inertia	J_x	J_y	J_z
Product of inertia	J_{yz}	J_{zx}	J_{xy}

§2.2 Euler Equations of the Missile Rigid Body Motion

The six-degree-of-freedom model of a missile motion in space comprises six dynamic equations (three for center of gravity motion and three for rotational dynamics) and six kinematic equations (three for center of gravity motion and three for rotational motion).

The coordinate systems used in missile guidance and control include the Earth coordinate system, the missile body coordinate system, the trajectory coordinate system, and the velocity coordinate system. The x-axis of the last two coordinate systems coincides with the missile velocity vector. However, the y-axis of the trajectory coordinate system is in the vertical plane, and the y-axis of the velocity coordinate system is in the longitudinal symmetrical plane of the missile body. The transformation between the four coordinate systems is achieved through a series of rotations (Fig. 2.2 – 1). For more detailed descriptions of these coordinate systems, consult general flight dynamics textbooks.

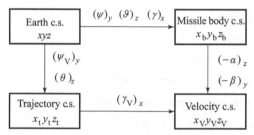

Fig. 2.2 – 1 Transformation from the Earth coordinate system to other coordinate systems

For instance, the rotation transformation from the Earth coordinate system to the missile body coordinate system is illustrated in Fig. 2.2 – 2.

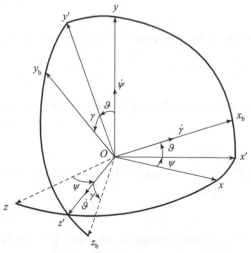

Fig. 2.2 – 2 Relationship between the Earth coordinate system
and the missile body coordinate system

In the study of coordinate transformation, it is necessary to know three basic coordinate system transformation matrixes about axes x, y, and z:

Rotation matrix that does rotation about the x-axis by angle φ_x,

$$L_x(\varphi_x) = \begin{bmatrix} 1 & 0 & 0 \\ 0 & \cos\varphi_x & \sin\varphi_x \\ 0 & -\sin\varphi_x & \cos\varphi_x \end{bmatrix};$$

Rotation matrix that does rotation about the y-axis by angle φ_y,

$$L_y(\varphi_y) = \begin{bmatrix} \cos\varphi_y & 0 & -\sin\varphi_y \\ 0 & 1 & 0 \\ \sin\varphi_y & 0 & \cos\varphi_y \end{bmatrix};$$

Rotation matrix that does rotation about the z-axis by angle φ_z,

$$L_z(\varphi_z) = \begin{bmatrix} \cos\varphi_z & \sin\varphi_z & 0 \\ -\sin\varphi_z & \cos\varphi_z & 0 \\ 0 & 0 & 1 \end{bmatrix}.$$

Define the following variables:

V——missile body velocity;

ψ_V, θ——missile flight path angle;

γ_V——missile symmetrical plane deflection angle;

ψ, ϑ, γ——missiles yaw angle, pitch angle, roll angle;

α, β——angle of attack, angle of sideslip;

V_x, V_y, V_z——velocity component (Earth coordinate system);

V_{x_b}, V_{y_b}, V_{z_b}——velocity component (missile body coordinate system);

ω_x, ω_y, ω_z——missile angular velocity component (missile body coordinate system);

F_{x_t}, F_{y_t}, F_{z_t}——resultant force component acting on the missile (trajectory coordinate system);

F_{x_b}, F_{y_b}, F_{z_b}——resultant force component acting on the missile (missile body coordinate system).

The total force F acting on the missile consists of aerodynamic force $R = \begin{bmatrix} X_b \\ Y_b \\ Z_b \end{bmatrix}$ (missile body coordinate system), thrust $P = \begin{bmatrix} P \\ 0 \\ 0 \end{bmatrix}$ (missile body coordinate system), and gravity $G = \begin{bmatrix} 0 \\ -G \\ 0 \end{bmatrix}$ (Earth coordinate system). The moment acting on the missile body is $M = \begin{bmatrix} M_x \\ M_y \\ M_z \end{bmatrix}$ (missile body coordinate system). The projections of related components in other coordinate systems are shown in

Table 2.2 –1.

Table 2.2 –1 Related projections in different coordinate systems

Items	Earth coordinate systems	Trajectory coordinate systems	Missile body coordinate systems
Aerodynamic force $\boldsymbol{R}$		$\boldsymbol{L}_x(-\gamma_V)\boldsymbol{L}_y(-\beta)\boldsymbol{L}_z(-\alpha)\begin{bmatrix} X_b \\ Y_b \\ Z_b \end{bmatrix}$	$\begin{bmatrix} X_b \\ Y_b \\ Z_b \end{bmatrix}$
Gravity $\boldsymbol{G}$	$\begin{bmatrix} 0 \\ -G \\ 0 \end{bmatrix}$	$\boldsymbol{L}_z(\theta)\boldsymbol{L}_y(\psi_V)\begin{bmatrix} 0 \\ -G \\ 0 \end{bmatrix}$	$\boldsymbol{L}_x(\gamma)\boldsymbol{L}_z(\vartheta)\boldsymbol{L}_y(\psi)\begin{bmatrix} 0 \\ -G \\ 0 \end{bmatrix}$
Thrust $\boldsymbol{P}$		$\boldsymbol{L}_x(-\gamma_V)\boldsymbol{L}_y(-\beta)\boldsymbol{L}_z(-\alpha)\begin{bmatrix} P \\ 0 \\ 0 \end{bmatrix}$	$\begin{bmatrix} P \\ 0 \\ 0 \end{bmatrix}$
Resultant force $\boldsymbol{F}$ ($\boldsymbol{F} = \boldsymbol{R} + \boldsymbol{G} + \boldsymbol{P}$)		$\begin{bmatrix} F_{x_t} \\ F_{y_t} \\ F_{z_t} \end{bmatrix}$	$\begin{bmatrix} F_{x_b} \\ F_{y_b} \\ F_{z_b} \end{bmatrix}$
Aerodynamic moment $\boldsymbol{M}$			$\begin{bmatrix} M_x \\ M_y \\ M_z \end{bmatrix}$
Velocity $\boldsymbol{V}$	$\begin{bmatrix} V_x \\ V_y \\ V_z \end{bmatrix} = \boldsymbol{L}_y(-\psi)\boldsymbol{L}_z(-\vartheta)$ $\boldsymbol{L}_x(-\gamma)\begin{bmatrix} V_{x_b} \\ V_{y_b} \\ V_{z_b} \end{bmatrix}$		$\begin{bmatrix} V_{x_b} \\ V_{y_b} \\ V_{z_b} \end{bmatrix}$

The six-degree-of-freedom missile model can be represented in the trajectory coordinate system or the missile body coordinate system. When the model is expressed in the trajectory coordinate system (Equation (2.2 – 1)), the state variables for the three dynamic translational and three rotational equations are taken as the velocity V, θ, and ψ_V (in the trajectory coordinate system) and the angular velocity ω_x, ω_y, and ω_z (in the missile body coordinate system). The state variables of the six kinematic equations are respectively taken as the position components x, y, and z (in the Earth coordinate system) and the Euler angle ϑ, ψ, and γ (in the missile body coordinate system). Other dependent derived parameters include α, β, and γ_V.

$$m\dot{V} = F_{x_t},$$
$$m V\dot{\theta} = F_{y_t},$$
$$-mV\cos\theta\,\dot{\psi}_V = F_{z_t},$$
$$J_x\dot{\omega}_x - (J_y - J_z)\omega_y\omega_z - J_{yz}(\omega_y^2 - \omega_z^2) - J_{zx}(\dot{\omega}_z + \omega_x\omega_y) - J_{xy}(\dot{\omega}_y - \omega_x\omega_z) = M_x,$$

$$J_y\dot{\omega}_y - (J_z - J_x)\omega_z\omega_x - J_{zx}(\omega_z^2 - \omega_x^2) - J_{xy}(\dot{\omega}_x + \omega_y\omega_z) - J_{yz}(\dot{\omega}_z - \omega_y\omega_x) = M_y,$$
$$J_z\dot{\omega}_z - (J_x - J_y)\omega_x\omega_y - J_{xy}(\omega_x^2 - \omega_y^2) - J_{yz}(\dot{\omega}_y + \omega_z\omega_x) - J_{zx}(\dot{\omega}_x - \omega_z\omega_y) = M_z,$$
$$\dot{x} = V\cos\theta\cos\psi_V,$$
$$\dot{y} = V\sin\theta,$$
$$\dot{z} = -V\cos\theta\sin\psi_V, \quad (2.2-1)$$
$$\dot{\vartheta} = \omega_y\sin\gamma + \omega_z\cos\gamma,$$
$$\dot{\psi} = (\omega_y\cos\gamma - \omega_z\sin\gamma)/\cos\vartheta,$$
$$\dot{\gamma} = \omega_x - \tan\vartheta(\omega_y\cos\gamma - \omega_z\sin\gamma),$$
$$\sin\beta = \cos\theta[\cos\gamma\sin(\psi - \psi_V) + \sin\vartheta\sin\gamma\sin(\psi - \psi_V)] - \sin\theta\cos\vartheta\sin\gamma,$$
$$\sin\alpha = \{\cos\theta[\sin\vartheta\cos\gamma\cos(\psi - \psi_V) - \sin\gamma\sin(\psi - \psi_V)] - \sin\theta\cos\vartheta\cos\gamma\}/\cos\beta,$$
$$\sin\gamma_V = (\cos\alpha\sin\beta\sin\vartheta - \sin\alpha\sin\beta\cos\gamma\cos\vartheta + \cos\beta\sin\gamma\cos\vartheta)/\cos\theta.$$

When the six-degree-of-freedom model of the missile is given in the missile body coordinate system (Equation (2.2-2)), aside from the state variables of the three translational dynamic equations changing to the velocity components V_{x_b}, V_{y_b}, and V_{z_b} (in the missile body coordinate system), the remaining state variables are the same as the trajectory system. That is, the state variables of the three dynamic rotational equations are taken as the angular velocity components ω_x, ω_y, and ω_z (in the missile body coordinate system). The state variables of the six kinematic equations are taken as the position components x, y, and z (in the Earth coordinate system) and the Euler angle ϑ, ψ, and γ (in the missile body coordinate system), respectively. Other useful dependent derived parameters are V_x, V_y, V_z, V, θ, ψ_V, α, β, and γ_V.

$$m(\dot{V}_{x_b} + V_{z_b}\omega_y - V_{y_b}\omega_z) = F_{x_b} = X_b - G\sin\vartheta + P,$$
$$m(\dot{V}_{y_b} + V_{x_b}\omega_z - V_{z_b}\omega_x) = F_{y_b} = Y_b - G\cos\vartheta\cos\gamma,$$
$$m(\dot{V}_{z_b} + V_{y_b}\omega_x - V_{x_b}\omega_y) = F_{z_b} = Z_b + G\cos\vartheta\sin\gamma,$$
$$J_x\dot{\omega}_x - (J_y - J_z)\omega_y\omega_z - J_{yz}(\omega_y^2 - \omega_z^2) - J_{zx}(\dot{\omega}_z + \omega_x\omega_y) - J_{xy}(\dot{\omega}_y - \omega_x\omega_z) = M_x,$$
$$J_y\dot{\omega}_y - (J_z - J_x)\omega_z\omega_x - J_{zx}(\omega_z^2 - \omega_x^2) - J_{xy}(\dot{\omega}_x + \omega_y\omega_z) - J_{yz}(\dot{\omega}_z - \omega_y\omega_x) = M_y,$$
$$J_z\dot{\omega}_z - (J_x - J_y)\omega_x\omega_y - J_{xy}(\omega_x^2 - \omega_y^2) - J_{yz}(\dot{\omega}_y + \omega_z\omega_x) - J_{zx}(\dot{\omega}_x - \omega_z\omega_y) = M_z,$$
$$\dot{x} = \cos\psi\cos\vartheta V_{x_b} - (\cos\psi\sin\vartheta\cos\gamma - \sin\psi\sin\gamma)V_{y_b} + (\cos\psi\sin\vartheta\sin\gamma + \sin\psi\cos\gamma)V_{z_b},$$
$$\dot{y} = \sin\vartheta V_{x_b} + \cos\vartheta\cos\gamma V_{y_b} - \cos\vartheta\sin\gamma V_{z_b}, \quad (2.2-2)$$
$$\dot{z} = -\sin\psi\cos\vartheta V_{x_b} + (\sin\psi\sin\vartheta\cos\gamma + \cos\psi\sin\gamma)V_{y_b} - (\sin\psi\sin\vartheta\sin\gamma - \cos\psi\cos\gamma)V_{z_b},$$
$$\dot{\vartheta} = \omega_y\sin\gamma + \omega_z\cos\gamma,$$
$$\dot{\psi} = (\omega_y\cos\gamma - \omega_z\sin\gamma)/\cos\vartheta,$$
$$\dot{\gamma} = \omega_x - \tan\vartheta(\omega_y\cos\gamma - \omega_z\sin\gamma),$$
$$V = \sqrt{V_{x_b}^2 + V_{y_b}^2 + V_{z_b}^2} = \sqrt{V_x^2 + V_y^2 + V_z^2} \text{ (Expressions of } V_x, V_y, \text{ and } V_z \text{ are shown in Table 2.2-1)},$$
$$\theta = \arctan(V_y/\sqrt{V_x^2 + V_z^2}),$$
$$\psi_V = \arctan(-V_z/V_x),$$
$$\alpha = \arctan(-V_{y_b}/V_{x_b}),$$
$$\beta = -\arcsin(V_{z_b}/V),$$

$$\gamma_V = \arcsin[(\cos\alpha\sin\beta\sin\vartheta - \sin\alpha\sin\beta\cos\gamma\cos\vartheta + \cos\beta\sin\gamma\cos\vartheta)/\cos\theta].$$

Typically, aerodynamic force $\boldsymbol{R}$ and moment $\boldsymbol{M}$ are functions of Mach number Ma, angle of attack α, angle of sideslip β, three-channel control surface deflection angles δ_x, δ_y, and δ_z, and three angular velocities ω_x, ω_y, and ω_z.

$$\boldsymbol{R} = \boldsymbol{R}(Ma, \alpha, \beta, \delta_x, \delta_y, \delta_z),$$
$$\boldsymbol{M} = \boldsymbol{M}(Ma, \alpha, \beta, \delta_x, \delta_y, \delta_z, \omega_x, \omega_y, \omega_z).$$

The precise expression for these functions and their reasonable simplifications can be determined through wind tunnel tests and analysis of test data.

Missile guidance and control is achieved through control surface deflection δ_x, δ_y, and δ_z commanded by guidance and control laws. Models incorporating guidance and control will involve additional equations. For instance, the complete system model will also include equations for the seeker dynamics, autopilot, command guidance radar, and control surface servo mechanisms.

These dynamic equations can often be simplified for specific mathematical simulations. For example, for axisymmetric missiles, their cross inertia moment J_{xy}, J_{yz}, and J_{zx} can be safely omitted. For three-channel control missiles, the related ω_x, ω_y, and ω_z are so small that their product $\omega_x\omega_y$, $\omega_y\omega_z$, $\omega_z\omega_x$, ω_x^2, ω_y^2, and ω_z^2 can also be omitted. Furthermore, since the projections of the velocity vectors on the missile body coordinate system, V_{y_b} and V_{z_b} are also of a small quantity, their product with the component of ω can also be omitted. Therefore, the dynamic equations represented in the missile body coordinate system can be simplified as

$$m\dot{V}_{x_b} = F_{x_b}, \tag{2.2-3}$$

$$m(\dot{V}_{y_b} + V_{x_b} \cdot \omega_z) = F_{y_b}, \tag{2.2-4}$$

$$m(\dot{V}_{z_b} - V_{x_b} \cdot \omega_y) = F_{z_b}, \tag{2.2-5}$$

$$J_x \cdot \dot{\omega}_x = M_x, \tag{2.2-6}$$

$$J_y \cdot \dot{\omega}_y = M_y, \tag{2.2-7}$$

$$J_z \cdot \dot{\omega}_z = M_z. \tag{2.2-8}$$

When expressed in the trajectory coordinate system, the above translational dynamic equations can be described as follows

$$m\dot{V} = F_{x_t}, \tag{2.2-9}$$

$$mV\dot{\theta} = F_{y_t}, \tag{2.2-10}$$

$$mV\cos\theta\dot{\psi}_V = -F_{z_t}. \tag{2.2-11}$$

§2.3 Configuration of the Control Surfaces

The sequential numbering of the control surface is shown in Fig. 2.3 – 1. The deflection angles δ_1, δ_2, δ_3, and δ_4 are considered positive when turning clockwise along each respective coordinate axis's positive direction. The definitions for these deflection angles are as follows.

Roll control deflection angle: $\delta_x = \dfrac{1}{4}(\delta_1 + \delta_2 + \delta_3 + \delta_4)$.

(When only a pair of actuators is moved, there is $\delta_x = (\delta_1 + \delta_3)/2$ or $\delta_x = (\delta_2 + \delta_4)/2$)

Pitch control deflection angle: $\delta_z = \dfrac{1}{2}(\delta_1 - \delta_3)$.

Yaw control deflection angle: $\delta_y = \dfrac{1}{2}(\delta_4 - \delta_2)$.

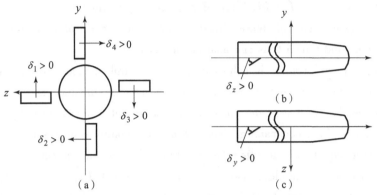

Fig. 2.3-1   Definition of control surface angles
(a) Front view; (b) Side view; (c) Top view

Readers can verify that,

A positive roll control deflection angle δ_x produces a negative moment around the x-axis. For normal control missiles, a positive pitch control deflection angle δ_z produces a negative pitch moment around the z-axis, and a positive yaw control deflection angle δ_y produces a negative yaw moment around the y-axis. A positive pitch control deflection angle δ_z produces a positive force along the y-axis, and a positive yaw control deflection angle δ_y produces a negative force along the z-axis.

For canard-controlled missiles, the pitch and yaw moments produced by the same control channel actuator deflection directions are opposite to those in conventional control missiles, though the direction of the force remains unchanged.

§2.4   Aerodynamic Derivatives and the Missile Control Dynamic Coefficients

To design missile control systems using established LTI system techniques, simplifications are necessary. To start, the small-disturbance nonlinear dynamic equations are linearized to derive their linearized time-varying differential equations. Subsequently, it is assumed that the system's time-varying parameters change slowly during transient periods, allowing them to be treated as constants. This approach facilitates the design of simplified LTI systems using various well-established control theories, such as frequency analysis, root locus design, optimal control, and robust control. A key aspect of this simplification process is the linearization of aerodynamic force and moment functions.

Suppose a certain fin-controlled missile flies at sea level with $Ma = 1.5$. The relation between the roll moment M_x and the roll actuator deflection angle δ_x and the total angle of attack α_T is shown in Fig. 2.4 – 1. It can be seen that here, M_x is not a strict linear function of δ_x. It can also be seen that the roll actuator produced M_x moment is slightly reduced with the increase of the total angle of attack α_T.

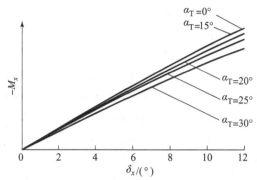

Fig. 2.4 – 1 Relationship between the roll moment M_x and the roll actuator deflection angle δ_x and the total angle of attack α_T

$M_x^{\delta_x}$ is defined as

$$M_x^{\delta_x} = \frac{\partial M_x}{\partial \delta_x}. \tag{2.4 – 1}$$

The moment increment ΔM_x caused by a small increment $\Delta \delta_x$ is

$$\Delta M_x = M_x^{\delta_x} \cdot \Delta \delta_x. \tag{2.4 – 2}$$

Among these, the value of the roll moment derivative $M_x^{\delta_x}$ is closely related to the chosen flight condition (the set point on the trajectory). It is noteworthy that since the value of $\Delta \delta_x$ in most applications is only a few degrees, $M_x^{\delta_x}$ is always regarded as a constant value at the selected flight condition.

$M_x^{\omega_x}$ is a roll damping derivative with dimensions of moment per unit roll angular velocity. Because this moment always prevents the rolling motion, its sign is always negative. For a given Mach number and flight altitude, $M_x^{\omega_x}$ is often considered as a constant. Aside from $M_x^{\delta_x}$ and $M_x^{\omega_x}$, there are no other important roll derivatives.

Let us now examine the aerodynamic derivatives associated with pitch and yaw. The lift Y caused by an angle of attack of the missile is usually expressed as

$$Y = \frac{1}{2}\rho V^2 S C_y. \tag{2.4 – 3}$$

Here, ρ is the atmosphere density; S is the characteristic area of the missile body, which is usually taken as the cross-sectional area of the missile body. $C_y(Ma, \alpha, \delta_z)$ is known as the lift coefficient, which is a function of the angle of attack and the actuator deflection angle δ_z for a given Mach number Ma. For symmetrically arranged missiles, $C_z(Ma, \beta, \delta_y)$ is related to the sideslip angle and the actuator deflection angle δ_y, and is equal to the lift coefficient C_y. The related derivatives are defined as follows

$$Y^\alpha = \frac{\partial Y}{\partial \alpha} = \frac{\partial C_y}{\partial \alpha} \cdot \frac{1}{2}\rho V^2 S = C_y^\alpha \cdot \frac{1}{2}\rho V^2 S, \qquad (2.4-4)$$

$$Y^{\delta_z} = \frac{\partial Y}{\partial \delta_z} = \frac{\partial C_y}{\partial \delta_z} \cdot \frac{1}{2}\rho V^2 S = C_y^{\delta_z} \cdot \frac{1}{2}\rho V^2 S, \qquad (2.4-5)$$

$$Z^\beta = \frac{\partial Z}{\partial \beta} = \frac{\partial C_z}{\partial \beta} \cdot \frac{1}{2}\rho V^2 S = C_z^\beta \cdot \frac{1}{2}\rho V^2 S, \qquad (2.4-6)$$

$$Z^{\delta_y} = \frac{\partial z}{\partial \delta_y} = \frac{\partial C_z}{\partial \delta_y} \cdot \frac{1}{2}\rho V^2 S = C_z^{\delta_y} \cdot \frac{1}{2}\rho V^2 S. \qquad (2.4-7)$$

In the design of most wings and control surfaces, lift generated by a small angle of attack is typically proportional to the angle of attack itself. However, for slender missile bodies, the lift consists of two components: one is proportional to α, and the other is proportional to α^2. This situation is commonly observed in fin-controlled supersonic missiles, as shown in Fig. 2.4-2.

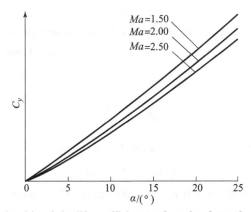

Fig. 2.4-2 Relationship of the lift coefficient and angle of attack and Mach number

It should be noted that if there are angle of attack α and actuator deflection δ_z at the same time, the actual aerodynamic incident angle of the pitch actuator is $\alpha + \delta_z$; however, the total aerodynamic force increment is not $Y^\alpha \cdot \alpha + Y^{\delta_z} \cdot (\alpha + \delta_z)$, but $Y^\alpha \cdot \alpha + Y^{\delta_z} \cdot \delta_z$. This is because the lift generated by the angle of attack α has already been included in Y^α.

Consider the expression for the pitch moment M_z as

$$M_z = \frac{1}{2}\rho V^2 SLm_z. \qquad (2.4-8)$$

In Equation (2.4-8), m_z (Ma, α, δ_z, ω_z) is called the pitch moment coefficient, L is the characteristic length of the missile (usually taken as the missile length), and the derivatives related to the pitch moment are defined as follows

$$M_z^\alpha = \frac{\partial M_z}{\partial \alpha} = \frac{1}{2}\rho V^2 SLm_z^\alpha, m_z^\alpha = \frac{\partial m_z}{\partial \alpha}, \qquad (2.4-9)$$

$$M_z^{\delta_z} = \frac{\partial M_z}{\partial \delta_z} = \frac{1}{2}\rho V^2 SLm_z^{\delta_z}, m_z^{\delta_z} = \frac{\partial m_z}{\partial \delta_z}, \qquad (2.4-10)$$

$$M_z^{\omega_z} = \frac{\partial M_z}{\partial \omega_z} = \frac{1}{2}\rho V^2 SLm_z^{\omega_z}, m_z^{\omega_z} = \frac{\partial m_z}{\partial \omega_z}. \qquad (2.4-11)$$

M_z^α is the product of the aerodynamic force derivative Y^α and the distance from the center of gravity to the center of pressure. This distance between the center of gravity and the center of pressure is referred to as the static stability. Essentially, it is an aerodynamic parameter that measures the missile's static stability. If the center of pressure is located behind the center of gravity, it generates an aerodynamic restoration moment that helps reduce the angle of attack and stabilize the missile body in response to disturbances. Conversely, if the center of pressure is in front of the center of gravity, disturbances will cause the angle of attack to increase continuously, resulting in a statically unstable missile body.

Generally, if the center of gravity is located approximately 50% of the missile's length from the nose, the center of pressure is typically positioned around 52%~55% of the missile's length from the nose. At subsonic and low supersonic speeds, the center of pressure is usually positioned further forward compared to high Mach numbers. Additionally, at low speeds, the position of the center of pressure is significantly affected by the angle of attack. This is primarily because, for a missile body without wings and fins, the center of pressure tends to shift backward with increasing angle of attack. In contrast, the center of pressure for control surfaces and missile wings remains relatively stable. Unfortunately, the position of the center of pressure is also a function of the angle of attack plane angle λ that characterizes the roll direction of the total angle of attack α_T plane.

$M_z^{\delta_z}$ is the moment derivative due to the pitch actuator deflection, which is equal to the actuator force derivative Y^{δ_z} multiplied by the distance from the center of gravity to the pitch actuator center of pressure ℓ_c. It is clear that if Y^{δ_z} is a constant, $M_z^{\delta_z}$ will only change when the center of gravity moves.

$M_z^{\omega_z}$ is the pitch damping moment derivative, which is equal to the aerodynamic moment produced by unit pitch angular velocity. This derivative is a small term that is not sensitive to the angle of attack.

To facilitate the control system design, the aerodynamic force derivative is usually divided by the product mV of the missile mass and speed, and the aerodynamic moment derivative is divided by its respective moment of inertia. This process yields all the dynamic coefficients relevant to missile control system design, which are detailed in Table 2.4-1 and Table 2.4-2. These coefficients include a_α, a_δ, and a_ω, which are associated with the missile's pitch and yaw rotation; b_α and b_δ, which pertain to the missile's translational motion; and the roll-related rotational dynamic coefficients c_δ and c_ω.

The physical meaning of the dynamic coefficients defined above is explained as follows. Firstly, to facilitate the use of these coefficients in the missile body transfer function analysis in the next section, the following conventions have been established:

(1) All aerodynamic coefficients (a_α, a_δ, a_ω, b_α, b_δ, c_δ, and c_ω) for a statically stable and normally controlled missile are considered positive;

(2) For canard-controlled missiles, only $a_\delta < 0$, the others are positive;

(3) For statically unstable missiles, only $a_\alpha < 0$.

The physical meaning of a_α is the missile pitch angular acceleration produced by the unit angle of attack. Its unit is $(\text{rad} \cdot \text{s}^{-2})/\text{rad}$ or s^{-2}. It reflects the level of the missile's static stability.

a_δ is the missile angular acceleration produced by unit control of surface deflection, and its unit is $(\text{rad} \cdot \text{s}^{-2})/\text{rad}$ or s^{-2}. It reflects the control surface's efficiency in controlling missile rotation.

a_ω is the missile angular acceleration produced by unit missile angular velocity. Its unit is $(\text{rad} \cdot \text{s}^{-2})/(\text{rad} \cdot \text{s}^{-1})$ or s^{-1}. It reflects the amount of the missile aerodynamic damping.

b_α is the missile velocity vector rotation angular velocity produced by the unit angle of attack, and its unit is $(\text{rad} \cdot \text{s}^{-2})/\text{rad}$ or s^{-1}. It is a very important aerodynamic derivative that characterizes the missile's maneuvering ability. The normal force produced by the angle of attack rotates the velocity vector.

b_δ is the missile velocity vector rotation angular velocity generated by unit control surface deflection and its unit is $(\text{rad} \cdot \text{s}^{-1})/\text{rad}$, that is s^{-1}. Since its value is relatively small compared with b_α, it has a limited contribution to missile velocity rotation. However, as the rotational control moment is equal to the product of the actuator force and the distance between the actuator's center of pressure and the missile center of gravity, it has a direct impact on a_δ value.

c_δ is the missile roll angle acceleration caused by unit roll actuator deflection, and its unit is $(\text{rad} \cdot \text{s}^{-2})/\text{rad}$ or s^{-2}. It reflects the control efficiency of the missile's roll actuator.

c_ω is the roll angle acceleration of the missile caused by unit missile roll angular velocity, and it is expressed in units of $(\text{rad} \cdot \text{s}^{-2})/(\text{rad} \cdot \text{s}^{-1})$ or s^{-1}. It reflects the size of the missile's roll damping.

Table 2.4-1 Symbols and dimensions of the main aerodynamic derivatives

Symbols	Algebraic signs	Dimensions	Symbols	Algebraic signs	Dimensions	Physical meanings
M_z^α	−ve Statically stable missile body; +ve Statically unstable missile body	$\text{N} \cdot \text{m}$	$a_\alpha = \dfrac{-M_z^\alpha}{J_z}$	+ve Statically stable missile body; −ve Statically unstable missile body	s^{-2}	$-\dfrac{\partial \dot{\omega}_z}{\partial \alpha}$
$M_z^{\delta_z}$	−ve, Fin-controlled; +ve, Canard-controlled	$\text{N} \cdot \text{m}$	$a_\delta = \dfrac{-M_z^{\delta_z}}{J_z}$	+ve, Fin-controlled; −ve, Canard-controlled	s^{-2}	$-\dfrac{\partial \dot{\omega}_z}{\partial \delta_z}$
$M_z^{\omega_z}$	−ve	$\text{N} \cdot \text{m} \cdot \text{s}$	$a_\omega = \dfrac{-M_z^{\omega_z}}{J_z}$	+ve	s^{-1}	$-\dfrac{\partial \dot{\omega}_z}{\partial \omega_z}$
Y^α	+ve	N	$b_\alpha = \dfrac{P + Y^\alpha}{mV}$	+ve	s^{-1}	$\dfrac{\partial \dot{\theta}}{\partial \alpha}$
Y^{δ_z}	+ve	N	$b_\delta = \dfrac{Y^{\delta_z}}{mV}$	+ve	s^{-1}	$\dfrac{\partial \dot{\theta}}{\partial \delta_z}$
$M_x^{\delta_x}$	−ve	$\text{N} \cdot \text{m}$	$c_\delta = \dfrac{-M_x^{\delta_x}}{J_x}$	+ve	s^{-2}	$-\dfrac{\partial \dot{\omega}_x}{\partial \delta_x}$
$M_x^{\omega_x}$	−ve	$\text{N} \cdot \text{m} \cdot \text{s}$	$c_\omega = \dfrac{-M_x^{\omega_x}}{J_x}$	+ve	s^{-1}	$-\dfrac{\partial \dot{\omega}_x}{\partial \omega_x}$

Table 2.4 – 2 Main dynamic coefficient expressions

Symbols	Expressions	Notes
$a_\alpha = \dfrac{-M_z^\alpha}{J_z}$	$\dfrac{-m_z^\alpha qSL}{J_z} = \dfrac{-m_z^\alpha \rho V^2 SL}{2J_z}$	$a_\alpha > 0$, $m_z^\alpha < 0$, Statically stable missile body; $a_\alpha < 0$, $m_z^\alpha > 0$, Statically unstable missile body
$a_\delta = \dfrac{-M_z^{\delta_z}}{J_z}$	$\dfrac{-m_z^{\delta_z} qSL}{J_z} = \dfrac{-m_z^{\delta_z} \rho V^2 SL}{2J_z}$	$a_\delta > 0$, $m_z^{\delta_z} < 0$, Fin-controlled; $a_\delta < 0$, $m_z^{\delta_z} > 0$, Canard-controlled
$a_\omega = \dfrac{-M_z^{\omega_z}}{J_z}$	$\dfrac{-m_z^{\omega_z} qSL \cdot \dfrac{L}{V}}{J_z} = \dfrac{-m_z^{\omega_z} \rho VSL^2}{2J_z}$	$a_\omega > 0$, $m_z^{\omega_z} < 0$
$b_\alpha = \dfrac{P + Y^\alpha}{mV}$	$\dfrac{P + C_y^\alpha qS}{mV} = \dfrac{2P + C_y^\alpha \rho V^2 S}{2mV}$	$b_\alpha > 0$, $C_y^\alpha > 0$
$b_\delta = \dfrac{Y^{\delta_z}}{mV}$	$\dfrac{C_y^{\delta_z} qS}{mV} = \dfrac{C_y^{\delta_z} \rho VS}{2m}$	$b_\delta > 0$, $C_y^{\delta_z} > 0$
$c_\delta = \dfrac{-M_x^{\delta_x}}{J_x}$	$\dfrac{-m_x^{\delta_x} qSL}{J_x} = \dfrac{-m_x^{\delta_x} \rho V^2 SL}{2J_x}$	$c_\delta > 0$, $m_x^{\delta_x} < 0$
$c_\omega = \dfrac{-M_x^{\omega_x}}{J_x}$	$\dfrac{-m_x^{\omega_x} qSL}{J_x} = \dfrac{-m_x^{\omega_x} \rho V^2 SL}{2J_x}$	$c_\omega > 0$, $m_x^{\omega_x} < 0$

§2.5 The Transfer Function of a Missile as the Controlled Object

As discussed in the previous section, the missile's dynamic equations can be simplified to a set of LTI differential equations for a given trajectory set point by employing small disturbance assumptions, linearization, and constant parameters. This simplification allows for the analysis of the missile body's transfer function as a controlled object. Although the differential equations are derived for small disturbance variables, the small disturbance symbol is often omitted in both state and control variables for brevity and to avoid excessive notation complexity.

For axisymmetric missiles, the pitch channel and the yaw channel are symmetric, allowing for a focus solely on the pitch channel transfer functions.

The simplified missile pitch channel differential equations are as follows.

$$\ddot{\vartheta} = -a_\omega \cdot \dot{\vartheta} - a_\alpha \cdot \alpha - a_\delta \cdot \delta_z, \quad (2.5-1)$$

$$\dot{\theta} = b_\alpha \cdot \alpha + b_\delta \cdot \delta_z, \quad (2.5-2)$$

$$\alpha = \vartheta - \theta. \quad (2.5-3)$$

This set of equations is a 3-state variable differential equation system. The system control variable is δ_z, the independent state variables are $\vartheta, \dot{\vartheta}$ and θ, and the derived dependent state variable is $\alpha = \vartheta - \theta$. Therefore, the equations above can be expressed as

$$\frac{d\vartheta}{dt} = \dot{\vartheta}, \quad (2.5-4)$$

$$\frac{d\dot{\vartheta}}{dt} = -a_\omega \cdot \dot{\vartheta} - a_\alpha \cdot (\vartheta - \theta) - a_\delta \cdot \delta_z, \qquad (2.5-5)$$

$$\frac{d\theta}{dt} = b_\alpha \cdot (\vartheta - \theta) + b_\delta \cdot \delta_z. \qquad (2.5-6)$$

That is

$$\frac{d}{dt}\begin{bmatrix}\vartheta\\\dot{\vartheta}\\\theta\end{bmatrix} = \begin{bmatrix}0 & 1 & 0\\-a_\alpha & -a_\omega & a_\alpha\\b_\alpha & 0 & -b_\alpha\end{bmatrix}\begin{bmatrix}\vartheta\\\dot{\vartheta}\\\theta\end{bmatrix} + \begin{bmatrix}0\\-a_\delta\\b_\delta\end{bmatrix}\delta_z, \qquad (2.5-7)$$

$$\alpha = \vartheta - \theta. \qquad (2.5-8)$$

The important transfer functions of the pitch channel obtained from the pitch state equations are,

(1) The transfer function from the pitch actuator to the missile's normal acceleration $a_y(s)/\delta_z(s)$. The transfer function of $a_y(s)/\delta_z(s)$ is derived with the help of the relation $a_y = V \cdot \dot{\theta}$.

$$\frac{a_y(s)}{\delta_z(s)} = -V \cdot \frac{-b_\delta s^2 - a_\omega b_\delta s + (a_\delta b_\alpha - a_\alpha b_\delta)}{s^2 + (a_\omega + b_\alpha)s + (a_\alpha + a_\omega b_\alpha)}, \qquad (2.5-9)$$

$$\frac{a_y(s)}{\delta_z(s)} = k_a \cdot \frac{A_2 s^2 + A_1 s + 1}{T_m^2 s^2 + 2\mu_m T_m s + 1}, \qquad (2.5-10)$$

where

$$k_a = -V \cdot \frac{a_\delta b_\alpha - a_\alpha b_\delta}{a_\alpha + a_\omega b_\alpha} \ ((\mathrm{m}\cdot\mathrm{s}^{-2})/\mathrm{rad}), \quad T_m = \frac{1}{\sqrt{a_\alpha + a_\omega b_\alpha}}\ (\mathrm{s}),$$

$$\omega_m = \frac{1}{T_m} = \sqrt{a_\alpha + a_\omega b_\alpha}\ (\mathrm{s}^{-1}), \qquad \mu_m = \frac{a_\omega + b_\alpha}{2\sqrt{a_\alpha + a_\omega b_\alpha}},$$

$$A_1 = -\frac{a_\omega b_\delta}{a_\delta b_\alpha - a_\alpha b_\delta}\ (\mathrm{s}), \qquad A_2 = -\frac{b_\delta}{a_\delta b_\alpha - a_\alpha b_\delta}\ (\mathrm{s}^2).$$

This is a typical second-order oscillation transfer function and its undamped natural frequency is $\omega_m = \sqrt{a_\alpha + a_\omega b_\alpha}$. Since the term $a_\omega b_\alpha$ for a common missile is much smaller than a_α, it has

$$\omega_m \approx \sqrt{a_\alpha} = \sqrt{\frac{\text{Restoring moment produced by unit angle of attack}}{\text{Rotational inertia}}}$$

$$= \sqrt{\frac{-M_z^\alpha}{J_z}} = \sqrt{\frac{-m_z^\alpha \rho V^2 SL}{2J_z}} = \sqrt{\frac{C_y^\alpha S \rho V^2 x^*}{2J_z}}. \qquad (2.5-11)$$

Here, x^* is the distance from the center of pressure to the center of gravity of the missile. It characterizes the static stability level of the missile. The higher the missile's static stability is, the greater the a_α and the missile's natural frequency are.

For a rear-controlled surface-to-air missile, the trajectory parameters for the chosen set point are as follows: the missile height from the ground is 1 500 m, and the Mach number is 1.4, which corresponds to $V = 467$ m/s. The dynamic coefficients of the missile at this specific set point are shown in Table 2.5-1.

Table 2.5-1 Dynamic coefficients of the missile at a specific set point

a_α/s^{-2}	a_δ/s^{-2}	a_ω/s^{-1}	b_α/s^{-1}	b_δ/s^{-1}
321	534	2.89	2.74	0.42

Among these, the missile length is 2 m, the moment of inertia around the z-axis is $J_z = 12.8 \text{ kg} \cdot \text{m}^2$, and the mass of the missile is $m = 53$ kg. Therefore, the static stability could be calculated as $x^* = -M_z^\alpha/Y^\alpha = a_\alpha J_z/b_\alpha mV = 321 \times 12.8 \times (2.74 \times 53 \times 467) = 61$ mm, which is approximately 3% of the missile's length. With all the related missile dynamic coefficients from the known Equation (2.5-9), the following results are obtained.

$$\frac{a_y(s)}{\delta_z(s)} = \frac{-1\,886 \times (-3.16 \times 10^{-4} s^2 - 9.14 \times 10^{-4} s + 1)}{0.003 s^2 + 0.0171 s + 1} \quad (2.5-12)$$

Based on this transfer function, it is known that the undamped natural frequency of this second-order oscillation is $\omega_m = 18.1$ rad/s $= 2.88$ Hz, and its damping coefficient is $\mu_m = 0.155$. In addition, $k_a = -1\,886$ m $\cdot$ s^{-2}/rad, that is to say, a pitch actuator deflection angle of 5° will produce a normal acceleration $a_y = 165$ m/s$^2 = 16.8g$. Fig. 2.5-1 illustrates the transient process of normal acceleration resulting from a 5° step change in the pitch actuator deflection angle.

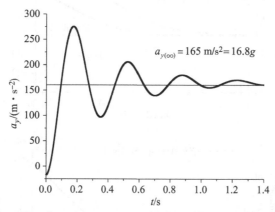

Fig. 2.5-1 Normal acceleration response generated by a 5° step pitch actuator deflection angle

Fig. 2.5-1 shows that the aerodynamic damping of a general missile body is very low (in this case, $\mu_m = 0.155$). The transient process of an uncontrolled missile body exhibits severe oscillations and significant overshoot. The transient behavior can only be improved by introducing artificial damping through an autopilot.

If the center of gravity of the above fin-controlled missile is moved forward to increase its static stability by a factor of four, the steady-state gain will decrease to about a quarter of its original value. Additionally, the oscillation frequency will double, and the damping coefficient will be halved. This shows that missile static stability is a critical design parameter, and a larger static stability will result in:

(a) Smaller steady-state gain (poor maneuverability) ($k_a \propto 1/x^*$) and good resistance to x^* change. That is to say, there is better robustness;

(b) Higher short-period oscillation frequency ($\omega_m \propto \sqrt{x^*}$);

(c) A smaller damping coefficient ($\mu_m \propto 1/\sqrt{x^*}$).

Similarly, lower static stability will result in:

(a) Larger steady state gains (higher maneuverability) and poor resistance to x^* change. That is to say, there is weak robustness;

(b) Lower short-period oscillation frequency;

(c) An improved, but still lower, damping coefficient.

Therefore, when designing control systems, it is crucial for designers to carefully consider the positions of the center of pressure and the center of gravity for the missile in the chosen design.

In this case, the transfer function for the numerator is

$$(A_2 s^2 + A_1 s + 1) = -0.000\,316 s^2 - 0.000\,914 s + 1,$$

that is, $(s/59.3 + 1)(-s/56.4 + 1)$.

The Bode diagram for the individual elements and the Bode diagram for the total second-order numerator transfer functions are presented separately in Fig. 2.5 – 2.

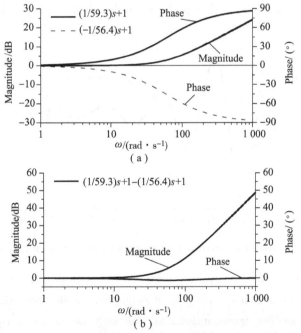

Fig. 2.5 – 2 Bode diagrams of the numerator transfer functions

(a) The Bode diagram for the individual elements; (b) The Bode diagram of the total second-order numerator transfer function

The s^2 term coefficient A_2 – of the transfer function numerator is given as

$$|A_2| = \frac{b_\delta}{a_\delta b_\alpha - a_\alpha b_\delta} = \left(\frac{b_\delta}{b_\alpha}\right)\left(\frac{1}{a_\delta - \frac{a_\alpha b_\delta}{b_\alpha}}\right) = \left(\frac{b_\delta}{b_\alpha}\right)\left(\frac{1}{a_\delta}\right)\frac{1}{1 - \left(\frac{a_\alpha}{b_\alpha}\right)\left(\frac{b_\delta}{a_\delta}\right)}.$$

The following relations are known:

$\dfrac{a_\alpha}{b_\alpha} = x^*$, x^* is the distance from the center of gravity to the missile's center of pressure;

$\dfrac{a_\delta}{b_\delta} = l_\delta$, l_δ is the distance from the center of gravity to the actuator center of pressure.

Therefore, $|A_2| = \left(\dfrac{b_\delta}{b_\alpha}\right)\left(\dfrac{1}{a_\delta}\right)\dfrac{1}{1-\left(\dfrac{x^*}{l_\delta}\right)}$, since x^* is smaller than l_δ, the last item can be simplified as 1. Thus, $|A_2| \approx \left(\dfrac{b_\delta}{b_\alpha}\right)\left(\dfrac{1}{a_\delta}\right)$. Therefore, the cutoff frequency of the numerator transfer function can be given as $\omega^* = \sqrt{\dfrac{1}{|A_2|}} = \sqrt{\dfrac{b_\alpha}{b_\delta}}\sqrt{a_\delta}$. When a_α and a_δ are in the same magnitude order, the following relation can be obtained, $\omega^* = \sqrt{\dfrac{b_\alpha}{b_\delta}}\sqrt{a_\alpha} = \sqrt{\dfrac{b_\alpha}{b_\delta}}\omega_m$. Since the lift force generated by the angle of attack is greater than the lift force of the actuator, it is commonly observed that $\dfrac{b_\alpha}{b_\delta} \approx 4 \sim 10$. Then, the cutoff frequency ω^* of the numerator transfer function of $\dfrac{a_y(s)}{\delta_z(s)}$ will be about φ_z times the natural frequency of the missile body ω_m. Here, ω_m is determined by the missile transfer function denominator.

It is known that the s item coefficient A_1 of the numerator transfer function is $A_1 = -\dfrac{a_\omega b_\delta}{a_\delta b_\alpha - a_\alpha b_\delta}$. Because the missile body damping a_ω is very small, A_1 is approximately 0. The results show that the two first-order cutoff frequencies of the missile numerator transfer function are nearly identical, and the resulting phase is almost zero. This is because the phases of these two elements are equal in magnitude but opposite in sign. It is more important that the same magnitudes of their two components will lead to an increase in the magnitude of the transfer function $\dfrac{a_y(s)}{\delta_z(s)}$, making its crossover frequency move to a higher frequency. The results indicate that the larger phase lag of the denominator at higher frequencies will complicate the autopilot design. However, the amplitudes of these two first-order transfer functions must be combined.

Since the value of ω^* is higher than the natural frequency of the missile ω_m and earlier missile autopilot design bandwidth is relatively low, the numerator transfer function was always omitted in past textbooks when discussing the normal acceleration transfer function of the missile body. However, with the higher bandwidth of modern autopilot designs, the numerator component of the normal acceleration transfer function should not be disregarded.

For axisymmetric missiles, the transfer functions a_z/δ_y and a_y/δ_z are essentially the same.

(2) Transfer function $\dot{\theta}(s)/\delta_z(s)$ from pitch actuator deflection angle δ_z to flight path angle angular velocity $\dot{\theta}$.

As $a_y(s) = V\dot{\theta}(s)$, so

$$\dfrac{\dot{\theta}(s)}{\delta_z(s)} = \dfrac{a_y(s)}{V} = k_{\dot{\theta}} \cdot \dfrac{A_2 s^2 + A_1 s + 1}{T_m^2 s^2 + 2\mu_m T_m s + 1}, \qquad (2.5-13)$$

where

$$k_{\dot{\theta}} = -\frac{a_\delta b_\alpha - a_\alpha b_\delta}{a_\alpha + a_\omega b_\alpha} \ ((\text{rad} \cdot \text{s}^{-1})/\text{rad}).$$

It should be noted that all the transfer functions that use the actuator deflection angle δ_z as input share the same denominator and have second-order oscillation characteristics.

(3) Transfer function $\dot{\vartheta}(s)/\delta_z(s)$ from actuator deflection angle δ_z to missile angular velocity $\dot{\vartheta}$.

It can be inferred from the Equation (2.5-4) ~ Equation (2.5-6) that

$$\frac{\dot{\vartheta}(s)}{\delta_z(s)} = -\frac{a_\delta s + (a_\delta b_\alpha - a_\alpha b_\delta)}{s^2 + (a_\omega + b_\alpha)s + (a_\alpha + a_\omega b_\alpha)}, \quad (2.5-14)$$

$$\frac{\dot{\vartheta}(s)}{\delta_z(s)} = k_{\dot{\vartheta}} \cdot \frac{T_\alpha s + 1}{T_m^2 s^2 + 2\mu_m T_m s + 1}. \quad (2.5-15)$$

It is noteworthy that

(a) The transfer function $\dot{\vartheta}(s)/\delta_z(s)$ and the transfer function $\dot{\theta}(s)/\delta_z(s)$ share the same gain, which means

$$k_{\dot{\vartheta}} = k_{\dot{\theta}}.$$

(b) From the previous derivation, it can be learned that $T_\alpha = \dfrac{a_\delta}{a_\delta b_\alpha - a_\alpha b_\delta} = \dfrac{1}{b_\alpha(1 - x^*/\ell_\delta)}$ $\approx \dfrac{1}{b_\alpha}$. Here, the unit of T_α is second, which is called the angle of attack time constant. Since the physical meaning of b_α is the flight path angle angular velocity $\dot{\theta}$ produced by the unit angle of attack α, the smaller the T_α, the higher the missile maneuverability. Because the air density and the value of b_α decrease with the increase in altitude, T_α will generally increase from a fraction of a second to several seconds with the increase in missile flight altitude.

(c) If the small term in the numerator of the transfer functions $a_y(s)/\delta_z(s)$ and $\dot{\theta}(s)/\delta_z(s)$ is omitted, assuming $b_\delta \approx 0$, the relationship between the missile's normal acceleration a_y, the flight path angle angular velocity $\dot{\theta}$, and the pitch angle angular velocity $\dot{\vartheta}$ is shown in Fig. 2.5-3.

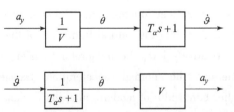

Fig. 2.5-3 The relationship between the normal acceleration, the flight path angle angular velocity, and the pitch angle angular velocity

Since $\dot{\vartheta}(s) = \dfrac{1}{V}(T_\alpha s + 1)a_y(s)$, the angular velocity feedback in the inner loop of an acceleration autopilot design is clearly equivalent to lead compensation for the normal acceleration.

(d) Since $\dot{\theta}(s) = \dfrac{1}{(T_\alpha s + 1)}\dot{\vartheta}(s)$ and $\theta(s) = \dfrac{1}{(T_\alpha s + 1)}\vartheta(s)$, $\dot{\theta}$ can be considered as the response of $\dot\theta$ to the pitch angle angular velocity $\dot\vartheta$ through dynamic lag $\dfrac{1}{(T_\alpha s + 1)}$ and the response of θ to the pitch angle ϑ through dynamic lag $\dfrac{1}{(T_\alpha s + 1)}$. Fig. 2.5–4 shows the time response curves of the missile body's angular velocity $\dot\vartheta$, flight path angle angular velocity $\dot\theta$, missile body attitude angle ϑ, and the flight path angle θ of a fin-controlled missile with unit step pitch actuator deflection. As can be clearly seen from Fig. 2.5–4, the $\dot\theta$ response lags behind the $\dot\vartheta$ response, and the θ response lags behind the ϑ response.

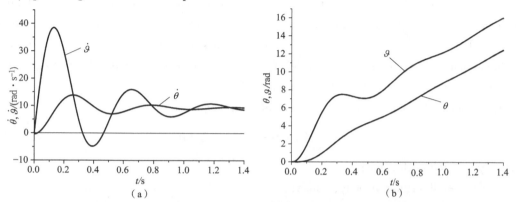

Fig. 2.5–4 The time response curves to unit step actuator input
(a) The time response curves of $\dot\vartheta$ and $\dot\theta$; (b) The time response curves of ϑ and θ

The physical mechanism of the flight path angle θ lagging behind the attitude angle ϑ is very simple. The missile's change in attitude results in an angle of attack. The normal acceleration produced by this angle of attack will alter the flight path angle's angular velocity, and its integral yields the flight path angle. Consequently, the response of the flight path angle will lag behind the missile's attitude angle, as illustrated in Fig. 2.5–5.

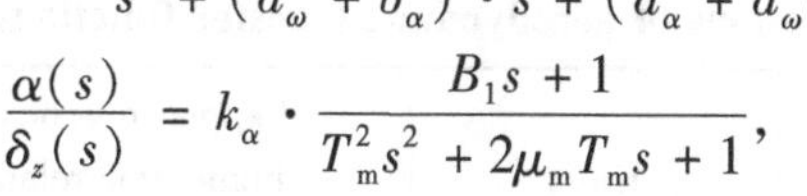

Fig. 2.5–5 Response of the flight path angle θ lagging behind the missile's attitude angle ϑ

(4) Transfer function from the pitch actuator deflection angle to the angle of attack $\alpha(s)/\delta_z(s)$

$$\dfrac{\alpha(s)}{\delta_z(s)} = -\dfrac{b_\delta \cdot s + (a_\omega b_\delta + a_\delta)}{s^2 + (a_\omega + b_\alpha) \cdot s + (a_\alpha + a_\omega b_\alpha)}, \qquad (2.5-16)$$

$$\dfrac{\alpha(s)}{\delta_z(s)} = k_\alpha \cdot \dfrac{B_1 s + 1}{T_m^2 s^2 + 2\mu_m T_m s + 1}, \qquad (2.5-17)$$

where

$$k_\alpha = -\dfrac{a_\omega b_\delta + a_\delta}{a_\alpha + a_\omega b_\alpha} \ (\text{rad/rad}), \ B_1 = \dfrac{b_\delta}{a_\omega b_\delta + a_\delta} \ (\text{s}).$$

Since the value of a_ω is very small, k_α can be expressed as $k_\alpha = -\dfrac{a_\delta}{a_\alpha}$. In missile design, $|k_\alpha| = \dfrac{a_\delta}{a_\alpha}$ is usually referred to as the control ratio of the missile, and it represents the steady angle of attack generated by the unit actuator deflection angle of the missile. It represents the actuator's capability for producing the angle of attack efficiency. Therefore, it seems that improving the control ratio of the missile can improve the efficiency of control. However, due to the limitation of the size of the actuator surface, the value of a_δ is unlikely to increase significantly. Therefore, the control ratio can only be improved by reducing the missile's static stability margin, that is, reducing the distance x^* between the center of gravity and the center of pressure. When x^* is very small, small changes in the position of the center of gravity or the center of pressure can cause large changes in the value of a_α, that is, the control ratio. This could cause k_α to deviate significantly from the designed value, potentially resulting in the designed autopilot failing to meet robustness evaluation standards. An efficient solution is to incorporate a pseudo angle of attack feedback loop into the missile body, creating a highly robust virtual missile model. This design approach will be detailed in Chapter 6.

(5) Transfer function of the roll channel $\dot{\gamma}(s)/\delta_x(s)$.

Roll channel dynamics is a second-order differential equation

$$\ddot{\gamma} = \frac{d}{dt}(\dot{\gamma}) = -c_\omega \cdot \dot{\gamma} - c_\delta \cdot \delta_x, \qquad (2.5-18)$$

where γ ——roll angle, rad;

$\dot{\gamma}$ ——roll angular velocity, rad/s;

$\ddot{\gamma}$ ——roll angular acceleration, rad/s^2.

Express it in the form of a state equation. take the state variables as γ and $\dot{\gamma}$, and the control variable as δ_x, that is

$$\frac{d}{dt}\begin{bmatrix} \gamma \\ \dot{\gamma} \end{bmatrix} = \begin{bmatrix} 0 & 1 \\ 0 & -c_\omega \end{bmatrix}\begin{bmatrix} \gamma \\ \dot{\gamma} \end{bmatrix} + \begin{bmatrix} 0 \\ -c_\delta \end{bmatrix}\delta_x. \qquad (2.5-19)$$

So, it can be derived that

$$\frac{\dot{\gamma}(s)}{\delta_x(s)} = \frac{-c_\delta}{s+c_\omega} = \frac{-c_\delta/c_\omega}{(1/c_\omega)s+1} = \frac{-c_\delta/c_\omega}{T_r s+1}, \qquad (2.5-20)$$

$$\frac{\dot{\gamma}(s)}{\delta_x(s)} = \frac{k_r}{T_r s+1}. \qquad (2.5-21)$$

Here $k_r = -c_\delta/c_\omega$ is the steady state gain from δ_x to $\dot{\gamma}$, and $T_r = 1/c_\omega$ is the aerodynamic time constant for this transfer function.

The detailed expressions of all the above transfer functions are shown in Table 2.5-2.

Table 2.5-2 **Expressions of aerodynamic transfer functions and their coefficients**

Aerodynamic transfer functions	Units	Expressions of the numerator related coefficients	Expressions of the denominator related coefficients
$\dfrac{a_y(s)}{\delta_z(s)} = k_a \cdot \dfrac{A_2 s^2 + A_1 s + 1}{T_m^2 s^2 + 2\mu_m T_m s + 1}$	$(m \cdot s^{-2})/\text{rad}$	$k_a = -V \cdot \dfrac{a_\delta b_\alpha - a_\alpha b_\delta}{a_\alpha + a_\omega b_\alpha}$	$T_m = \dfrac{1}{\sqrt{a_\alpha + a_\omega b_\alpha}}$
		$A_1 = -\dfrac{a_\omega b_\delta}{a_\delta b_\alpha - a_\alpha b_\delta}$	$\mu_m = \dfrac{a_\omega + b_\alpha}{2\sqrt{a_\alpha + a_\omega b_\alpha}}$

Continued

Aerodynamic transfer functions	Units	Expressions of the numerator related coefficients	Expressions of the denominator related coefficients
$\dfrac{a_y(s)}{\delta_z(s)} = k_a \cdot \dfrac{A_2 s^2 + A_1 s + 1}{T_m^2 s^2 + 2\mu_m T_m s + 1}$	$(\text{m} \cdot \text{s}^{-2})/\text{rad}$	$A_2 = -\dfrac{b_\delta}{a_\delta b_\alpha - a_\alpha b_\delta}$	
$\dfrac{\dot{\theta}(s)}{\delta_z(s)} = k_{\dot{\theta}} \cdot \dfrac{A_2 s^2 + A_1 s + 1}{T_m^2 s^2 + 2\mu_m T_m s + 1}$	$(\text{rad} \cdot \text{s}^{-1})/\text{rad}$	$k_{\dot{\theta}} = -\dfrac{a_\delta b_\alpha - a_\alpha b_\delta}{a_\alpha + a_\omega b_\alpha}$ $A_1 = -\dfrac{a_\omega b_\delta}{a_\delta b_\alpha - a_\alpha b_\delta}$ $A_2 = -\dfrac{b_\delta}{a_\delta b_\alpha - a_\alpha b_\delta}$	
$\dfrac{\dot{\vartheta}(s)}{\delta_z(s)} = k_{\dot{\vartheta}} \cdot \dfrac{T_\alpha s + 1}{T_m^2 s^2 + 2\mu_m T_m s + 1}$	$(\text{rad} \cdot \text{s}^{-1})/\text{rad}$	$k_{\dot{\vartheta}} = -\dfrac{a_\delta b_\alpha - a_\alpha b_\delta}{a_\alpha + a_\omega b_\alpha}$ $T_\alpha = \dfrac{a_\delta}{a_\delta b_\alpha - a_\alpha b_\delta}$	
$\dfrac{\alpha(s)}{\delta_z(s)} = k_\alpha \cdot \dfrac{B_1 s + 1}{T_m^2 s^2 + 2\mu_m T_m s + 1}$	rad/rad	$k_\alpha = -\dfrac{a_\omega b_\delta + a_\delta}{a_\alpha + a_\omega b_\alpha}$ $B_1 = \dfrac{b_\delta}{a_\omega b_\delta + a_\delta}$	
$\dfrac{\dot{\gamma}(s)}{\delta_x(s)} = \dfrac{k_r}{T_r s + 1}$	$(\text{rad} \cdot \text{s}^{-1})/\text{rad}$	$k_r = -\dfrac{c_\delta}{c_\omega}$	$T_r = \dfrac{1}{c_\omega}$

3

Basic Missile Control Component Mathematical Models

This chapter provides an overview of the mathematical models for basic missile control components.

§ 3.1 Seeker

The seeker is a crucial control component of the missile. Given its importance, it will be discussed in detail in Chapter 5.

§ 3.2 Actuator

Another critical missile component is the actuator. Actuators come in various types, including electrical, hydraulic, hot gas, and cold gas actuators. Hydraulic actuators offer the highest load capacity and broadest frequency bandwidth. However, advancements in electrical actuators are rapidly improving their load capacity and frequency bandwidth. As a result, electrical actuators are now the most commonly used. Hot gas and cold gas actuators, though limited in load capacity, have reasonable frequency bandwidths and are primarily used in low-end, short-range missiles.

The actuator is a key element that restricts the bandwidth of the missile autopilot. Its dynamic model is typically represented using first-order or second-order models. The first-order model is expressed as follows

$$\frac{\delta(s)}{\delta_c(s)} = \frac{1}{Ts+1}e^{-\tau s}, \qquad (3.2-1)$$

where T is the actuator time constant, $\omega = \frac{1}{T}$ is the first-order bandwidth, and τ is the time delay.

The expression of the second-order model is

$$\frac{\delta(s)}{\delta_c(s)} = \frac{\omega_n^2}{s^2 + 2\mu\omega_n s + \omega_n^2}e^{-\tau s}, \qquad (3.2-2)$$

where ω_n is the undamped natural frequency of the actuator, μ is the damping coefficient, and τ is the time delay.

In practical engineering applications, it is often necessary to determine several commonly used characteristic parameters of the actuator by testing and identifying its transfer function.

(1) Transfer function -3 dB bandwidth $\omega_{-3\,\mathrm{dB}}$.

(2) Transfer function $-90°$ bandwidth $\omega_{90°}$.
(3) Damping coefficient μ.
(4) Time delay τ.
(5) Angular velocity capability of the actuator, both with and without load (Fig. 3.2 – 1)

$$\dot{\delta} = \frac{\delta^*}{\Delta t}. \qquad (3.2-3)$$

Typically, actuators have a broad frequency band, allowing them to respond to high-frequency noise signals. This not only exacerbates the actuator's friction effect but, more importantly, when the high-frequency noise is superimposed on a normal actuator response to a command δ_c, it could hit the bidirectional saturation zone of the actuator. After the actuator output δ effects have been filtered by the missile body, the effective δ angle will be different from δ_c, and an actuator response error will occur.

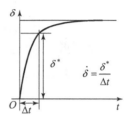

Fig. 3.2 – 1 Actuator output δ in response to unit command

To mitigate this issue, a notch filter is commonly incorporated in the autopilot design before the actuator to eliminate the impact of high-frequency noise.

§ 3.3 Gyroscope

Gyroscopes can be broadly categorized into three main types: mechanical rotor gyroscopes, optical gyroscopes, and vibrating gyroscopes.

Mechanical rotor gyroscopes, primarily dynamically tuned gyroscopes, use a high-speed rotating rotor as the central sensing component, combined with precise mechanical parts to measure angular velocity. These gyroscopes tend to have complex structures, larger sizes, higher costs, and limited measurement ranges.

Optical gyroscopes, including laser and fiber optic gyroscopes, are solid-state devices that operate based on the Sagnac effect. They are known for their high accuracy and wide dynamic range but come with limitations due to the complexity of their optical design.

Vibrating gyroscopes include piezoelectric vibrating gyroscopes, quartz vibrating gyroscopes (such as tuning fork gyroscopes and hemispherical resonator gyroscopes), MEMS (micro-electro-mechanical system) gyroscopes, and metal-shell resonator gyroscopes. These gyroscopes operate by detecting changes in the vibration pattern of a resonator caused by Coriolis forces to measure angular velocity. Unlike traditional gyroscopes that use a high-speed rotating rotor, vibrating gyroscopes use a vibrating element and small-scale vibrations. This makes them particularly suitable for measuring angular velocity in high-overload environments.

MEMS gyroscopes use the Coriolis effect to convert the angular velocity of a rotating object into a direct current voltage signal that is proportional to the angular velocity. The core components of these gyroscopes are produced in bulk using techniques like doping, photolithography, etching, LIGA (lithography, electroplating, and molding), and packaging. Typically, low-precision MEMS sensors are used in consumer electronics, medium-precision sensors in industrial and automotive applications, and high-precision sensors in military and aerospace applications. Typical MEMS navigation module is shown in Fig. 3. 3 – 1.

Fig. 3. 3 – 1   Typical MEMS navigation module

The effect caused by radial motion in the rotating system is known as the Coriolis effect, which can be expressed by the following equation

$$\vec{a}_A = a_r \vec{r}_0 - 2v_r \omega \vec{r}_0 \times \vec{\omega}_0 - \omega^2 r \vec{r}_0. \tag{3.3-1}$$

Equation 3. 3 – 1 shows that the acceleration of point A is composed of three parts: the first part is the radial acceleration along the $\vec{r}_0$ direction, the second part is the Coriolis acceleration, and the third part is the centripetal acceleration.

MEMS vibrating gyroscopes typically include a proof mass, springs, a support frame, a driving mechanism, and a signal detection system. In operation, the proof mass vibrates along the x-axis due to electrostatic forces from the driving mechanism. When an angular velocity is applied along the z-axis, the Coriolis force causes the proof mass to vibrate along the y-axis (Sense Mode). The amplitude of this vibration is directly proportional to the applied angular velocity. By measuring the vibration along the y-axis, the input angular velocity can be accurately determined. Its equivalent model is shown in Fig. 3. 3 – 2.

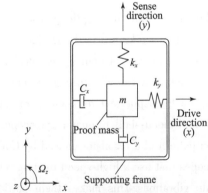

Fig. 3. 3 – 2   Schematic and equivalent model of a
frame-type MEMS vibrating gyroscope

This can be represented as a standard second-order model

$$m \frac{d^2 x}{dt^2} + c_x \frac{dx}{dt} + k_x x = F_0 \sin \omega t. \tag{3.3-2}$$

Metal-shell resonator gyroscopes are a significant subset of vibrating gyroscopes. Their sensitive component is a metal shell called a metal resonator. When the resonator rotates with the carrier, the Coriolis effect causes a "shift" in the vibration pattern of the resonator, which is the fundamental principle behind its sensitivity to rotation. Metal-shell resonator gyroscopes not only have the inertial qualities of traditional gyroscopes but also offer unique advantages, such as high overload resistance, a wide measurement range, and compact size-features not found in other types of gyroscopes.

In 1980, renowned British physicist Bryan studied the vibration characteristics of axisymmetric shell resonators, laying the theoretical foundation for metal-shell resonator gyroscopes. Notable gyroscope expert Lynch, through extensive research on Hemispherical Resonator Gyroscopes, developed a universal model for axisymmetric shell resonator gyroscopes using averaging methods. This advancement significantly accelerated the development of metal-shell resonator gyroscopes. Innalabs has since introduced a series of these products, which are now widely used in drones, autonomous vehicles, and guided missiles.

In China, early research on metal-shell resonator gyroscopes was conducted by scholars such as Fan Shangchun from Beihang University, who initially focused on shell model analysis. Inspired by traditional large clocks, Su Zhong and others from the Beijing Key Laboratory of High-Dynamic Navigation Technology innovatively developed a bell-shaped vibrating resonator gyroscope (BVG) based on existing metal-shell resonator designs. This new design features excellent high overload resistance, a wide measurement range, compact size, and low cost.

The transfer function of the angular rate gyro is typically represented by a second-order transfer function.

$$\frac{\dot{\vartheta}_s(s)}{\dot{\vartheta}(s)} = \frac{\omega_n^2}{s^2 + 2\mu\omega_n s + \omega_n^2} \qquad (3.3-3)$$

Here, $\dot{\vartheta}(s)$ is the angular velocity of the missile, and $\dot{\vartheta}_s(s)$ is the output of the angular rate gyro. ω_n is the undamped natural frequency of the angular rate gyro, and μ is its damping ratio. Current angular rate gyros typically feature a wide frequency bandwidth, often reaching up to 80 Hz. This broad bandwidth has minimal impact on the autopilot design.

§ 3.4 Accelerometer

Fig. 3.4 – 1 depicts the working mechanism of an accelerator. Fig. 3.4 – 2 shows the force model of the accelerator proof mass.

The force kx_A (where x_A is the displacement) exerted by the spring, when compressed or stretched, causes the proof mass to move with the same acceleration a relative to an inertial frame. In a steady state, the relative displacement x_A of the proof mass is directly proportional to the acceleration a of the carrier.

Assuming the proof mass of the accelerometer is subjected to a gravitational force mG (where G is the acceleration due to gravity) along the sensitive axis (ignoring the acceleration of the carrier), the following relationship holds in a steady state.

$$kx_G = mG \qquad (3.4-1)$$

or

$$x_G = \frac{m}{k}G \qquad (3.4-2)$$

In a steady state, the relative displacement x_G of the proof mass is directly proportional to the gravitational acceleration G.

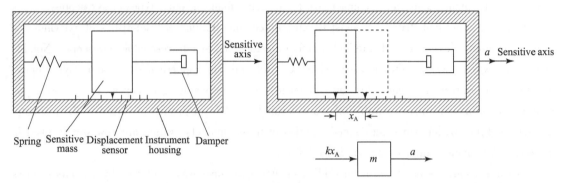

Fig. 3.4-1 The operating principle of an accelerometer based on Newton's laws of motion

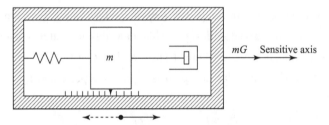

Fig. 3.4-2 Force model of the accelerometer proof mass

As illustrated, the displacement direction of the proof mass caused by the acceleration vector a and the gravitational vector G along the same axis is exactly opposite. Taking into account both the carrier's acceleration and the gravitational acceleration, the relative displacement of the proof mass in a steady state is given by

$$x = \frac{m}{k}(a - G). \qquad (3.4-3)$$

This means that, in a steady state, the relative displacement x of the proof mass is proportional to $(a - G)$. A damper is used to minimize oscillations of the proof mass as it settles into a stable position. The displacement can be converted into an electrical signal by a displacement sensor, making the accelerometer's output directly proportional to $(a - G)$. In inertial technology, the input $(a - G)$ is known as "specific force f", which represents the external force acting on a unit mass and is also referred to as "non-gravitational acceleration". Therefore, accelerometers are sometimes called specific force sensors (since specific force shares the same dimension as acceleration).

For typical precision inertial systems, ignoring the effects of gravitational accelerations from the Moon and other celestial bodies, the specific force measured by the accelerometer can be expressed as follows.

$$f = \left.\frac{d^2 r}{dt^2}\right|_e + 2\Omega \times \left.\frac{dr}{dt}\right|_e + \Omega \times (\Omega \times r) - G_e, \qquad (3.4-4)$$

where $\left.\frac{dr}{dt}\right|_e$ is the velocity of the carrier relative to the Earth, represented by v.

$\Omega \times (\Omega \times r)$ includes both the Earth's gravitational acceleration G_e and the centripetal acceleration due to Earth's rotation;

These two components together constitute the Earth's gravitational acceleration: $g = G_e - \Omega \times (\Omega \times r)$.

Thus, the specific force measured by the accelerometer can be expressed as

$$f = \left.\frac{dv}{dt}\right|_e + 2\Omega \times v - g. \qquad (3.4-5)$$

In an inertial system, the accelerometer is installed in a measurement coordinate system on the carrier. It can be directly mounted within the carrier's coordinate system (as in a strap-down inertial system) or within the platform's coordinate system (as in a platform-based inertial system). If the accelerometer is placed in the measurement coordinate system p, and the angular velocity of p relative to the Earth coordinate system is w_{ep}, then

$$\left.\frac{dv}{dt}\right|_e = \frac{dv}{dt} + \omega_{ep} \times v. \qquad (3.4-6)$$

Therefore, the specific force measured by the accelerometer can be further expressed as

$$f = \left.\frac{dv}{dt}\right|_p + \omega_{ep} \times v + 2\Omega \times v - g, \qquad (3.4-7)$$

or

$$f = \dot{v} + \omega_{ep} \times v + 2\Omega \times v - g. \qquad (3.4-8)$$

This expression represents the specific force measured by the accelerometer when the carrier moves relative to the Earth and is commonly known as the specific force equation.

Recently, advancements in MEMS technology have led to the development of navigation-grade and even strategic-grade silicon micromechanical vibrating accelerometers. These accelerometers use silicon-on-insulator (SOI) micromechanical processes to replace traditional quartz resonators.

With the progress of MEMS technology, MEMS-based inertial measurement units (MIMUs) have become mainstream, offering significant benefits such as smaller size, lighter weight, lower power consumption, and higher precision.

Similarly, the transfer function of an accelerometer is usually represented as a second-order system.

$$\frac{a_s(s)}{a(s)} = \frac{\omega_n^2}{s^2 + 2\mu\omega_n s + \omega_n^2} \qquad (3.4-9)$$

Here, $a(s)$ is the normal acceleration of the missile, $a_s(s)$ is the output of the accelerometer, ω_n is the undamped natural frequency of the accelerometer, and μ is the damping ratio. In practice, the accelerometers typically have a very wide frequency bandwidth, often reaching up to 80 Hz. Consequently, their impact on autopilot design is minimal.

§ 3.5 Inertial Navigation Components and Integrated Inertial Navigation Module

An inertial navigation module, or integrated inertial navigation module, can simultaneously provide high-rate outputs for missile acceleration, velocity, position, angular velocity, and orientation (e.g., 100 – 200 Hz). These modules significantly enhance the information available for missile guidance and autopilot design, enabling the application of advanced guidance laws and more sophisticated autopilot structures.

4

Guidance Radar

§ 4.1 Introduction

Guidance radar is a crucial component of command guidance systems, commonly used in short- to medium-range ground-to-air missile guidance to precisely track targets and determine their line-of-sight (LOS) direction. After launch, the missile is initially directed towards the guidance radar's narrow tracking beam. Once within this beam, the radar uses the missile's transponder echo to calculate the missile's deviation from the beam center. This deviation is then communicated to the missile as a command, steering it to align with the beam center until it intercepts the target. Given that guidance radar can be installed on ships or launch vehicles, it is defined as a precise angle-tracking system mounted on either a low-speed moving carrier or a stationary platform.

§ 4.2 Motion Characteristic of the Target LOS

Fig. 4.2 – 1 illustrates the geometric relationship between the target and the guidance radar.

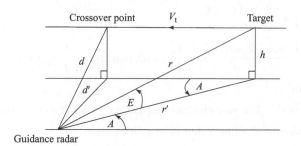

Fig. 4.2 – 1 Geometric relationship between the target and the guidance radar
(the target moves at a constant speed along a straight line)

Suppose that the target follows along a line parallel to the ground at a constant speed of V_t. The target flight height is h, and the minimum distance from the target to the guidance radar is d' when the target flies over the guidance radar. The distance between the target and the radar is r at the moment of time t. The tracking angles of the radar are A in the yaw channel direction and E in the pitch channel direction, respectively.

In the following analysis, the focus will be on the angular velocities of the guidance radar yaw and pitch tracking channels $\dot{A}$ and $\dot{E}$, as well as their angular accelerations $\ddot{A}$ and $\ddot{E}$. Take V_t, h, and d' (or d) as constant parameters in this analysis, while the variables to be studied are A, $\dot{A}$, $\ddot{A}$, E, $\dot{E}$, and $\ddot{E}$. According to the geometric relationship shown in Fig. 4.2-1, the expression of the angular velocity for the yaw channel $\dot{A}$ with the variation of A is

$$\dot{A} = \frac{dA}{dt} = \frac{V_t \sin A}{r'} = \frac{V_t \sin^2 A}{d'} = \frac{V_t}{d'} f_1(A), \qquad (4.2-1)$$

where $f_1(A) = \sin^2 A$. The function $f_1(A)$ is shown in Fig. 4.2-2.

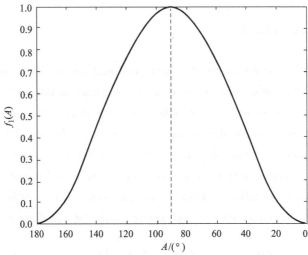

Fig. 4.2-2 $f_1(A)$

Taking A in Equation (4.2-1) as an intermediate variable and taking the derivative of $\dot{A}$ with respect to time t to get the expression of $\ddot{A}$ with the variation of A

$$\ddot{A} = \frac{d}{dt}\left(\frac{dA}{dt}\right) = \frac{d}{dA}\left(\frac{dA}{dt}\right) \cdot \frac{dA}{dt}$$

$$= \frac{2V_t \sin A \cos A}{d'} \cdot \frac{V_t \sin^2 A}{d'} = \frac{V_t^2}{(d')^2} \sin 2A \sin^2 A = \frac{V_t^2}{(d')^2} \cdot f_2(A), \qquad (4.2-2)$$

where $f_2(A) = \sin 2A \sin^2 A$.

For a given value of d', $\ddot{A}$ in yaw channel has the maximum absolute value when $A = 60°$ or $120°$ (Fig. 4.2-3). Its maximum absolute value is as follows

$$|\ddot{A}_{max}| = \frac{V_t^2}{(d')^2} \sin 120° \cdot \sin^2 60° = \frac{3\sqrt{3}}{8} \cdot \frac{V_t^2}{(d')^2} = 0.65 \frac{V_t^2}{(d')^2}. \qquad (4.2-3)$$

The maximum value of $\ddot{A}$ is positive when $A = 60°$, and negative when $A = 120°$. That is, there is a positive maximum angular acceleration for a coming target $A = 60°$ and a negative maximum angular acceleration for a departing target $A = 120°$.

Similarly, for a given d' and h (or $k = h/d'$) in the pitch channel, the expression of E with the variation of A is as follows

$$\tan E = \frac{h}{d'/\sin A} = \frac{h \sin A}{d'} = k \sin A. \qquad (4.2-4)$$

Take d' as a constant and A and k as variables. The expressions of $\dot{E}$ and $\ddot{E}$ with the variation of A are as follows

$$\dot{E} = \frac{V_t}{r}\sin E\cos A = \frac{V_t}{d'}\sin E\cos E\sin A\cos A$$

$$= \frac{V_t \cdot k \cdot \sin^2 A\cos A}{d'(1 + k^2\sin^2 A)} = \frac{V_t}{d'}f_3(A,k), \qquad (4.2-5)$$

where $f_3(A,k) = \dfrac{k \cdot \sin^2 A\cos A}{(1 + k^2\sin^2 A)}$; and

$$\ddot{E} = -\frac{kV_t^2\sin^3 A(\sin^2 A + k^2\sin^4 A - 2\cos^2 A)}{(d')^2(1 + k^2\sin^2 A)^2}$$

$$= -\frac{V_t^2}{d'^2} \cdot f_4(A,k), \qquad (4.2-6)$$

where $f_4(A,k) = \dfrac{k\sin^3 A(\sin^2 A + k^2\sin^4 A - 2\cos^2 A)}{(1 + k^2\sin^2 A)^2}$. Fig. 4.2 – 4 shows the variation of the function $f_4(A,k)$ with respect to A and k.

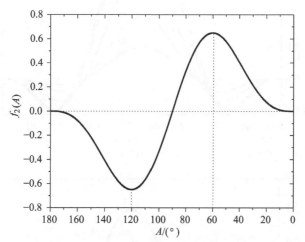

Fig. 4.2 – 3 Change of $f_2(A)$ with the variation of A

It can be proven that $\ddot{E}$ reaches its maximum value when d' and k are given and $A = 90°$ (Fig. 4.2 – 4), that is

$$\ddot{E}_{max} = -\frac{kV_t^2}{(d')^2(1 + k^2)}. \qquad (4.2-7)$$

It can be seen from Equation (4.2 – 7) that $\ddot{E}_{max}$ changes with the variation of k for a given d'. Furthermore, $(\ddot{E}_{max})_{max}$ exists as follows when $k = 1$ (that is, $A = 90°$ and $E = 45°$).

$$(\ddot{E}_{max})_{max} = -0.5\frac{V_t^2}{d'^2}(E = 45°, A = 90°).$$

If the target passes directly above the guidance radar (when $d = 0$), h can be taken as a constant and E can be regarded as a variable. The expressions in the horizontal plane can be converted to vertical plane ones as

$$\dot{E} = \frac{V_t \sin^2 E}{h}, \tag{4.2-8}$$

$$\ddot{E} = \frac{V_t^2 \sin 2E \sin^2 E}{h^2}, \tag{4.2-9}$$

$$\ddot{E}_{max} = \frac{3\sqrt{3}}{8} \cdot \frac{V_t^2}{h^2} = 0.65 \frac{V_t^2}{h^2} (E = 45°),$$

$$\ddot{E}_{max} = \frac{3\sqrt{3}}{8} \cdot \frac{V_t^2}{h^2} = 0.65 \frac{V_t^2}{h^2} (E = 45°). \tag{4.2-10}$$

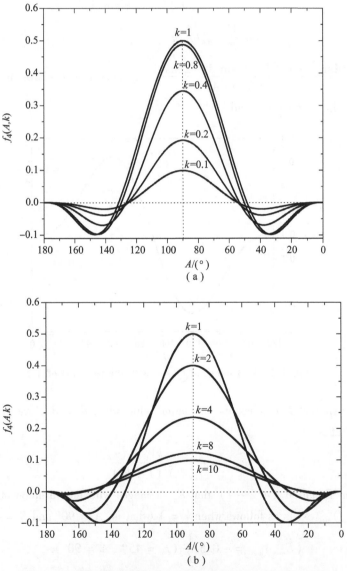

Fig. 4.2-4 Variation of $f_4(A, k)$ with respect to A and k

(a) $k = 0.1, 0.2, 0.4, 0.8, 1$; (b) $k = 1, 2, 4, 8, 10$

The analysis reveals that even with a target moving at a constant speed, the yaw and pitch angular accelerations of the tracking system are present. This necessitates designing the guidance radar tracking system as a type II system with two integrators to ensure tracking accuracy.

§4.3 Control Loop of the Guidance Radar

According to the analysis in Section 4.2, even if the target is flying at a constant speed, the angular accelerations of both $(A)_t$ and $(E)_t$ are not constant. Therefore, the control loop of the guidance radar must be designed as a type II system at a minimum. The system should also meet accuracy requirements when tracking slow variations in angular acceleration.

Fig. 4.3 − 1 illustrates the control block diagram of a typical guidance radar tracking system. The feedback element of the high-gain stabilization loop is usually a rate gyro, ensuring that the antenna axis remains stabilized in inertial space, even if the radar is mounted on a slowly moving platform. The high gain in the stabilization loop helps maintain a small angular tracking error despite various disturbances. The tracking loop's compensation network typically employs a proportional-intgeral (PI) compensator scheme, which realizes a type II system capable of stably tracking the angular acceleration of the LOS.

Fig. 4.3 − 2 and Fig. 4.3 − 3 show the block diagram of the tracking loop of the guidance radar. In Fig. 4.3 − 3, $k = k_1 k_2 k_3$.

To ensure a minimal phase shift in the stability loop at the tracking loop's crossover frequency, the stability loop bandwidth is typically designed to be at least five times greater than that of the tracking loop. Fig. 4.3 − 4 shows the Bode diagram of the PI compensator networks for different T_i values. It can be seen that the compensator position in the frequency domain is different depending on the T_i value. The compensator Bode diagram is shifted toward a high frequency when the T_i value is small. When the T_i value is large, it moves to a low frequency. This compensator structure has two possible applications.

1) For a type I system

This structure enhances the gain at low frequencies to reduce the system's steady-state error. In this application, the PI compensator network is positioned to operate at low frequencies, resulting in a minimal phase shift at the system's crossover frequency (e.g., 5° − 10°). Consequently, the introduction of this network has a negligible impact on the overall bandwidth of the final design. Under these conditions, the crossover frequency of the system can be estimated as $\omega_c \approx k$ rad/s (Fig. 4.3 − 3).

2) For a type II system

In this application, it is best to modify Fig. 4.3 − 3 to Fig. 4.3 − 5 when this scheme is used. The function of the PI compensator $\dfrac{(T_i s + 1)}{T_i s}$ here is to provide an integrator for the type II design requirement and use the numerator $(T_i s + 1)$ term to introduce a lead compensator for system stability consideration because the two integrators will have a −180° phase lag in the tracking loop.

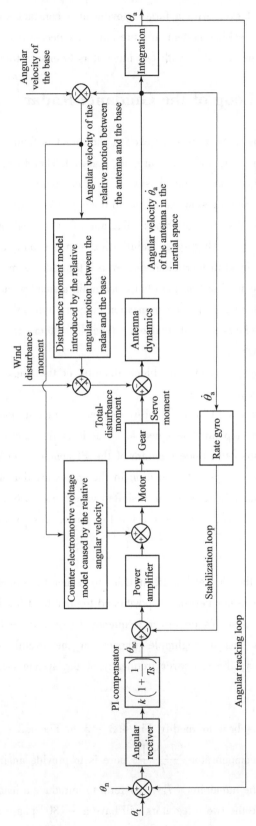

Fig.4.3-1 Control block diagram of a typical guidance radar tracking system

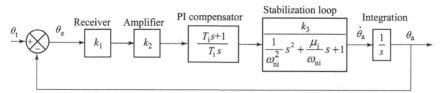

Fig. 4.3 – 2 Block diagram of the tracking toop of the guidance radar

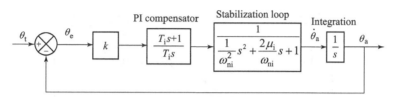

Fig. 4.3 – 3 Model of the tracking toop of the guidance radar control system

When the PI compensator is used in a type II system, the open-loop gain of the system should be $K = \dfrac{k}{T_i}$, and the system crossover frequency is about $\omega_c \approx \sqrt{K}$ rad/s (Fig. 4.3 – 5).

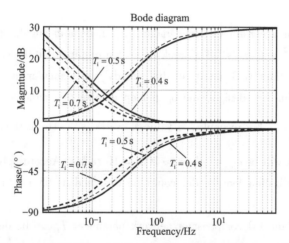

Fig. 4.3 – 4 Bode diagram of the PI compensator networks for different values of parameter T_i

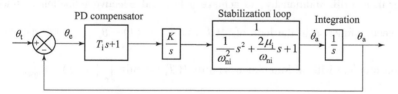

Fig. 4.3 – 5 Another structure of the system radar control model

Fig. 4.3 – 6 shows the Bode diagram of the proportional-derivative (PD) compensator $T_i s + 1$ for different T_i values. The proper time constant T_i should be chosen to give a good phase compensator at the loop crossover frequency and, at the same time, guarantee a shorter tracking error reduction transient. Thus, identifying the appropriate position for PD compensator in the mid-frequency domain requires a separate process for optimizing parameters.

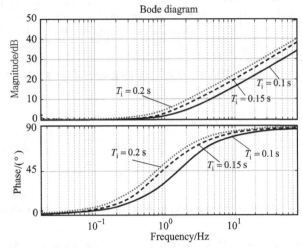

Fig. 4.3-6 Bode diagram of PD compensator for different T_i values

This example illustrates key considerations that should be taken into account when designing a tracking system.

It is known that for a certain guidance radar, the maximum angular acceleration of the LOS given is $\ddot{\theta}_t = 28 \text{ mrad/s}^2 = 1.6°/\text{s}^2$. The tracking design requires that the steady state tracking error of the system be less than 0.3 mrad at this steady state angular acceleration. To meet this requirement in the preliminary design, the open loop gain of this radar should be chosen as $K \geqslant \dfrac{28 \text{ mrad} \cdot \text{s}^{-2}}{0.3 \text{ mrad}} = 93.3 \text{ s}^{-2}$. For example, a value $K = 100 \text{ s}^{-2}$. The estimated open loop crossover frequency of this system will be $\omega_c = \sqrt{K} = \sqrt{100} = 10 \text{ rad/s}$. Accordingly, the bandwidth of the second-order inner stabilization loop model could be designed as 65 rad/s, with a damping ratio of 0.5.

Next, the value of the PD compensator network parameter T_i can be designed according to the following optimal strategy.

Suppose that, for stability considerations, the phase margin must exceed 40°, and the gain margin must be greater than 6 dB. Additionally, to achieve a fast and effective reduction in tracking error, an optimization objective function in the form of $J(T_i) = \int_0^1 \left| \theta_e(t) - \theta_{\text{steady state}} \right| dt$ can be chosen, and the optimization problem will be formulated as $\min J(T_i) = \min \int_0^T \left| \theta_e(t) - \theta_{\text{steady state}} \right| dt$, in which

$\theta_{\text{steady state}} = \dfrac{\ddot{\theta}_{\max}}{K} = \dfrac{28}{100} = 0.28 \text{ mrad}$, subject to $\Delta\phi(T_i) > 40°$ and $\Delta L(T_i) > 6 \text{ dB}$.

Fig. 4.3-7 shows the variation of the gain margin ΔL and phase margin $\Delta\phi$ with the change of T_i. Fig. 4.3-8 shows the variation of the objective function $J(T_i) = \int_0^T \left| \theta_e(t) - \theta_{\text{steady state}} \right| dt$ with the change of T_i.

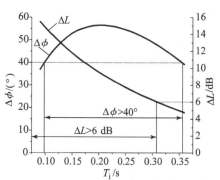

Fig. 4.3-7 Variation of the gain margin and phase margins with the change of T_i

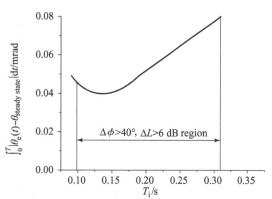

Fig. 4.3-8 Variation of the objective function J with the change of T_i

It can be seen from Fig. 4.3-7 that the range of T_i, which meets the requirements of both the stability ΔL and $\Delta \phi$ constraints, is $0.10 \text{ s} < T_i < 0.31 \text{ s}$. In this region, the optimal value of T_i can be obtained according to the minimum value of the object function J as $T_i = 0.15 \text{ s}$. It is known from Fig. 4.3-8 that this object function J changes smoothly in the vicinity of the optimum solution, that is to say, this optimum solution is also very robust. Fig. 4.3-9 gives the tracking error transient for the optimum solution $T_i = 0.15 \text{ s}$ together with the transients from the stability boundaries $T_i = 0.10 \text{ s}$ and 0.30 s when the input is a step angular acceleration $\ddot{\theta}_t = 28 \text{ mrad/s}^2$. It can be seen that $T_i = 0.10 \text{ s}$ gives a fast response and $T_i = 0.30 \text{ s}$ gives a slow response at the expense of stability margins.

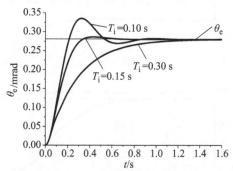

Fig. 4.3-9 Tracking error transients for different T_i values with a step input of angular acceleration of $\ddot{\theta}_t = 28 \text{ mrad/s}^2$

Fig. 4.3-10 illustrates the block diagram of the design results. Fig. 4.3-11 presents the open-loop Bode diagram of the tracking system. From this, it is observed that the system crossover frequency is $\omega_c = 2.65 \text{ Hz}$, the phase margin is $\Delta \varphi = 52.9°$, and the gain margin is $\Delta L = 11.8 \text{ dB}$. The PD compensator provides a phase lead of $68.2°$, while the inner stability loop exhibits a phase lag of $-15.3°$ at the system crossover frequency. Consequently, the system's phase margin is $\Delta \varphi = 68.2° - 15.3° = 52.9°$. The closed-loop Bode diagram of the tracking system is depicted in Fig. 4.3-12.

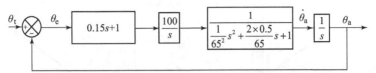

Fig. 4.3 – 10 Block diagram of the design results

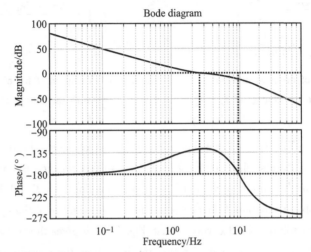

Fig. 4.3 – 11 Open-loop Bode diagram of the tracking system

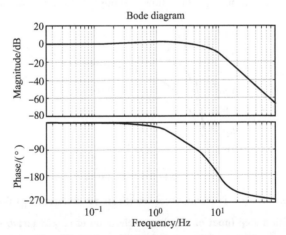

Fig. 4.3 – 12 Closed-loop Bode diagram of the tracking system

Fig. 4.3 – 13 shows the system's response to a unit step input of $\theta_t = 50$ mrad for the LOS angle. Fig. 4.3 – 14, Fig. 4.3 – 15, and Fig. 4.3 – 16 show the tracking angle error curves of the system for inputs of $\theta_t = 50$ mrad, $\dot{\theta}_t = 50$ mrad/s, and $\ddot{\theta}_t = 28$ mrad/s^2, respectively. It is known that a type-II tracking system has zero steady-state error when tracking a constant target angle θ_t and the angular velocity $\dot{\theta}_t$, but there is a steady state error $\theta_e = \dfrac{\ddot{\theta}_t}{K}$ when tracking a constant $\ddot{\theta}_t$ input. In this case, $\ddot{\theta}_t = 28$ mrad/s², $K = 100$ s⁻², and the steady state angular error is $\theta_e = \dfrac{\ddot{\theta}_t}{K} = 0.28$ mrad.

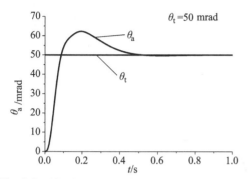

Fig. 4.3 – 13 Response of the radar antenna to a unit step input of $\theta_t = 50$ mrad for the LOS angle

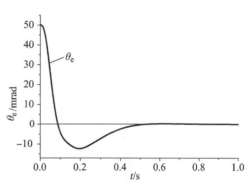

Fig. 4.3 – 14 Tracking angle error curve with a step input $\theta_t = 50$ mrad

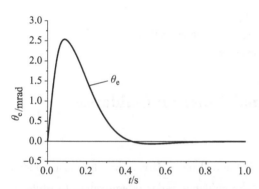

Fig. 4.3 – 15 Tracking angle error curve with a step input $\dot{\theta}_t = 50$ mrad/s

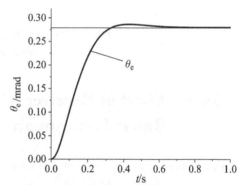

Fig. 4.3 – 16 Tracking angle error curve with a step input $\ddot{\theta}_t = 28$ mrad/s²

This example demonstrates the rationale behind designing a type Ⅱ angular tracking system using the maximum LOS angular acceleration, $(\ddot{\theta}_t)_{max}$, as its input. Consider a scenario where the target's flight speed is $V_t = 250$ m/s, and the minimum slant range is $d' = 1\,300$ m. It is known from Equation (4.2 – 2) that the maximum LOS angular acceleration will be $(\ddot{\theta}_t)_{max} = 28$ mrad/s² when $A = 60°$. Fig. 4.2 – 3 in Section 4.2 shows the variation curve of $\ddot{A}_t$ with the change of A when the target is flying past, and the $\ddot{A}_t$ is not even a constant in a real scenario. Fig. 4.3 – 17 shows the actual tracking error and estimated tracking error $\left(\theta_{e\text{ estimated}} = \dfrac{\ddot{\theta}}{K}\right)$ curves for the real LOS variation with a changing $\ddot{\theta}_t$. It can be seen that when $\ddot{\theta}_t$ input changes slowly, the estimation equation can predict the tracking error nicely.

From Fig. 4.3 – 17, it can be observed that:

(1) In a real tracking scenario, the tracking error can be estimated quite accurately using $\theta_{e\text{ estimated}} = \dfrac{\ddot{\theta}}{K}$;

(2) The target will have a relatively large third-order derivative $\dddot{\theta}_t$ value over the overhead flight region, but its time duration is very short. The target typically moves out of the guidance radar's field of view before the tracking error accumulates significantly. Therefore, a type II system structure is sufficient for designing the guidance radar angle tracking system rather than a type III system.

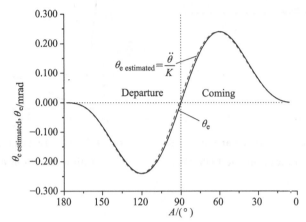

Fig. 4.3 – 17　Actual tracking error and estimated tracking error curves

§ 4.4　Effect of Receiver Thermal Noise on Guidance Radar Performance

The primary source of noise in a radar receiver is thermal noise. Given that the noise bandwidth far exceeds that of the tracking system, its effect on the guidance radar is equivalent to white noise interference. According to the basic radar equation, the echo signal strength of the target is inversely proportional to the fourth power of the target distance. Consequently, the received signal strength varies significantly with changes in target distance. To ensure the subsequent circuitry operates normally, the guidance radar receiver includes an automatic gain control circuit. This circuit adjusts its gain value inversely with the signal strength, maintaining the signal within the designed operational range.

Fig. 4.4 – 1 shows the schematic of the guidance radar automatic gain control. In Fig. 4.4 – 1, P_s and P_n are the signal strengths of the target and the thermal noise, respectively, before the gain control, while P_{s0} and P_{n0} are the target signal strengths and the thermal noise signal strength after the gain control. Generally, the thermal noise strength P_n before the gain adjustment can be considered as a constant. The thermal noise after the gain control $P_{n0} = KP_n$ could be considered as an angular noise by subsequent processing circuits. With this design, as K is proportional to the fourth power of the echoed target signal strength, the angular noise caused by the thermal noise will be proportional to the fourth power of the target distance. In other words, the effect of thermal noise increases as the signal weakens at greater distances from the target. Assuming the white noise power spectral density (angular noise) caused by thermal noise remains constant,

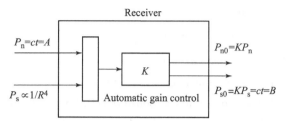

Fig. 4.4 – 1 Schematic of the guidance radar automatic gain control

Assuming the white noise power spectral density $S(\omega)$ caused by thermal noise remains constant,
$$S(\omega) = K_s^2 \ (\text{rad}^2/\text{Hz}).$$
In computer simulation, a white noise could be approximated by a random series of normally distributed digital signals of standard deviation σ and width h (Fig. 4.4 – 2). The autocorrelation function of the random sequence $R(\tau)$ is
$$R(\tau) = \sigma^2 \quad (\tau = 0 \sim h), \tag{4.4-1}$$
$$R(\tau) = 0 \quad (\tau > h). \tag{4.4-2}$$
Its power spectrum density is given by
$$S(\omega) = \int_{-\infty}^{\infty} e^{-j\omega\tau} R(\tau) \, d\tau = \sigma^2 h = K_s^2 \ (\text{rad}^2/\text{Hz}). \tag{4.4-3}$$

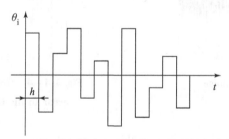

Fig. 4.4 – 2 The white noise input simulation

Given the thermal noise power spectrum density K_s^2 and the simulation step size h, the variance of the normal distribution random number σ^2 should be taken as
$$\sigma^2 = \frac{K_s^2}{h}. \tag{4.4-4}$$

Suppose that the variation of the power spectrum density in the power spectral density of the guidance radar's thermal noise with respect to changes in target distance is given by
$$S(\omega) = K_s^2 = 2.51 \times 10^{-14} \left(\frac{R_t^4}{R_0^4} \right) \text{rad}^2/\text{Hz}, \tag{4.4-5}$$
where R_0 is the reference distance ($R_0 = 1$ km). Suppose that the maximum working distance of the radar is $R_t = 32$ km. Then, the maximum thermal noise power spectrum density will be
$$S(\omega) = 2.51 \times 10^{-14} \times 32^4 = 2.63 \times 10^{-8} \, \text{rad}^2/\text{Hz}.$$
If the computer simulation step size is taken $h = 0.001$ s, the random number variance will be

$$\sigma^2 = \frac{K_s^2}{h} = 2.63 \times 10^{-5} \text{ rad, then}$$

$$\sigma = 0.005\ 13 \text{ rad} = 5.13 \text{ mrad}. \qquad (4.4-6)$$

Fig. 4.4 – 3 displays the thermal noise angle input used in this computer simulation. Additionally, Fig. 4.4 – 4 presents the response of the rader angular error to the thermal noise disturbance. The standard deviation of this response σ is 0.53 mrad, and its value is reduced by about an order of magnitude in comparison with the white noise σ. This is because the tracking system has a low-frequency bandwidth, which effectively filters out high-frequency white noise input. It is indicated in Section 4.2 that $\ddot{A}$ and $\ddot{E}$ values will decrease as the target distance Rincreases. Since the radar tracking error is proportional to $\ddot{A}/K$ and $\ddot{E}/K$, the system can be designed to automatically reduce its open-loop gain K as the target distance increases. This adjustment compensates for the increased impact of thermal noise at greater distances. This self-adaptive design effectively mitigates the negative effects of thermal noise.

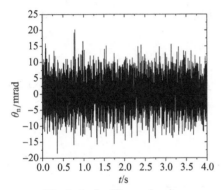

Fig. 4.4 – 3 Thermal noise angle input

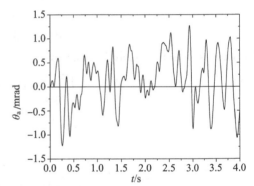

Fig. 4.4 – 4 Response of the radar angular error to the thermal noise disturbance

Fig. 4.4 – 5 shows the implementation of this design scheme.

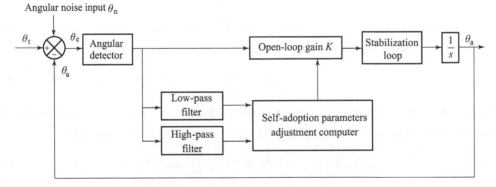

Fig. 4.4 – 5 Block diagram of the adaptive radar tracking system

The output of the low-pass filter in Fig. 4.4 – 5 represents the low-frequency tracking error, while the high-pass filter output indicates the system's response to noise input. When the relative value of the low-pass filter output is high, the system gain will be automatically increased. Conversely, if the high-pass filter output is high, the system gain will be reduced. This adaptive control strategy effectively minimizes the overall tracking error of the guidance radar across different target distances.

§ 4.5 Effect of Target Glint on Guidance Radar Performance

When the radar illuminates the target, the phase shifts of echoes from different parts of the target vary because the target is not a perfect sphere. As a result, the energy center of the target echo can deviate from the geometric center in both pitch and yaw. This phenomenon is known as target glint. The glint effect can be simulated using the output of a first-order colored noise filter, as shown in Fig. 4.5 – 1.

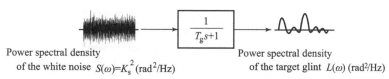

Fig. 4.5 – 1 Target glint model

In Fig. 4.5 – 1, the target glint characteristic can be expressed with two parameters, in which K_g^2 is the power spectrum density of the white noise input (m^2/Hz), T_g is the time constant of the colored noise filter, and $L(\omega)$ is the power spectrum density of the target glint. For a certain target, the two parameters of the target glint are known as $K_g^2 = 3.14 \ m^2/Hz$, $T_g = 0.25 \ s$. Fig. 4.5 – 2 displays the tracking error when the guidance radar, as designed in Section 4.3, is tracking a target with the glint model shown in Fig. 4.5 – 1. Since glint is a low-frequency disturbance with a frequency bandwidth comparable to that of the tracking radar, it can be effectively followed by the radar. This is why target glint is considered one of the primary sources of tracking error in guidance radar systems.

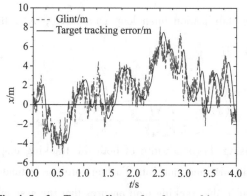

Fig. 4.5 – 2 Target glint and radar tracking error

§ 4.6 Effect of Other Disturbances on Guidance Radar Performance

4.6.1 Effect of Disturbance Moment on Tracking Radar Performance

Firstly, since the antenna and its base are connected by bearings and various cables, relative movement between them can generate spring and damping disturbance torques. Additionally, the tracking antenna's physical size means that wind can exert significant disturbance moments on it. It is crucial that even with substantial wind disturbances, the radar tracking error remains within the system specifications. Fig. 4.6 – 1 illustrates a simplified block diagram of the guidance radar tracking system accounting for moment disturbances.

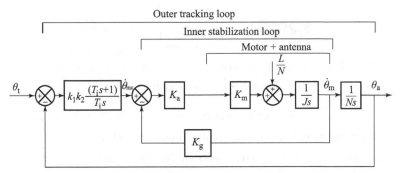

Fig. 4.6 – 1 Simplified block diagram of the guidance radar tracking system with moment disturbances

In Fig. 4.6 – 1, J is the moment of inertia of the antenna, and N is the transmission ratio of the motor gears. The transfer function $\dfrac{\theta_a(s)}{L(s)}$ from the disturbance moment L to the antenna angle error θ_a is given in Equation (4.6 – 1). It is known from Equation (4.6 – 1) that the antenna error is inversely proportional to the stabilization open loop gain of K_a, K_m, the tracking loop gain $\dfrac{k_1 k_2}{T_i}$, and the transmission ratio N.

$$\frac{\theta_a(s)}{L(s)} = \frac{\dfrac{T_i}{N K_a K_m k_1 k_2} s}{\dfrac{T_i N J}{k_a k_m k_1 k_2} s^3 + \dfrac{T_i N K_g}{k_1 k_2} s^2 + T_i s + 1} \quad (4.6-1)$$

In other words, increasing the bandwidth of both the tracking loop and the stability loop will help mitigate the effects of disturbance moments. Furthermore, as indicated by Equation (4.6 – 1), the inclusion of PI correction in the tracking loop ensures that the system's steady-state error remains zero even in the presence of a steady-state moment disturbance.

Suppose the parameters of an example tracking system are as follows: $k_1 k_2 = 1\,300$, $T_i =$

0. 13 s, $N=100$, $K_a K_m = 80$ N·m/V, $J = 1.2$ kg·m², $K_g = 1$ V/(rad·s⁻¹). The Bode diagram of $\theta_a(s)/L(s)$ is shown in Fig. 4.6-2, where responses to both low-frequency and high-frequency interferences can be neglected. However, there may be significant response errors for alternating moment disturbances near the radar system's bandwidth due to resonance effects, such as those caused by wind gusts.

Fig. 4.6-3 shows the angular error response curve of the above system under $\dfrac{L}{N} = 31$ N·m constant moment disturbance.

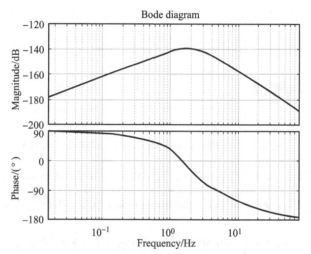

Fig. 4.6-2 Bode diagram of $\dfrac{\theta_a(s)}{L(s)}$

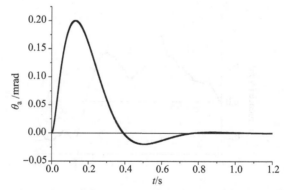

Fig. 4.6-3 Angular error response curve with constant moment disturbance ($L/N = 31$ N·m)

Generally, the power spectrum density of the wind disturbance moment model is taken as

$$S(\omega) = \frac{1}{2\pi} \cdot \frac{cV^4}{(1 + \omega^2 T_1^2)(1 + \omega^2 T_2^2)} \quad ((\text{N}^2 \cdot \text{m}^2)/\text{Hz}), \qquad (4.6-2)$$

where the value of c depends on the shape and the size of the antenna and the relative direction of the wind. V is the average wind speed; the time constant T_1 is about 8.5 s and T_2 is about 0.5 s.

It is known that for a LTI system, its input and output power spectrum densities $S_i(\omega)$ and $S_o(\omega)$ have the following relation

$$S_o(\omega) = G(j\omega)G(-j\omega)S_i(\omega), \quad (4.6-3)$$

where $S_i(\omega)$ is the input power spectrum density, $S_o(\omega)$ is the output power spectrum density and $G(s)$ is the system transfer function.

When $S_i(\omega)$ is taken as a white noise in Equation (4.6-3) and its power spectrum density is $K_s^2 = \dfrac{1}{2\pi} \cdot cV^4$, it can be seen from Equation (4.6-2) that the expression of the filter $G(s)$ in Equation (4.6-3) is $G(s) = \dfrac{1}{T_1 s + 1} \cdot \dfrac{1}{T_2 s + 1}$. Therefore, the model illustrated in the Fig. 4.6-4 can be used to simulate the time domain output characteristics of the wind disturbance moment.

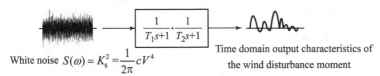

Fig. 4.6-4 The wind disturbance simulation model

Set that $c = 1$, $V = 5$ m/s, $T_1 = 8.5$ s, $T_2 = 0.5$ s, and the power spectrum density of the white noise is $K_s^2 = \dfrac{1}{2\pi} cV^4 = 99.5$ m²/Hz, the integration step $h = 0.001$ s, and the random number variance of the white noise $\sigma^2 = \dfrac{K_s^2}{h} = 9.95 \times 10^4$ rad. Fig. 4.6-5 displays the time-domain response of the radar system under the influence of the wind disturbance moment. It is evident that the gust is characterized by a low-frequency disturbance moment model.

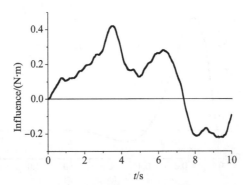

Fig. 4.6-5 Time domain response of the radar system under the influence of the wind disturbance moment

4.6.2 Effect of Target Maneuvers

If the target has a maneuvering acceleration a_t (m/s²) perpendicular to the direction of the missile-target LOS (Fig. 4.6-6), it corresponds to an angular acceleration $\ddot{\theta} = \dfrac{a_t}{R}$ of the LOS, and its effect can be managed in accordance with the model for tracking the angular acceleration target in Section 4.3. For example, if the targe is making a 4g maneuver ($a_t = 39.2$ m/s^2), and at the maneuvering time the target distance is $R = 4$ km, then the corresponding angular acceleration will be $\ddot{\theta} = \dfrac{39.2}{4\,000} \times 1\,000 = 9.8$ mrad/s^2, which has a value of about 35% of the maximum angular acceleration 28 mrad/s^2 in the example of Section 4.3. Clearly, if such a situation arises, its impact cannot be ignored.

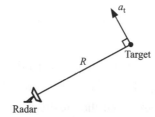

Fig. 4.6 – 6 Target linear acceleration motion effect on the guidance radar tracking

5

Seekers

§ 5.1 Overview

Currently, most missile guidance systems employ proportional guidance or its variants. The main information required for this kind of guidance law is the inertial LOS angular velocity $\dot{q}$. Fig. 5.1 – 1 illustrates the angles related to the seeker and the missile during guidance:

ϑ——Angle of the missile's x-axis with respect to the inertial space;

θ——Angle of the missile velocity vector with respect to the inertial space;

q——LOS angle of the missile and the target relative to the inertial space;

q_d——Angle of the seeker antenna axis with respect to the inertial space;

ε——Angular error of the missile-target line with respect to the seeker axis measured by the seeker detector (optical, radio, laser, etc.).

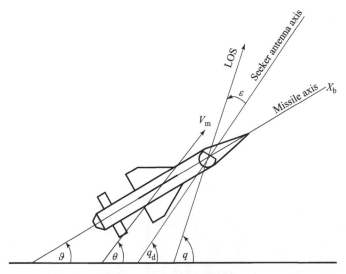

Fig. 5.1 –1   Angle definitions

The seeker itself is an angular tracking device capable of tracking the missile-target line in inertial space. As the target is tracked by the seeker, the $\dot{q}$ signal required for guidance can be obtained by measuring the seeker's angular velocity in the inertial space.

Since the seeker is mounted on the moving base of the missile, it is essential for the seeker to isolate the influence of the missile's angular motion on the $\dot{q}$ measurement of the seeker.

§5.2 Electromechanical Structure of Commonly Used Seekers

Seekers used in current missiles can be categorized into dynamic gyro-stabilized seekers, stabilized platform-based seekers, semi-strap-down seekers, strap-down seekers, and roll-pitch seekers.

5.2.1 Dynamic Gyro Seeker

This type of seeker features two angular degrees of freedom relative to the missile body, achieved through an outer gimbal and an inner gimbal. The seeker detector is mounted on the inner gimbal, while a dynamic gyro with high angular momentum rotates around the detector's load via a high-speed bearing (Fig. 5.2-1).

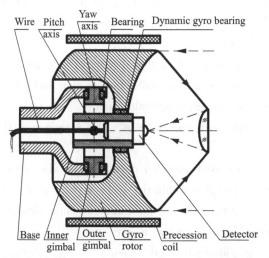

Fig. 5.2-1 Structure of the dynamic gyro-stabilized seeker

The function of the seeker in measuring the LOS angular velocity $\dot{q}$ and isolating the missile's motion can be illustrated by the block diagram in Fig. 5.2-2.

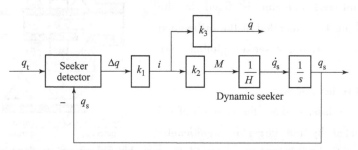

Fig. 5.2-2 Block diagram of the dynamic gyro-stabilized seeker system

The angular tracking error signal Δq output by the detector will produce a corresponding precession current i, a precession moment M, and a precession angular velocity of the dynamic gyro-stabilized seeker $\dot{q}$. Therefore, the seeker precession current can be taken as the rotational angular velocity $\dot{q}$ of the seeker in the inertial space. The open loop gain of the dynamic gyro-stabilized seeker tracking loop (usually referred to as the quality factor of the seeker) is denoted as $D = k_1 k_2 / H$ (rad · s^{-1}/rad), and its physical meaning is the seeker steady state angular velocity D (°/s) per unit seeker detector error 1°. The transfer function of the seeker closed-loop tracking system is:

$$\frac{q_s(s)}{q_t(s)} = \frac{1}{T_s s + 1}\left(T_s = \frac{1}{D}\right). \qquad (5.2-1)$$

The momentum moment of the dynamic gyro-stabilized seeker $H = J\omega$ is usually designed to be very large. Therefore, when the angular motion of the missile body creates a disturbance moment M_d, the corresponding error of the seeker precession angular velocity $\Delta \dot{q} = M_d / H$ will be very small. Since the dynamic gyro provides excellent disturbance rejection capability, the seeker decoupling performance testing is often omitted during the production process for this type of seeker.

Here are some examples of commonly seen dynamic gyro-stabilized seekers.

1) Infrared seeker (IR seeker)

The infrared seeker is a passive system that detects weak target infrared signals. To minimize background clutter, these seekers require a smaller field of view and longer focal length. As a result, most infrared seekers employ a Cassegrain optical design.

(1) Early infrared dynamic gyro seeker (e.g., the American AIM – 9B Sidewinder and the Russian Arrow – 2M).

Early infrared seekers often used uncooled detectors. The absence of a cooling device allowed the detector to be smaller and installed within the seeker's inner gimbal. Consequently, the detector did not need to be positioned at the center of the inner and outer gimbal axes.

(2) Strap-down-cooled infrared dynamic gyro seeker (e.g., the American Sidewinder and Israeli Python3 missile).

The cooled infrared detector assembly, which includes a cooling device, has a larger volume. Typically, the infrared detector is fixed to the missile body and must be installed at the orthogonal center of the inner and outer seeker gimbal axes (Fig. 5.2 – 3).

2) Semi-active laser seeker

In a semi-active laser seeker, the strength of the laser signal reflected by the target is significantly stronger than the background clutter, enabling a larger field of view. Typically, the field of view can range from 15° to 30°, eliminating the need for a Cassegrain structure.

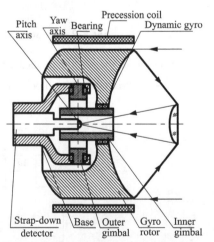

Fig. 5.2 – 3 **Strap-down-cooled infrared detector dynamic gyro seeker**

(1) The laser semi-active dynamic gyro seeker of Hellfire missiles.

The American Hellfire, a helicopter-mounted laser semi-active air-to-surface guided missile, features a four-quadrant laser detector. This detector is mounted in front of the inner gimbal of the seeker's dynamic gyro (Fig. 5.2 −4).

(2) The semi-active dynamic gyro laser seeker of the Russian guided projectile (Fig. 5.2 −5).

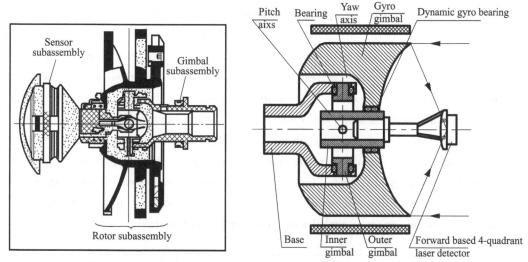

Fig. 5.2 −4 Semi-active dynamic gyro laser seeker of a Hellfire missile

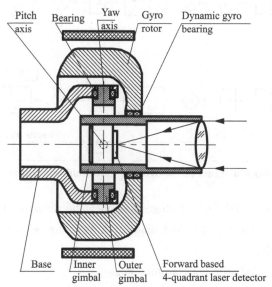

Fig. 5.2 −5 Semi-active dynamic gyro laser seeker
of the Russian guided projectile

3) Dynamic gyro seeker driven by the motor

In this system, the precession moment of the dynamic gyro is generated by two electrical motors connected to the two gimbals of the dynamic gyro. Since the precession direction of the dynamic gyro is oriented 90° from the precession moment, the actual implementation of the dynamic gyro seeker is illustrated in Fig. 5.2 – 6.

4) Dynamic gyro-stabilized seeker with additional tracking gimbal

One disadvantage of the dynamic gyro seeker is that when the seeker has a gimbal angle ϕ, the effective precession moment generated by the precession coil is only $\cos\theta$ times the precession moment when $\phi = 0$. As a result, the seeker cannot work under large gimbal angles (usually the upper limit of the gimbal angle is 40°). To address this issue, a dynamic gyro seeker has been designed with an additional tracking gimbal that aligns with the central axis of the seeker while the precession coil is mounted on the tracking gimbal. This configuration minimizes the angle between the precession moment and the momentum axis of the dynamic gyro, effectively reducing precession moment loss due to large gimbal angles. Consequently, the seeker can operate with a larger gimbal angle, up to 60°. Additionally, the small angle between the seeker load and the tracking gimbal significantly reduces the impact of gimbal angle disturbances on the seeker.

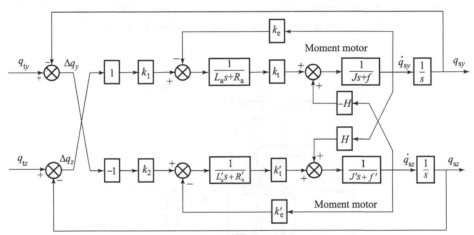

Fig. 5.2 – 6 Block diagram of the dynamic gyro-stabilized seeker control system

Fig. 5.2 – 7(a), (b), and (c) illustrate the schematic diagram of the dynamic gyro seeker with an additional tracking gimbal, the block diagram of its control system, and the definitions of its related angles, respectively.

In Fig. 5.2 – 7, ϑ——Missile body attitude angle;
q_s——Seeker inertial space angle;
q_g——Tracking gimbal inertial space angle;
ϕ——Angle between the optic axis of the seeker and the missile-axis;
ϕ_s——Angle between the tracking gimbal and the missile-axis;
$\Delta\phi$——Gimbal angle of the seeker's optical axis with respect to the tracking gimbal.

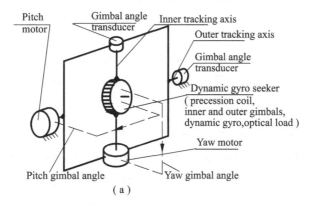

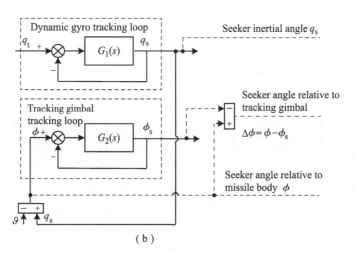

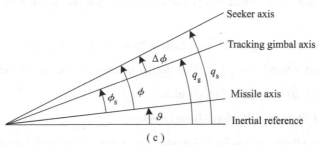

Fig. 5.2-7 Dynamic gyro seeker with an additional tracking gimbal
(a) Schematic diagram of the dynamic gyro seeker with an additional tracking gimbal;
(b) Block diagram of the dynamic gyro seeker control system with an additional tracking gimbal;
(c) Definitions of angles relative to the dynamic gyro seeker with an additional tracking gimbal

5.2.2 Stabilized Platform-Based Seeker

1) Two-gimbal stabilized platform-based seeker

The platform-based seeker is designed with two control loops. The inner high-gain inertial space angular velocity stabilization loop mitigates angular motion disturbances caused by missile body movements and provides the angular velocity output for the missile-target line of sight in inertial space. The outer tracking loop ensures accurate tracking of the missile-target line, as shown in Fig. 5.2 – 8. Compared to the dynamic gyro-stabilized seeker, the stabilized platform-based seeker offers a faster response, accommodates larger gimbal angles, and supports heavier seeker loads. The American GBU – 15 guided missile, the AGM – 130 air-to-ground missile, and the Polyphem fiber optic guided missile (a joint development by France, Germany, and Italy) all utilize image guidance seekers with this configuration. Additionally, the US Sparrow air-to-air missile employs a two-gimbal stabilized platform radar seeker.

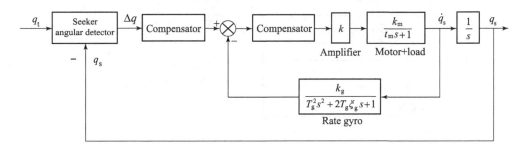

Fig. 5.2 – 8 Block diagram of a stabilized platform-based seeker

Fig. 5.2 – 9 depicts a two-gimbal stabilized platform-based radar seeker designed by Marconi Space Defense System (MSDS) Co., Ltd. This design incorporates a conventional permanent magnet DC motor paired with a 20∶1 gear reduction unit.

2) Three-gimbal Stabilized Platform-Based Seeker

To counteract the effects of missile roll motion on seeker image stabilization, an additional roll gimbal can be integrated with the two-gimbal platform to provide roll stabilization, as illustrated in Fig. 5.2 – 10. Literature indicates that this structure has been utilized in some Russian air-to-surface image-guided missiles and bombs.

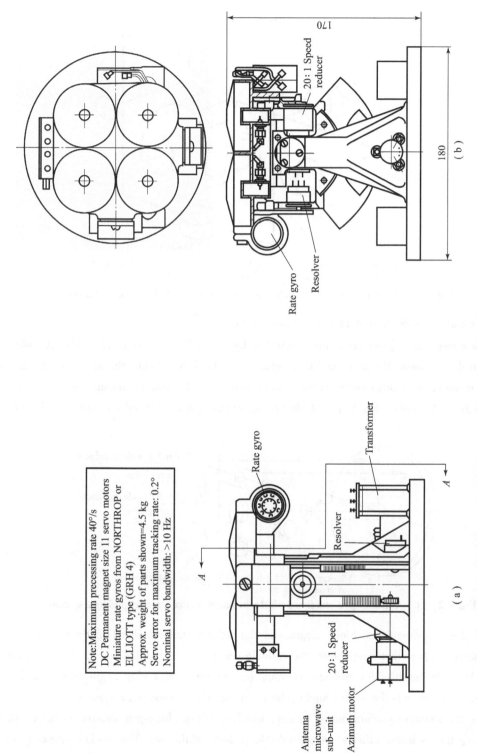

Fig. 5.2-9 Seeker designed by Marconi Space Defense System (MSDS) Co., Ltd (DC motor and angular rate gyro)
(a) View with gyro choke removed; (b) Section A—A with transformer removed

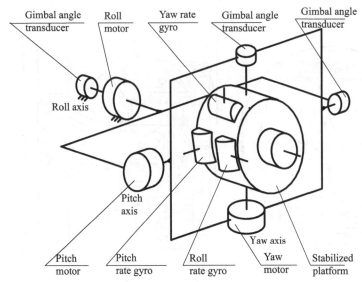

Fig. 5.2 – 10 Schematic of the three-gimbal stabilized platform-based seeker

3) Attitude Gyro-Stabilized Platform-Based Seeker

In this configuration, the error angle detected by the radar is used to drive the attitude gyro mounted on the platform. The inner and outer gimbals of the seeker follow the motion of the attitude gyro axis, meaning the gimbal angles of the attitude gyro serve as control commands for the inner and outer gimbals of the seeker. Fig. 5.2 – 11 shows the attitude gyro-stabilized platform-based seeker.

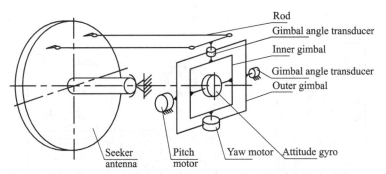

Fig. 5.2 – 11 Attitude gyro-stabilized platform-based seeker (radar guidance system)

Fig. 5.2 – 12 presents the block diagram of the attitude gyro-stabilized platform-based seeker control system. Suppose that x represents the variable relative to the inertial space, $\tilde{x}$ represents the variable relative to the missile coordinate system, q_g represents the seeker angle relative to inertial space, and $\tilde{q}_g$ represents the seeker angle relative to the missile coordinate system.

Unlike the standard platform-based seeker, which employs a high-gain angular velocity loop for stabilization, this scheme utilizes an attitude loop for stabilization. The seeker tracking loop's frequency bandwidth is limited by the low sampling frequency and delay of the target detector. However, the attitude gyro gimbal angle sensor and the seeker detector platform gimbal angle sensor in this angular stabilization loop can have high sampling frequencies. This allows them to

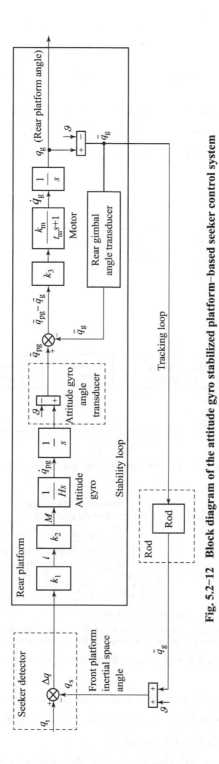

Fig. 5.2-12 Block diagram of the attitude gyro stabilized platform-based seeker control system

create an angular tracking loop with a wide bandwidth, effectively eliminating angular disturbances caused by missile motion and stabilizing the seeker's optical axis.

5.2.3 Detector Strap-down Stabilized Optic Seeker

Unlike the traditional method of stabilizing the detector on a platform, this approach involves strapping the seeker detector directly onto the missile body. The optical axis of the detector is stabilized by rotating a moving mirror, as depicted in Fig. 5.2 – 13. The angle definitions for this design are shown in Fig. 5.2 – 14.

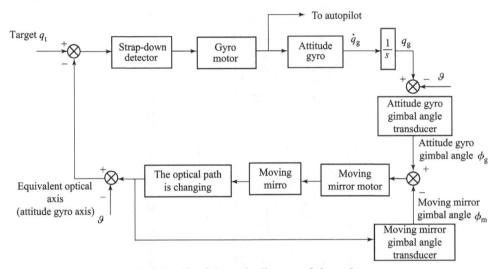

Fig. 5.2 – 13 Schematic diagram of the seeker

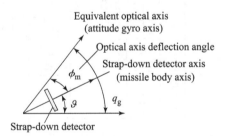

Moving mirror rotation produces optical axis deflection angle ϕ_m to make the optical axis coincide with the detector axis

Fig. 5.2 – 14 Angle definitions of the seeker

In this scheme, the attitude gyro axis serves as the reference axis. The mirror rotates to compensate for changes in the optical axis caused by missile angle variations, ensuring that the optical axis remains aligned with the detector axis.

This seeker uses an angular stabilization loop to decouple missile body motion disturbances, similar to the attitude stabilization loop used in the attitude gyro-stabilized platform-based seeker described in Part 3) of Section 5.2.2.

5.2.4 Semi-strap-down Platform Seeker

In a semi-strap-down platform seeker, the angular velocity gyro is mounted on the inner gimbal of the seeker platform. To reduce the load on the seeker platform and lower the cost, the angular velocity gyro on the seeker platform can be removed. In this design, the angular velocity $\dot{q}_s$ of the seeker platform in the inertial space is obtained by adding the angular velocity $\dot{\vartheta}$ of the missile measured by the onboard inertial navigation system and the derivative $\dot{\phi}$ of the seeker gimbal angle ϕ. This approach significantly reduces the load and size of the seeker platform, allowing the missile to have a more streamlined nose. This reduction in drag enhances the missile's flight speed.

Since only the gyro is strapped down while the seeker detector remains on the platform, this configuration is commonly referred to as a semi-strap-down platform seeker, as illustrated in Fig. 5.2 – 15.

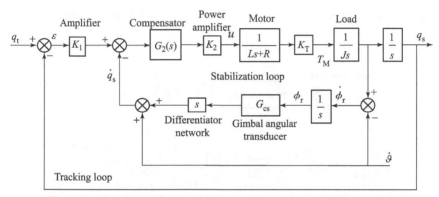

Fig. 5.2 – 15 Block diagram of the semi-strap-down platform seeker

5.2.5 Strap-down Seeker

1) Phased array strap-down seeker

In this design, this phased array antenna is fixed to the missile body (Fig. 5.2 – 16). Here, $x_b y_b z_b$ is the missile coordinate system, $x_p y_p z_p$ is the beam coordinate system, and $x_i y_i z_i$ is the inertial coordinate system. Taking the pitch channel as an example, the relationship between the angles involved is shown in Fig. 5.2 – 17, where θ_B is the beam angle formed by the beam direction x_p and the missile axis x_b, ϑ is the pitch attitude angle of the missile, and ε_p is the phase array antenna measured target LOS error.

The scheme that is used for the phased array strap-down seeker to measure the LOS angular velocity $\dot{q}$ is shown in Fig. 5.2 – 18. Similar to the ordinary platform-based seeker, the seeker output $\dot{q}_s$ is still proportional to the angular error ε_p, between the beam direction and the target direction. However, since the beam pointing control is in relation to the missile coordinate system, the angular velocity $\dot{\vartheta}$ of the missile should be subtracted from the $\dot{q}_s$ output, and then integration should be made to obtain the required phased array beam command θ_{BC} relative to the missile body. The expected beam angle θ_B is obtained by the phase array beam control design. As a result, closed-loop tracking of the target can be effectively achieved.

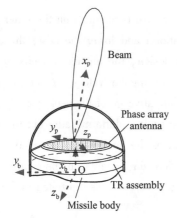

Fig. 5.2 – 16 Structure of the phased array strap-down seeker

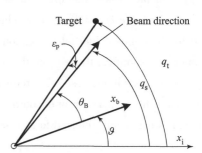

Fig. 5.2 – 17 Relationship of the phased array antenna angles

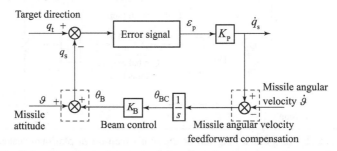

Fig. 5.2 – 18 Control system of the phased array strap-down seeker

It should be noted that the control from the phased array beam command θ_{BC} to the actual beam direction θ_B is an open-loop control. The accuracy of this control is entirely dependent on the calibration process during the phased array antenna production. Given the stringent requirements for seeker decoupling performance, this calibration must be highly precise and involve a significant amount of work. At present, it is known that it is required during production to carry out two-dimensional angular calibration of the relationship between the beam command θ_{BC} and the beam deflection angle θ_B for every phase array operational frequency.

2) Image strap-down seeker

In an image strap-down seeker, the image detector is a highly accurate linear detector. However, when strapped to the missile body, it must have a field of view large enough to cover the gimbal angle range of a conventional seeker. This requirement ensures that the detector can adequately capture the necessary imagery despite the missile's movement. If it can be ensured that the target always remains in the image detector field of view during the whole guidance phase, the angular velocity $\dot{q}$ of the missile-target LOS can be acquired by adding the derivative of the angular error signal ε measured by the detector onboard to the rate gyro output

$\dot{v}$ of the missile (Fig. 5.2-19).

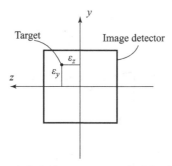

Fig. 5.2-19 Target deviation error acquired by the image detector

That is: $\dot{q} = \dot{v} + \dot{\varepsilon}$.

Fig. 5.2-20 shows the US extra-atmospheric interceptor, which utilizes an infrared image strap-down seeker for its final stage. This missile employs direct force attitude control and trajectory control schemes. Similarly, the Israeli short-range guided missile Spike-SR also uses an infrared strap-down image seeker.

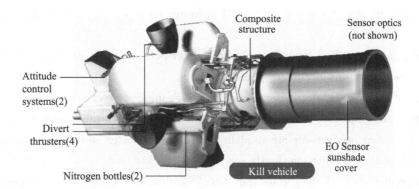

Fig. 5.2-20 EKV extra-atmospheric interceptor used by KEI intercept missile

3) Laser strap-down seeker

Typically, a laser seeker detector employs a four-quadrant scheme to measure target angular error. To ensure accurate angle measurement, the size of the diffused laser image is kept very small, resulting in a limited linear measurement range. However, because the reflected energy of the laser signal is much stronger than background clutter, the detector's nonlinear range can be quite large, allowing for a larger field of view design. Theoretically, increasing the size of the diffused laser image can extend the linear measurement range, making the strap-down seeker scheme feasible for missile guidance, as illustrated in Fig. 5.2-21. However, it should be noted that the accuracy of the angle measurement in this solution is low. That is, the output accuracy of the seeker angular velocity $\dot{q} = \dot{v} + \dot{\varepsilon}$ of the missile-target line depends entirely on the angular error measurement calibration accuracy of each specific seeker detector and its level of sensitivity to environmental change.

However, it is known that this scheme has been applied to some guided bombs and guided rockets that do not require extremely high guidance accuracy. Clearly, the accuracy of angle measurement can be enhanced by increasing the number of detector pixels. For example, the US APKWS 70 mm guided rocket utilizes a seven-pixel laser detector scheme.

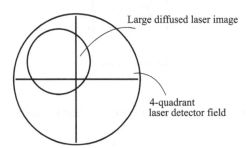

Fig. 5.2-21   Laser strap-down seeker solution

5.2.6　Roll-pitch Seeker

The United States AIM-9X and the European "Meteor" missiles require a 90° gimbal angle to expand their interception boundary. To minimize flight drag and extend operational range, it is crucial to reduce the size of the seeker. Several measures have been implemented to achieve these objectives.

(1) Adoption of the Roll-Pitch Two-Gimbal Seeker Scheme: This design allows for a gimbal angle of 90°, facilitating a wider interception capability.

(2) Use of Strap-down Infrared Image Detectors: To minimize seeker size, strap-down infrared image detectors are employed. Additionally, a missile body-fixed Stirling refrigerator and flexible optical path deflection mechanisms are used to relocate the detector and the J-T refrigerator away from the seeker platform, which previously served as the seeker load.

(3) Implementation of the Semi-Strap-down Stabilization Scheme: This approach involves removing the angular velocity gyro from the seeker platform, further reducing the overall size and weight of the seeker.

If the seeker has a roll gimbal angle ϕ_R and a pitch gimbal angle ϕ_P when using the Kurd optical path, it can be known that if ε_y and ε_z are the pitch and yaw angle errors on the seeker optical lens. ε_y^* and ε_z^* are the pitch and yaw angle errors on the strap-down detector. Then, the following relation exists:

$$\begin{bmatrix} \varepsilon_y^* \\ \varepsilon_z^* \end{bmatrix} = \begin{bmatrix} \cos(\phi_R + \phi_P) & -\sin(\phi_R + \phi_P) \\ \sin(\phi_R + \phi_P) & \cos(\phi_R + \phi_P) \end{bmatrix} \begin{bmatrix} \varepsilon_y \\ \varepsilon_z \end{bmatrix}. \quad (5.2-2)$$

That is to say, the image of the target on the strap-down detector will generate an $(\phi_R + \phi_P)$ angular rotation around the optical axis.

Fig. 5.2-22 shows the image of the target seen from the seeker's optical lens and the same image on the strap-down detector when $\phi_R = -45°$ and $\phi_R = -25°$.

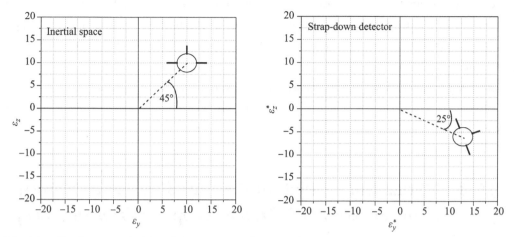

Fig. 5.2-22 Image variations of the target on the seeker's optical lens and the strap-down detector

The traditional seeker stabilization loop aims to stabilize the optical axis and counteract disturbances from missile body motion. In a standard two-gimbal seeker, this decoupling function is achieved through the pitch and yaw gimbal bearings, which reject the missile's pitch and yaw movements. However, in the case of a pitch-roll seeker, the system must perform a roll rotation to align the pitch gimbal with the composite direction of the missile's angular motion. This alignment allows the pitch bearing alone to isolate the missile's two-axis motions, as illustrated in Fig. 5.2-23.

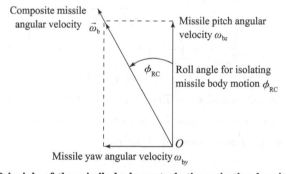

Fig. 5.2-23 Principle of the missile body perturbation rejection by pitch gimbal alone

When the missile undergoes random pitch and yaw angular motions, the roll gimbal must rotate rapidly, which can degrade the seeker's decoupling performance compared to a traditional two-gimbal stabilized seeker. As a result, meeting the design specifications can be challenging even with this approach.

The key challenge is that the primary function of a seeker is target tracking, and the roll gimbal angles needed for effective tracking do not align with those required for missile body motion decoupling. As a result, a pitch-roll seeker cannot perform as effectively as a traditional seeker in practical applications.

As the pitch-roll seeker cannot track the target accurately, a different approach has to be taken to obtain the guidance required for LOS angular velocity $\dot{q}$ output. Suppose that with a proper balance of the target tracking and base decoupling requirements, the seeker optical axis can be controlled to maintain the target in the seeker field view. Then, from the navigation system supplied, missile angular velocity $\dot{\vartheta}$ and $\dot{\phi}$, seeker gimbal angle ϕ_P and ϕ_R and seeker tracking error ε_P and ε_y, the required LOS angular velocity could be derived as below.

As a simple example, in just one plane, the $\dot{q}$ expression can be given as

$$\dot{q} = \omega_i + \dot{\phi}_i + \dot{\varepsilon}_i. \tag{5.2-3}$$

With this scheme, the target image on the strap-down detector may continuously translate and rotate, significantly increasing the complexity of image processing.

§ 5.3 Mechanism Analysis of the Anti-disturbance Moment of the Seeker's Stabilization Loop and Tracking Loop

The platform-based seeker features two main loops: the angular tracking loop and the stabilization loop, as illustrated in Fig. 5.3 – 1. When the seeker has a disturbance moment input M_d, a signal $\dot{q}_s$ will appear in the stabilization loop feedback, and a signal $\dot{q}_c$ will appear in the tracking loop. The sum of the two ($\dot{q}_c - \dot{q}_s$) will produce the required anti-disturbance moment $k_2(\dot{q}_c - \dot{q}_s)$ to balance the disturbance of M_d. With the disturbance M_d as the input, the output transfer functions of $\dot{q}_s$ and $\dot{q}_c$ are respectively expressed as

$$\frac{\dot{q}_c(s)}{M_d(s)} = -\left(\frac{1}{k_2}\right)\frac{1}{\frac{J_y s^2}{k_2 k_1} + \frac{s}{k_1} + 1}, \tag{5.3-1}$$

$$\frac{\dot{q}_s(s)}{M_d(s)} = \left(\frac{1}{k_1 k_2}\right) \times \frac{s}{\frac{J_y s^2}{k_2 k_1} + \frac{s}{k_1} + 1}. \tag{5.3-2}$$

It can be seen that the transmission coefficient for $\dot{q}_c$ is $1/k_2$, and for $\dot{q}_s$ is $1/k_1 k_2$. Therefore, a stabilization loop with high open loop gain k_2 can simultaneously reduce the negative effect of the disturbance moment on the guidance command $\dot{q}_s$ or $\dot{q}_c$.

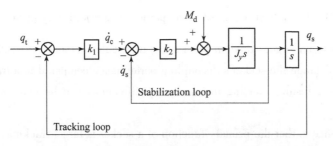

Fig. 5.3 – 1 The platform-based seeker tracking loop

For convenience, the disturbance moment can be converted into a disturbance angular velocity $\dot{q}_d$, $\dot{q}_d = M_d/k_2$. At the same time, take $K_2 = k_2/J_y$, $K_1 = k_1$, and Fig. 5.3 – 1 can now be simplified as Fig. 5.3 – 2.

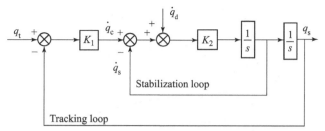

Fig. 5.3-2 Equivalent seeker block diagram with disturbance input

Under the action of disturbance $\dot{q}_d$, the output transfer functions of $\dot{q}_s$ and $\dot{q}_c$ can be expressed as

$$\frac{\dot{q}_c(s)}{\dot{q}_d(s)} = -\frac{1}{\frac{s^2}{K_2 K_1} + \frac{s}{K_1} + 1}. \tag{5.3-3}$$

$$\frac{\dot{q}_s(s)}{\dot{q}_d(s)} = \left(\frac{1}{K_1}\right) \times \frac{s}{\frac{s^2}{K_2 K_1} + \frac{s}{K_1} + 1}. \tag{5.3-4}$$

Since the missile guidance signal $\dot{q}$ can be taken as $\dot{q}_c$ or $\dot{q}_s$, it is clear that the disturbance effect for different guidance signal extraction points is quite different.

It can be seen from the two transfer functions that there is always the following relationship between the stabilization loop command $\dot{q}_c(s)$ and the stabilization loop feedback $\dot{q}_s(s)$.

$$\dot{q}_c(s) = -\frac{K_1}{s}\dot{q}_s(s), \tag{5.3-5}$$

where $\dot{q}_c(s)$ is the anti-disturbance signal from the outer loop, and $-\dot{q}_s(s)$, is the anti-disturbance signal from the stabilization loop. From the relation $\dot{q}_c(s) = (K_1/s)(-\dot{q}_s(s))$ it is clear that the components $\dot{q}_c(s)$ and $-\dot{q}_s(s)$ are always perpendicular to each other, and $\dot{q}_c(s)$ lags behind $-\dot{q}_s(s)$ by 90°. Fig. 5.3-3 gives the vector diagram of $\dot{q}_c(s) - \dot{q}_s(s)$ and the $\dot{q}_d(s)$ disturbance.

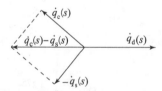

Fig. 5.3-3 Vector synthesis of the control moment

Take the parameters of a typical platform seeker, $K_1 = 10$ and $K_2 = 120$, for example. Fig. 5.3-4 shows the Bode diagram of its $\dot{q}_c(s)/\dot{q}_d(s)$ and $-\dot{q}_s(s)/\dot{q}_d(s)$ transfer functions.

It is known from $\dot{q}_c(s) = -K_1\dot{q}_s(s)/s$ that when $\omega < K_1$ rad/s (for low-frequency disturbance), the decoupling level of $\dot{q}_s(s)$ is superior to $\dot{q}_c(s)$, when $\omega \approx K_1$ rad/s, $\dot{q}_s(s)$ and $\dot{q}_c(s)$ will have equivalent decoupling levels; when $\omega > K_1$ rad/s (for high-frequency disturbance), the decoupling level of $\dot{q}_c(s)$ is better than that of $\dot{q}_s(s)$. The above conclusions can be clearly seen from the decoupling levels of the two at three frequencies of low frequency ($\omega = K_1/2$ rad/s), medium frequency ($\omega = K_1$ rad/s), and high frequency ($\omega = 2K_1$ rad/s) (Table 5.3-1) and the anti-disturbance vector diagram of $\dot{q}_c - \dot{q}_s$ (Fig. 5.3-5).

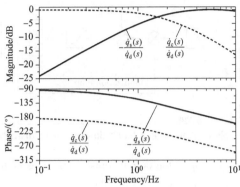

Fig. 5.3-4 Bode diagram of the disturbance effect for different guidance signal extraction points

Table 5.3-1 Coupling level for different disturbance frequencies

Frequency/(rad · s^{-1})	Coupling level	
	$|\dot{q}_c/\dot{q}_d|$	$|\dot{q}_s/\dot{q}_d|$
Low frequency $\omega = K_1/2 = 5$	0.909	0.455
Medium frequency $\omega = K_1 = 10$	0.737	0.737
High frequency $\omega = 2K_1 = 20$	0.474	0.949

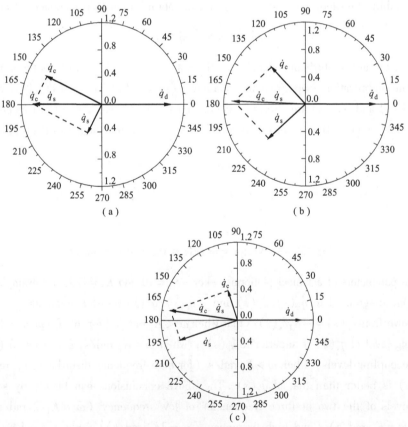

Fig. 5.3-5 Vector synthesis of the disturbance moment decoupling mechanism
(a) $\omega = 5$ rad/s; (b) $\omega = 10$ rad/s; (c) $\omega = 20$ rad/s

Given that there is no disturbance moment input and the seeker is tracking a target $\dot{q}_t$ with a high-bandwidth stabilization loop, the condition will be $\dot{q}_c \approx \dot{q}_s$. That is to say that there is no difference in taking $\dot{q}_s$ or $\dot{q}_c$ as the guidance command, but if disturbance exists, taking $\dot{q}_s$ or $\dot{q}_c$ as the guidance signal will be totally different, so its effects on the missile guidance performance are different. This will be one of the main topics addressed in the following section.

§5.4 Transfer Function of Body Motion Coupling and the Parasitic Loop

5.4.1 Transfer Function of Body Motion Coupling

The block diagram of the seeker, considering the influence of missile body motion coupling disturbance moments, is depicted in Fig. 5.4 – 1.

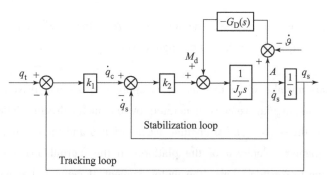

Fig. 5.4 – 1 Seeker block diagram under the influence of missile
body motion coupling disturbance moments

In Fig. 5.4 – 1, $G_D(s)$ is the transfer function of the disturbance moment M_d caused by the seeker-missile body relative motion. Usually the decoupling level of the platform-based seeker is very high, and the output signal $\dot{q}_s$ of the stabilization loop caused by missile angular velocity is small compared with missile angular velocity $\dot{\vartheta}$, which is usually not more than 5%. Therefore, the influence of the change of $\dot{q}_s$ at point A in Fig. 5.4 – 1 can be neglected, and a basic model for studying the effect of the missile body disturbance is shown in Fig. 5.4 – 2.

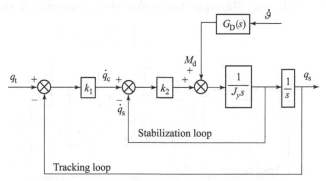

Fig. 5.4 – 2 Equivalent seeker block diagram

It is important to note that the seeker block diagram model, as shown in Fig. 5.4-3, has been widely adopted in many published guidance control books and reference papers, particularly those from early Russian publications.

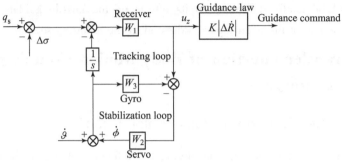

Fig. 5.4-3 Model of taking the motor output angular velocity as the relative angular velocity

In Fig. 5.4-3, $\Delta\sigma$ is the radome refraction error, $\dot{\vartheta}$ is the missile angular velocity, q_s is the LOS angle of the target, u_z is the seeker LOS angular velocity output, and ϕ is the antenna gimbal angle.

There is a serious mistake when this model is applied to a motor-driven platform. Here, the angular acceleration and angular velocity generated by the motor-driven platform are respectively mistakenly regarded as the seeker gimbal angular acceleration $\ddot{\phi}$ and the angular velocity $\dot{\phi}$, and it is considered that the angular velocity $\dot{q}_s$ of the platform in the inertial space is the sum of $\dot{\phi}$ and $\dot{\vartheta}$. Actually, according to Newton's Law, the motor moment M generated seeker platform's angular acceleration and angular velocity are relative to the inertial space rather than to the missile body coordinate system. It is important to note that the missile body's motion affects the seeker only through wire moments, bearing moments, and other forces arising from the relative motion between the seeker and the missile. In the absence of these moment disturbances, the seeker is unaffected by the missile's motion due to the decoupling function of the gimbal bearings.

Based on this analysis, the seeker block diagram in Fig. 5.4-2 accurately reflects the missile body motion coupling characteristics. For simplification, the influence of the disturbance moment M_d can be changed to a disturbance angular acceleration input $\ddot{q}_d$, i.e., $\ddot{q}_d = M_d/J_y$, $\overline{G}_D(s) = G_D(s)/J_y$, $K_2 = k_2/J_y$ and $K_1 = k_1$. The equivalent seeker block diagram is shown in Fig. 5.4-4.

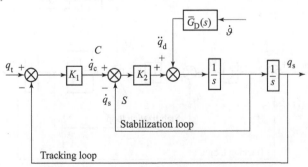

Fig. 5.4-4 Equivalent seeker block diagram

According to this model, the transfer function from $\dot{\vartheta}$ to $\Delta \dot{q}_s$ is

$$-\frac{\Delta \dot{q}_s(s)}{\dot{\vartheta}(s)} = \left(\frac{\bar{G}_D(s)}{K_1 K_2}\right) \times \frac{s}{\dfrac{s^2}{K_2 K_1} + \dfrac{s}{K_1} + 1}, \qquad (5.4-1)$$

and the transfer function from $\dot{\vartheta}$ to $\Delta \dot{q}_c$ is

$$\frac{\Delta \dot{q}_c(s)}{\dot{\vartheta}(s)} = -\left(\frac{\bar{G}_D(s)}{K_2}\right) \times \frac{1}{\dfrac{s^2}{K_2 K_1} + \dfrac{s}{K_1} + 1}. \qquad (5.4-2)$$

From the difference between the two transfer functions, it is indicated that regardless of the disturbance moment model $\bar{G}_D(s)$, There is always the following relationship between $\dfrac{\Delta \dot{q}_c(s)}{\dot{\vartheta}(s)}$ and $\dfrac{\Delta \dot{q}_s(s)}{\dot{\vartheta}(s)}$

$$\frac{\Delta \dot{q}_c(s)}{\dot{\vartheta}(s)} = -\frac{K_1}{s} \cdot \frac{\Delta \dot{q}_s(s)}{\dot{\vartheta}(s)}.$$

That is, the two have an equivalent order of magnitude decoupling level when $\omega \approx K_1$ rad/s, and the decoupling level for the stabilization loop command $\Delta \dot{q}_c$ is better than that of the stabilization loop feedback $\Delta \dot{q}_s$ at a high-frequency range (when $\omega > K_1$ rad/s). While the conclusion is the opposite in the low-frequency range (when $\omega < K_1$ rad/s), and the $\Delta \dot{q}_s$ always lags behind the $\Delta \dot{q}_c$ by 90°.

For an actual seeker, the disturbance moment model varies greatly depending on the different seeker structure designs, and it can be known only by specific design identification. In the following, the damping moment will be taken as the disturbance model to analyze the decoupling problem and its influence on the missile guidance parasitic loop when the seeker takes the stabilization loop command $\dot{q}_c$ or stabilization loop feedback $\dot{q}_s$ as its output. The damping disturbance is associated with the damping moment produced by the seeker bearings and the connecting wires as the missile moves relative to the seeker. Generally, this disturbance moment can be expressed as

$$\ddot{q}_d(s) = K_\omega \dot{\vartheta}(s). \qquad (5.4-3)$$

In Equation (5.4-3), $K_\omega = k_\omega / J_y$, and the equivalent diagram of the seeker control system is given in Fig. 5.4-5.

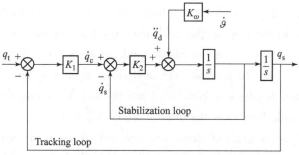

Fig. 5.4-5 Equivalent diagram of the seeker control system

Therefore, the transfer function from $\dot{\vartheta}(s)$ to $\Delta \dot{q}_s$ becomes

$$\frac{\Delta \dot{q}_s(s)}{\dot{\vartheta}(s)} = \left(\frac{K_\omega}{K_1 K_2}\right) \times \frac{s}{\frac{s^2}{K_2 K_1} + \frac{s}{K_1} + 1}. \quad (5.4-4)$$

The transfer function from $\dot{\vartheta}(s)$ to $\Delta \dot{q}_c$ is

$$\frac{\Delta \dot{q}_c(s)}{\dot{\vartheta}(s)} = -\left(\frac{K_\omega}{K_2}\right) \times \frac{1}{\frac{s^2}{K_2 K_1} + \frac{s}{K_1} + 1}. \quad (5.4-5)$$

Take the tracking loop parameter $K_1 = 10$ and the stabilization loop parameter $K_2 = 169$. Suppose $K_\omega = 1$ and the Bode diagram of the above second-order transfer function can be obtained (Fig. 5.4-6). It can be seen that when the disturbance moment is the damping moment, the relationship between the amplitude and the phase of the two is still consistent with our conclusions about the general disturbance moment $\bar{G}_D(s)$.

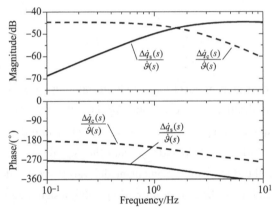

Fig. 5.4-6　Bode diagram of the coupling level transfer function at different output points for a seeker under the damping moment disturbance

5.4.2　Seeker-missile Coupling Introduced Guidance Parasitic Loop

It is known that in guidance loop operation, the seeker output $\dot{q}$ is used to generate an acceleration command a_c for the missile autopilot through the guidance law $a_c = NV_c \dot{q}$. For any seeker without disturbance moments, its output is independent of the missile's angular motion. However, when a disturbance moment is produced by the missile motion $\dot{\vartheta}$, it generates a seeker disturbance output $\Delta \dot{q}$ through its coupling model. This seeker-missile coupling output $\Delta \dot{q}$ will produce an inner parasitic loop in the guidance loop. The presence of this parasitic loop can significantly impact the performance of the missile guidance system. Fig. 5.4-7 illustrates the block diagram of this parasitic loop. To facilitate the analysis of control system stability using negative feedback theory, negative feedback is introduced into the loop. Consequently, the coupling transfer function within the parasitic loop takes the form $-\Delta \dot{q}(s)/\dot{\vartheta}(s)$.

In Fig. 5.4-7, T_α is the angle of attack time constant, N is the effective navigation ratio, V_c is the missile-target relative velocity, V_m is the missile flight velocity, a_c is the acceleration command of the autopilot, and a_m is the acceleration response of the missile. According to the previous analysis, the coupling transfer function exhibits different characteristics depending on the extraction points of the seeker guidance signal. Fig. 5.4-8 and Fig. 5.4-9 illustrate the seeker parasitic loop, with each figure representing the loop composed of different coupling transfer functions corresponding to various output points of the seeker.

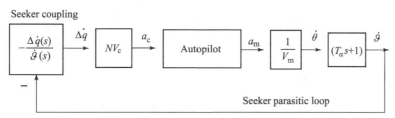

Fig. 5.4-7 Parasitic loop of the seeker coupling

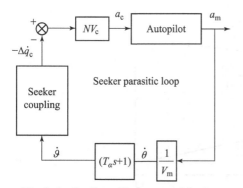

Fig. 5.4-8 Parasitic loop model when extracting the seeker guidance signal at point C

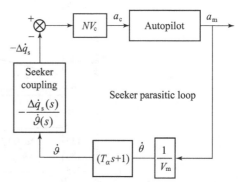

Fig. 5.4-9 Parasitic loop model when extracting the seeker guidance signal at point S

In Fig. 5.4-8 and Fig. 5.4-9, $\Delta\dot{q}_c$ is the coupling LOS angular velocity generated when the guidance signal is extracted from point C (the seeker stabilization loop command) and $\Delta\dot{q}_s$ is the coupling LOS angular velocity generated when the guidance signal is extracted from point S (the seeker stabilization loop feedback).

From the previous analysis, it is known that as $\dot{q}_c(s) = -(K_1/s)\dot{q}_s(s)$, point C and point S have equivalent decoupling levels under the disturbance only when $\omega = K_1 \text{rad/s}$.

For the coupling model (Fig. 5.4-10), as an example, take the loop parameters as $K_1 = 10$ and $K_2 = 169$. When $\omega = K_1 = 10$ (1.6 Hz), the coupling level or magnitude is the same no matter whether the guidance signal is extracted from point C or point S. As the damping moment coefficient $K_\omega = 7.86$, at the 1.6 Hz, both of the coupling levels are 3%. The coupling level transfer function for different extraction points C, S, and different frequency ω are shown in Fig. 5.4-11.

Comparison of the Bode diagrams for transfer functions $-\Delta\dot{q}_c(s)/\dot{\vartheta}(s)$ and $-\Delta\dot{q}_s(s)/\dot{\vartheta}(s)$ shows that, in terms of magnitude, the decoupling performance is superior for signals extracted from point S in the low-frequency region, while point C performs better in the high-frequency region. Additionally, the phase characteristics indicate that the signal from point S has a 90° phase lag compared to the signal from point C. This phase lag can negatively impact the stability of the parasitic loop.

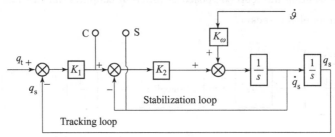

Fig. 5.4 – 10   Seeker coupling block diagram

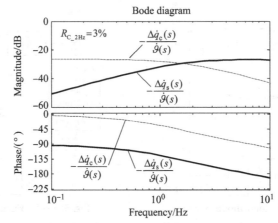

Fig. 5.4 – 11   Coupling transfer function for the given example

Based on the above coupling transfer function model, suppose the seeker disturbance moment parameter is R_ω ($R_\omega = K_\omega/K_2$), and the parasitic loop non – dimensional parameters are taken as $\overline{T}_\alpha$ and $NV_c R_\omega/V_m$ (Fig. 5.4 – 12). Fig. 5.4 – 13 shows the stable region and unstable region of the parasitic loop, taking non – dimensional $\overline{T}_\alpha$ and $NV_c R_\omega/V_m$ as parameters. Here $\overline{s} = T_g s$, $\overline{T}_\alpha = \dfrac{T_\alpha}{T_g}$ and T_g is the system guidance time constant.

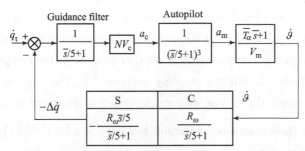

Fig. 5.4 – 12   Seeker parasitic loop with damping moment coupling

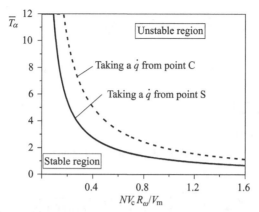

Fig. 5.4-13 Stable region and unstable region of the parasitic loop with damping moment coupling

§5.5 A Real Seeker Model and Testing Methods

5.5.1 A Real Seeker Model

The impact of the seeker parasitic loop on missile guidance performance can be effectively illustrated using a real-world seeker model that incorporates parasitic loops. When there is no parasitic loop in the seeker, the guidance block diagram from the target $\dot{q}_t$ to the autopilot output a_m is shown in Fig. 5.5-1. When a parasitic loop exists, the modified block diagram is shown in Fig. 5.5-2.

Fig. 5.5-1 Guidance block diagram from $\dot{q}_t$ to a_m with no parasitic loop

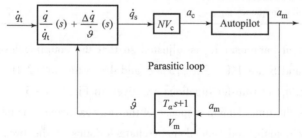

Fig. 5.5-2 Modified guidance block diagram from $\dot{q}_t$ to a_m with parasitic loop

The real seeker model $\dot{q}_s(s)/\dot{q}_t(s)$ or $\dot{q}_c(s)/\dot{q}_t(s)$ can be derived through an equivalent transformation of Fig. 5.5-2, as shown in Fig. 5.5-3.

It can be observed from Fig. 5.5-4 that when the real seeker model is used to replace the seeker model with no coupling disturbance, the block diagram from $\dot{q}_t$ to a_m will be the same as before.

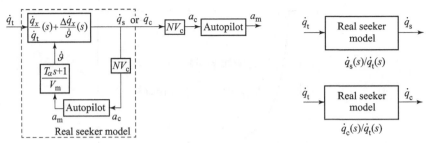

Fig. 5.5 – 3 Real seeker model

Fig. 5.5 – 4 Guidance model after introducing a real seeker model

The next step is to analyze the effect of the parasitic loop on the real seeker model, using damping disturbance as an example. The parameters for the seeker, guidance law, and autopilot are detailed in Table 5.5 – 1.

Table 5.5 – 1 Typical seeker model parameters

Parameters	Values
Tracking loop K_1	10 rad/s (2 Hz)
Stabilization loop K_2	169 rad/s (27 Hz)
Relative velocity V_c	1200 m/s
Missile velocity V_m	800 m/s
Angle of attack time constant T_α	1 s
Navigation ratio N	4
Autopilot damping ratio ζ_b	0.65
Autopilot frequency ω_b	2 Hz

The damping moment parameter K_ω is adjusted so that the coupling levels $\Delta\dot{q}/\dot{\vartheta}$ of the two signal output points C and S are 1%, 2%, 3%, and 4% when $\omega = 2$ Hz. The Bode diagrams of $\dot{q}_c(s)/\dot{q}_t(s)$ and $\dot{q}_s(s)/\dot{q}_t(s)$ transfer functions are given in Fig. 5.5 – 5.

Fig. 5.5 – 6 shows the time domain response of the two real seeker models to a step input signal $\dot{q}_t$. From the frequency domain and time domain characteristics of the two, it is evident that the seeker maintains good stability when the signal is taken from point C. However, the presence of a parasitic loop leads to a decrease in low-frequency gain. This phenomenon does not occur when the $\dot{q}$ signal is taken from point S, but the seeker will have less stability.

Based on the analysis, it is clear that the parasitic loop of the seeker at different output extraction points affects both the time domain and frequency domain responses differently, thus impacting guidance performance in varying ways.

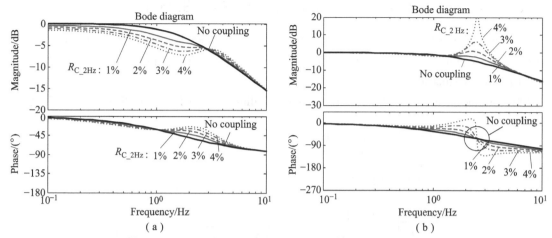

Fig. 5.5-5 Bode diagrams of transfer functions under damping moment parameter K_ω adjustment

(a) $\dot{q}_c(s)/\dot{q}_t(s)$ at output point C; (b) $\dot{q}_s(s)/\dot{q}_t(s)$ at output point S

In practical seeker design, the real coupling moment model can be highly complex depending on the specific platform design, making it nearly impossible to simulate accurately with a simple model. Consequently, hardware-in-the-loop simulation is commonly employed in real guidance system design to assess the missile-seeker coupling effects on overall system performance.

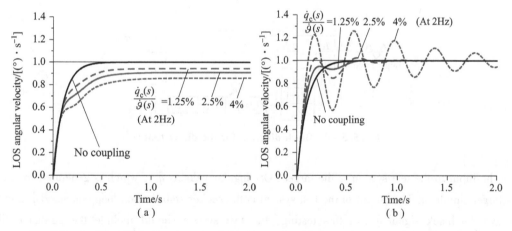

Fig. 5.5-6 Time domain response of the two real seeker models to a step input signal $\dot{q}_t$

(a) Extraction point C; (b) Extraction point S

Owing to the restricted resources, only a limited number of hardware-in-the-loop simulations can be performed. But if a real seeker model $\dfrac{\dot{q}_s(s)}{\dot{q}_t(s)}$ or $\dfrac{\dot{q}_c(s)}{\dot{q}_t(s)}$ can be obtained via test identification, the guidance system performance evaluation can then be performed by way of mathematical simulation. Furthermore, the mathematical simulation results can help to determine which seeker output $\dot{q}_s(s)$ or $\dot{q}_c(s)$ should be chosen in the final guidance system implementation.

5.5.2 Testing Methods for Modeling the Real Seeker

Based on the preceding analysis, two testing methods can be employed to obtain the real seeker's transfer function, incorporating the effects of missile-seeker coupling and parasitic loops.

1) Direct testing method

The arrangement for the direct testing method is illustrated in Fig. 5.5-7 and Fig. 5.5-8.

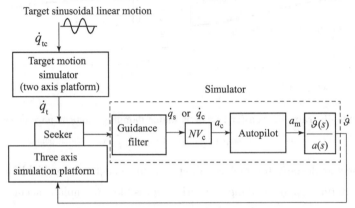

Fig. 5.5-7 Schematic of the direct testing method arrangement

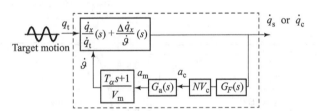

Fig. 5.5-8 Block diagram for the direct testing

The input to the test system is the target LOS angular velocity $\dot{q}_t$, which is generated by a two-axis target simulator. The output of the test system is the seeker stabilization loop command $\dot{q}_c$ and the rate gyro feedback signal $\dot{q}_s$. In this testing, the test system should include the guidance filter model, the guidance law model, and the autopilot model installed in the simulation computer, and the simulation computer output (the missile attitude angular velocity $\dot{\vartheta}$ and angle ϑ) should be used to drive the three-axis simulation platform.

2) Indirect testing method

This method tests both the seeker model with no coupling $\dfrac{\dot{q}_x(s)}{\dot{q}_t(s)}$ and the coupling model $\dfrac{\Delta \dot{q}_c(s)}{\dot{\vartheta}(s)}$ and $\dfrac{\Delta \dot{q}_s(s)}{\dot{\vartheta}(s)}$ separately, and combining the two will give the required result $\dfrac{\Delta \dot{q}_c(s)}{\dot{q}_t(s)}$ or $\dfrac{\Delta \dot{q}_s(s)}{\dot{q}_t(s)}$ (Fig 5.5-9).

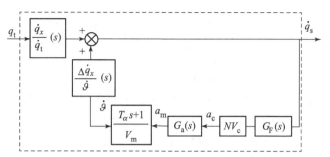

Fig. 5.5 – 9 Schematic of the indirect testing method

Through the seeker transfer function testing, the seeker transfer function without coupling can be obtained as

$$G_s(s) = \frac{\dot{q}(s)}{\dot{q}_t(s)}. \tag{5.5-1}$$

Through the missile seeker coupling testing, the coupling transfer function can be obtained as

$$R_{\text{DRE}}(s) = \frac{\Delta\dot{q}_c(s)}{\dot{\vartheta}(s)} \text{ or } \frac{\Delta\dot{q}_s(s)}{\dot{\vartheta}(s)}. \tag{5.5-2}$$

Finally, the real seeker transfer function, including the coupling parasitic loop, can be expressed as

$$R_{\text{seeker}}(s) = \frac{\dfrac{\dot{q}_x(s)}{\dot{q}_t(s)}}{1 + \dfrac{\Delta\dot{q}(s)}{\dot{\vartheta}(s)} G_a(s) G_F(s)(T_\alpha s + 1)\dfrac{NV_c}{V_m}}, \tag{5.5-3}$$

where $G_F(s)$ is the guidance filter model and $G_a(s)$ is the autopilot model.

§ 5.6 Other Parasitic Loop Models

5.6.1 Parasitic Loop Model for a Phase Array Strap-down Seeker

Using the block diagram of the phased array strap-down seeker shown in Fig. 5.6 – 1, the seeker output can be expressed as follows

$$\dot{q}_s(s) = \frac{\dfrac{1}{K_B}}{\dfrac{1}{K_B K_P}s + 1}\dot{q}_t + \frac{1 - \dfrac{1}{K_B}}{\dfrac{1}{K_P K_B}s + 1}\dot{\vartheta}. \tag{5.6-1}$$

According to Equation (5.6 – 1), when the beam control gain is $K_B = 1$, the missile angular velocity $\dot{\vartheta}$ disturbance to the seeker can be completely eliminated from the seeker output $\dot{q}_s$. At this time, the transfer function of the phased array strap-down seeker will be simply as follows

$$\frac{\dot{q}_s(s)}{\dot{q}_t(s)} = \frac{\frac{1}{K_B}}{\frac{1}{K_B K_P}s + 1} \approx \frac{1}{T_s s + 1} \quad \left(T_s = \frac{1}{K_P}\right), \quad (5.6-2)$$

where K_B is the beam control gain. Since the beam control from its command θ_{BC} to the beam reflection angle θ_B is an open loop control, the gain K_B can only be obtained through the strict calibration for different beam angles and operation frequencies. K_B can be adjusted to 1 in theory by computer compensation after calibration. When the gain K_B cannot be perfectly compensated to 1, a parasitic loop will be generated, and its effect on the seeker output is given by

$$\frac{\Delta \dot{q}_s(s)}{\dot{\vartheta}(s)} = \frac{1 - \frac{1}{K_B}}{T_s s + 1}. \quad (5.6-3)$$

When an error exists in the gain K_B, $K_B = 1 + \Delta K_B$, in which ΔK_B is the calibration error, that is $\Delta K_B = \frac{\theta_B - \theta_{BC}}{\theta_{BC}}$. Therefore

$$1 - \frac{1}{K_B} = 1 - \frac{1}{1 + \Delta K_B} \approx 1 - (1 - \Delta K_B) = \Delta K_B. \quad (5.6-4)$$

So, the coupling transfer function will be

$$\frac{\Delta \dot{q}_s(s)}{\dot{\vartheta}(s)} = \frac{\Delta K_B}{T_s s + 1}. \quad (5.6-5)$$

Since this decoupling model is analogous to the radome slope error model, its influence is analyzed in the following section.

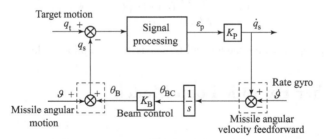

Fig. 5.6-1 Block diagram of the phase array strap-down seeker

5.6.2 Parasitic Loop Due to Radome Slope Error

When the target echo passes through the missile radome, variations in the dielectric constant and the radome's conical shape can cause deviations from a straight-line path. As this happens, the seeker will track a false target with a tracking error Δq, as shown in Fig. 5.6-2.

Usually, the error of the Δq model is set to be proportional to the seeker gimbal angle ϕ, that is $\Delta q = R_{dom} \phi$, where R_{dom} is the random error coefficient.

When the seeker tracks the target in the inertial space, and the missile has a certain angular velocity $\dot{\vartheta}$, the seeker gimbal angular will be $\phi = -\vartheta$ and $\dot{\phi} = -\dot{\vartheta}$.

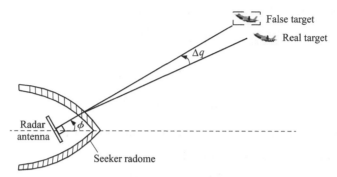

Fig. 5.6−2 Mechanism diagram of radome beam pointing error

Therefore, considering the beam pointing error caused by the radome, the seeker model is illustrated in Fig. 5.6−3, with a simplified version shown in Fig. 5.6−4.

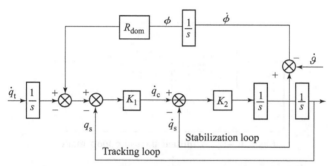

Fig. 5.6−3 Seeker model with radome slope error effect

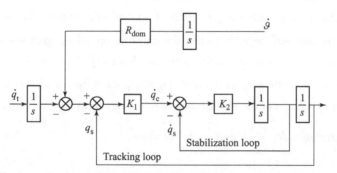

Fig. 5.6−4 Simplified seeker model with radome slope error effect

Therefore, the coupling transfer function caused by the radome error R_{dom} will be

$$\frac{\Delta \dot{q}_c(s)}{\dot{\vartheta}(s)} \approx \frac{\Delta \dot{q}_s(s)}{\dot{\vartheta}(s)} = \frac{-R_{\text{dom}}}{\frac{1}{K_1}s + 1} = \frac{-R_{\text{dom}}}{T_s s + 1}. \qquad (5.6-6)$$

5.6.3 Beam Control Gain Error ΔK_B of the Phased Array Seeker and the Radome Slope Error R_{dom} Effect on the Seeker's Performance

It is seen from the above two subsections that the seeker coupling caused by ΔK_B of the phased array and R_{dom} of the radome slope error are respectively

$$\frac{\Delta \dot{q}_c(s)}{\dot{\vartheta}(s)} = \frac{\Delta K_B}{T_s s + 1}, \qquad (5.6-7)$$

$$\frac{\Delta \dot{q}_c(s)}{\dot{\vartheta}(s)} = \frac{-R_{dom}}{T_s s + 1}. \qquad (5.6-8)$$

It can be seen that both transfer functions share the same form. Therefore, their influence on the parasitic loop can be represented by a unified transfer function

$$\frac{\Delta \dot{q}_c(s)}{\dot{\vartheta}(s)} = \frac{-R}{T_s s + 1}. \qquad (5.6-9)$$

In the above-unified transfer function, $R = R_{dom} = -\Delta K_B$.

After incorporating this coupling model, the parasitic loop model is depicted in Fig. 5.6-5.

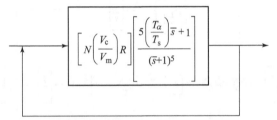

Fig. 5.6-5   A unified parasitic loop model

From Fig. 5.6-5, it can be seen that when the coupling coefficient R is negative, the parasitic loop functions as a negative feedback, and when R is positive, the parasitic loop operates as a positive feedback. Due to the lead compensation of the missile transfer function ($T_\alpha s + 1$), the positive feedback case may still be stable when the parasitic open loop gain is low. However, for the same absolute value $|R|$, the parasitic loop is definitely less stable as the feedback is positive. When R is taken both as positive and negative, its different effects on the parasitic loop stability can be clearly seen from the stabilization region diagram in Fig. 5.6-6, which is given with non-dimensional parameters $N\left(\dfrac{V_c}{V_m}\right)R$ and $\dfrac{T_\alpha}{T_g}$ as variables.

Fig. 5.6-6 shows that the parasitic loop stability is poor when the parameter $\dfrac{T_\alpha}{T_g}$ is high. At high altitudes, the value of the angle of attack time constant T_α is large. To avoid the instability problem of the parasitic loop, some missiles adopt an approach of reducing the autopilot bandwidth at high altitudes to increase the value of T_g, so that the value of $\dfrac{T_\alpha}{T_g}$ is not too large when the missile enters high altitude. Another parameter, $N\left(\dfrac{V_c}{V_m}\right)R$, shows that the increase of the proportional navigation constant N, the relative velocity V_c to the missile velocity V_m ratio $\dfrac{V_c}{V_m}$ as well as the coupling coefficient R will all have a negative effect on the seeker parasitic loop stability.

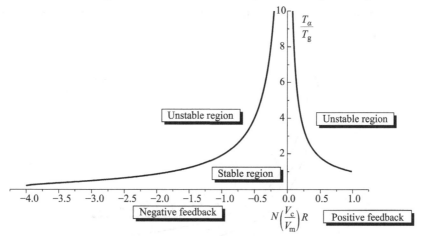

Fig. 5.6-6 Stabilization region diagram of the parasitic loop with $N\left(\dfrac{V_c}{V_m}\right)R$ and $\dfrac{T_\alpha}{T_g}$ as variables

The above analysis of the parasitic loop is based on the assumption that $t_{go} = \infty$ and the gain of the guidance loop is zero (Fig. 5.6-7). This means that the guidance loop does not affect the stability of the parasitic loop. However, when the missile is approaching the target and t_{go} is not large, the presence of the guidance loop can indeed reduce the stability of the parasitic loop. Fig. 5.6-7 shows the block diagram for the stability analysis of the parasitic loop in the presence of a guidance loop.

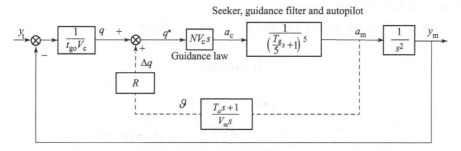

Fig. 5.6-7 Block diagram of the stabilization region analysis of the parasitic loop with a guidance loop

Taking the same non-dimensional parameters $N\left(\dfrac{V_c}{V_m}\right)R$ and $\dfrac{T_\alpha}{T_g}$ as variables and the non-dimensional time-to-go $\dfrac{t_{go}}{T_g}$ as parameters, the new stabilization region diagram is given as Fig. 5.6-8 (hereby taking $\dfrac{t_{go}}{T_g} = \infty, 10, 5$). From Fig. 5.6-8, it can be observed that the stability of the parasitic loop deteriorates as the missile approaches the target (i.e., as $\dfrac{t_{go}}{T_g}$ gets smaller).

It should be noted that as t_{go} changes, the guidance loop becomes a time-varying system, and it is wrong to analyze the stability of the parasitic loop with a time-invariant system analysis method. However, for qualitative analysis, the conclusion that small t_{go} will reduce the parasitic loop stability is still qualitatively correct.

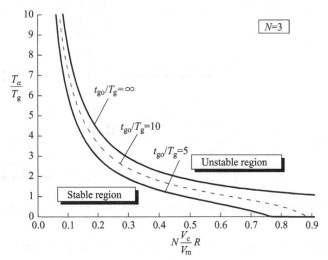

Fig. 5.6-8 Influence of guidance loops on the stabilization region of the parasitic loop

5.6.4 A Novel Online Estimation and Compensation Method for Strap-down Phased Array Seeker Disturbance Rejection Effect Using Extended State Kalman Filter

5.6.4.1 Mathematical model

The basic missile-target geometry is illustrated in Fig. 5.6-9.

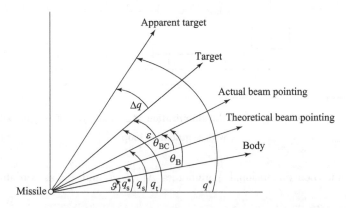

Fig. 5.6-9 The basic missile-target geometry model of strap-down phased array radar seeker

q_t—The true LOS angle; q_s—The LOS angular rate; q^*—The apparent LOS angle; θ_B—The beam pointing angle; θ_{BC}—The beam pointing angle instruction; ε—The tracking error angle; $\dot{\vartheta}$—The body moving rate; ϑ—The attitude angle

The apparent LOS angle q^* can be expressed as

$$q^* = q_t + (q_s - \vartheta)R. \tag{5.6-10}$$

In the stable tracking situation, the tracking error angle ε can be ignored, and the radome error

slope R is far less than 1 in practice. Hence, Equation (5.6-10) can be substituted into
$$q^* \cong q_t + (q_t - \vartheta)R = q_t(1 + R) - \vartheta R = q_t - \vartheta R. \qquad (5.6-11)$$

5.6.4.2 Designing the online estimation and compensation method for disturbance rejection effect (DRE) in the parasitic loop

1) ESKF (Error-State Kalman Filter) design

Consider the following class of nonlinear time-varying uncertain systems,
$$\begin{cases} X_{k+1} = \bar{A}_k X_k + \bar{B}_k F(X_k, k) + w_k \\ Z_k = \bar{C}_k X_k + n_k, k = 0, 1, \ldots, \end{cases} \qquad (5.6-12)$$

where $X_k \in \mathbb{R}^n$ is the state; $\bar{A}_k$, $\bar{B}_k$, and $\bar{C}_k$ are known time-varying matrixes with $\bar{A}_k \in \mathbb{R}^{n \times n}$, $\bar{B}_k \in \mathbb{R}^{n \times l}$, $\bar{C}_k \in \mathbb{R}^{m \times n}$; $F(\bar{X}_k, k) \in \mathbb{R}^l$ is the nonlinear uncertain dynamics in the system, and its nominal model is the known function $\bar{F}(X_k, k)$; $w_k \in \mathbb{R}^n$ and $n_k \in \mathbb{R}^m$ are the process noise and measurement noise respectively; $Z_k \in \mathbb{R}^m$ is the measurement output.

In the model, the uncertain dynamics are divided into three parts: the known linear part $\bar{A}_k X_k$, the time-varying nonlinear uncertain dynamics $F(X_k, k)$, and the noise (w_k, n_k). The ESKF method suggested using different approaches to deal with different kinds of uncertainties. $F(X_k, k)$ is treated as an extended state to be estimated and compensated for, while (w_k, n_k) is attenuated by the optimization technique of the Kalman filter. Therefore, the system can be equivalently transformed to

$$\begin{cases} \begin{bmatrix} X_{k+1} \\ F_{k+1} \end{bmatrix} = A_k \begin{bmatrix} X_k \\ F_k \end{bmatrix} + B_k G_k + \begin{bmatrix} w_k \\ 0 \end{bmatrix}, \\ Z_k = C_k \begin{bmatrix} X_k \\ F_k \end{bmatrix} + n_k, \end{cases} \qquad (5.6-13)$$

where
$$F_k \triangleq F(X_k, k), G_k = F_{k+1} - F_k, A_k = \begin{bmatrix} \bar{A}_k & \bar{B}_k \\ 0 & I \end{bmatrix}, B_k = \begin{bmatrix} 0 \\ I \end{bmatrix}, C_k = [\bar{C}_k \quad 0].$$

Design of the extended-state observer (ESO) is based on the extended model in Equation (5.6-14):
$$\begin{bmatrix} \hat{X}_{k+1} \\ \hat{F}_{k+1} \end{bmatrix} = A_k \begin{bmatrix} \hat{X}_k \\ \hat{F}_k \end{bmatrix} + B_k \hat{G}_k - K_k \left(Z_k - C_k \begin{bmatrix} \hat{X}_k \\ \hat{F}_k \end{bmatrix} \right), \qquad (5.6-14)$$

where $\hat{G}_k$, the estimate of G_k, is used to correct the estimation error of the state and the uncertainty by making full use of the model information. Thus, use the nominal model of G_k, $\bar{G}_k = \bar{F}(X_{x+1}, k+1) - \bar{F}(X_k, k)$. Then, the estimate is denoted as $\hat{\bar{G}}_k = \bar{F}(\bar{A}_k \hat{X}_k + \bar{B}_k \hat{F}_k, k+1) - \bar{F}(X_k, k)$. According to the estimate of the nominal model $\bar{G}_k$, $\hat{G}_k$ is designed as $\hat{G}_{k,i} = \text{sat}(\hat{\bar{G}}_{k,i}, \sqrt{\bar{q}_{k,i}})$, $i = 1, 2, \ldots$, where sat($\cdot$) is the saturation function defined by sat(f, b) = max{min{f, b}, $-b$}, $b > 0$. The saturation function sat($\cdot$) is used to ensure the boundedness of $\hat{G}_{k,i}$. Then

$$Q_{1,k} = \begin{bmatrix} 0_{n \times n} & 0_{n \times l} \\ 0_{l \times n} & 4\bar{Q}_k \end{bmatrix}, \qquad (5.6-15)$$

where $\bar{Q}_k \triangleq l \cdot \text{diag}([\bar{q}_{k,1} \quad \bar{q}_{k,2} \quad \cdots \quad \bar{q}_{k,l}])$.

Design θ to decouple the cross terms of estimation error and the uncertainties,

$$\theta = \sqrt{\frac{\mathrm{tr}(Q_{1,0})}{\mathrm{tr}(P_0)}} \tag{5.6-16}$$

and

$$Q_{2,k} = \begin{bmatrix} S_k & 0_{n \times l} \\ 0_{l \times n} & 0_{l \times l} \end{bmatrix}. \tag{5.6-17}$$

Let P_k satisfy the iteration equation,

$$P_{k+1} = (1+\theta)(A_k + K_k C_k) P_k (A_k + K_k C_k)^{\mathrm{T}} + K_k R_k K_k^{\mathrm{T}} + \left(1 + \frac{1}{\theta}\right) Q_{1,k} + Q_{2,k}, \tag{5.6-18}$$

with the initial value P_0.

Define $K_k^* = \arg\min_{K_k} \{(1+\theta)(A_k + K_k C_k) P_k (A_k + K_k C_k)^{\mathrm{T}} + K_k R_k K_k^{\mathrm{T}}\}$, since P_k is a positive semi-definite matrix and R_k is a positive definite matrix, $C_k P_k C_k^{\mathrm{T}} + \frac{1}{1+\theta} R_k$ is positive definite. It is straightforward to obtain,

$$K_k^* = -A_k P_k C_k^{\mathrm{T}} \left(C_k P_k C_k^{\mathrm{T}} + \frac{1}{1+\theta} R_k\right)^{-1}. \tag{5.6-19}$$

As a consequence, the ESKF can be designed as follows

$$\begin{bmatrix} \hat{X}_{k+1} \\ \hat{F}_{k+1} \end{bmatrix} = A_k \begin{bmatrix} \hat{X}_k \\ \hat{F}_k \end{bmatrix} + B_k \hat{G}_k - K_k \left(Z_k - C_k \begin{bmatrix} \hat{X}_k \\ \hat{F}_k \end{bmatrix}\right), \tag{5.6-20}$$

$$K_k = -A_k P_k C_k^{\mathrm{T}} \left(C_k P_k C_k^{\mathrm{T}} + \frac{1}{1+\theta} R_k\right)^{-1}, \tag{5.6-21}$$

$$P_{k+1} = (1+\theta)(A_k + K_k C_k) P_k (A_k + K_k C_k)^{\mathrm{T}} + K_k R_k K_k^{\mathrm{T}} + \left(1 + \frac{1}{\theta}\right) Q_{1,k} + Q_{2,k}, \tag{5.6-22}$$

where

$$Q_{1,k} = \begin{bmatrix} 0_{n \times n} & 0_{n \times l} \\ 0_{l \times n} & 4\bar{Q}_k \end{bmatrix}, Q_{2,k} = \begin{bmatrix} S_k & 0_{n \times l} \\ 0_{l \times n} & 0_{l \times l} \end{bmatrix}, \bar{Q}_k \triangleq l \cdot \mathrm{diag}([\bar{q}_{k,1} \quad \bar{q}_{k,2} \quad \cdots \quad \bar{q}_{k,l}]),$$

$$\theta = \sqrt{\frac{\mathrm{tr}(Q_{l,0})}{\mathrm{tr}(P_0)}}, \hat{G}_{k,i} \triangleq \mathrm{sat}(\bar{G}_{k,i}, \sqrt{\bar{q}_{k,i}}), \bar{G}_k \triangleq \bar{G}(\hat{X}_k, k),$$

$$\mathrm{sat}(f,b) = \max\{\min\{f,b\}, -b\}, b > 0.$$

For the system with uncertainty and noise, the tuning of K_k becomes a tradeoff between the disturbance rejection and the noise sensitivity. This is because higher K_k leads to faster tracking, but also results in higher levels of noise. Thus, different from ESO, K_k is no longer static and manually tuned here. Instead, K_k is optimized to ensure the mean square estimation error is minimal at each step. For more details about ESKF, please refer to the literature [1].

2) Online estimation and compensation method for DRE parasitic loop based on ESKF

The beam pointing error slope R^* and the radome error slope R (which affect the DRE parasitic loop of the strap-down phased array radar seeker) are set as the state variables. The beam pointing error slope estimate value $\hat{R}^*$, the radome error slope estimate value $\hat{R}$, and the true LOS angular rate estimate value $\hat{\dot{q}}_t$ are directly obtained with ESKF. Thereby, the online compensation for the DRE parasitic loop is completed.

Setting the system state variables of ESKF to
$$X = [X_1, X_2, X_3, X_4, X_5]^T = [q_t, \dot{q}_t, \theta_B, R, R^*]^T, \quad (5.6-23)$$
the LOS angular rate $\dot{q}_s$ is a measured output and is expressed as
$$Z = \dot{q}_s. \quad (5.6-24)$$

The attitude angle ϑ_m and the attitude angular rate $\dot{\vartheta}_m$ are measured by the inertial navigation system onboard the missiles. They are the input quantities for the ESKF, and the corresponding noise standard deviations are σ_{u1} and σ_{u2}, respectively. Thus, the ESKF model for the DRE parasitic loop is illustrated in Fig. 5.6 – 10.

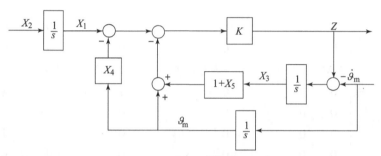

Fig. 5.6 – 10 The block diagram of the ESKF model for the DRE parasitic loop

According to Fig. 5.6 – 10, the state equation and the measurement equation of ESKF are obtained as follows
$$\dot{X}(t) = f[X(t)] + mw(t), \quad (5.6-25)$$
$$Z(t) = h[X(t)] + v(t), \quad (5.6-26)$$
where
$$f[X(t)] = \begin{bmatrix} X_2 \\ 0 \\ KX_1 - K\vartheta_m - KX_5\vartheta_m - KX_3 - KX_3X_4 - \dot{\vartheta}_m \\ 0 \\ 0 \end{bmatrix},$$
$$h[X(t)] = KX_1 - KX_5\vartheta_m - KX_3 - KX_3X_4 - K\vartheta_m - \dot{\vartheta}_m,$$
$$w(t) = [0 \quad w_2(t) \quad 0 \quad w_4(t) \quad w_5(t)]^T,$$
where $f[X(t)]$ represents the state matrix and $h[X(t)]$ represents the measurement matrix. $w(t)$ and $v(t)$ are the process noise and measurement noise, respectively. Design S_{w2}, S_{w4}, and S_{w5} are the process noise power spectral densities. S_k is the system noise covariance matrix. $v(t)$ is the zero-mean Gauss white-noise and its standard deviation is represented as σ_v. R_k is the measurement noise covariance matrix.

The nonlinear uncertain dynamics F_k can be set as

$$F_k = \begin{bmatrix} \dot{R} \\ R^* \end{bmatrix}, \qquad (5.6-27)$$

Then, the state equation and measurement equation can be substituted, as shown in Equation(5.6-28)

$$\begin{cases} X_{k+1} = \begin{bmatrix} 1 & \Delta t & 0 & 0 & 0 \\ 0 & 1 & 0 & 0 & 0 \\ K\Delta t & 0 & 1-K\Delta t - K\hat{X}_4\Delta t & -K\hat{X}_3\Delta t & -K\vartheta_m\Delta t \\ 0 & 0 & 0 & 1 & 0 \\ 0 & 0 & 0 & 0 & 1 \end{bmatrix} X_k + \begin{bmatrix} 0 & 0 \\ 0 & 0 \\ 0 & 0 \\ \Delta t & 0 \\ 0 & \Delta t \end{bmatrix} \begin{bmatrix} \dot{R} \\ R^* \end{bmatrix} + mw_k, \\ Z_k = \begin{bmatrix} K & 0 & -K-K\hat{X}_4 & -K\hat{X}_3 & -K\vartheta_m \end{bmatrix} \hat{X}_k + n_k, k=0,1,\ldots, \end{cases}$$

$$(5.6-28)$$

where Δt is sample time. Then, the discrete fundamental matrix $\bar{A}[\hat{X}(t)]$ is shown in Equation (5.6-29)

$$\bar{A}[\hat{X}(t)] = \begin{bmatrix} 1 & \Delta t & 0 & 0 & 0 & 0 & 0 \\ 0 & 1 & 0 & 0 & 0 & 0 & 0 \\ K\Delta t & 0 & 1-K\Delta t - K\hat{X}_{4,k-1}\Delta t & -K\hat{X}_{3,k-1}\Delta t & -K\vartheta_m\Delta t & 0 & 0 \\ 0 & 0 & 0 & 1 & 0 & \Delta t & 0 \\ 0 & 0 & 0 & 0 & 1 & 0 & \Delta t \\ 0 & 0 & 0 & 0 & 0 & 1 & 0 \\ 0 & 0 & 0 & 0 & 0 & 0 & 1 \end{bmatrix},$$

$$(5.6-29)$$

and the disturbance matrix $\bar{B}[\hat{X}(t)]$ and the discrete measurement matrix $\bar{C}[\hat{X}(t)]$ are obtained in Equation (5.6-30) and Equation (5.6-31), respectively.

$$\bar{B}[\hat{X}(t)] = \begin{bmatrix} 0 & 0 \\ 0 & 0 \\ 0 & 0 \\ 0 & 0 \\ 0 & 0 \\ 1 & 0 \\ 0 & 1 \end{bmatrix}, \qquad (5.6-30)$$

$$\bar{C}[\hat{X}(t)] = \begin{bmatrix} K & 0 & -K-KX_{4,k/k-1} & -KX_{3,k/k-1} & -K\vartheta & 0 & 0 \end{bmatrix}. \qquad (5.6-31)$$

The unknown disturbance nominal matrix G_k can be expressed as

$$G_k = \begin{bmatrix} \dot{R} + w_4(k) \\ \dot{R}^* + w_5(k) \end{bmatrix}. \qquad (5.6-32)$$

The extended-state observer (ESO) method is applied to estimate and compensate for the unknown disturbance. Hence, the nominal model of G_k is used as

$$\bar{G}_k = \begin{bmatrix} X_{4,k} - \hat{X}_{4,k-1} \\ X_{5,k} - \hat{X}_{5,k-1} \end{bmatrix}. \qquad (5.6-33)$$

The extended model disturbance estimation covariance matrix Q_{1k} is

$$Q_{1k} = \begin{bmatrix} \mathbf{0}_{5\times 5} & \mathbf{0}_{5\times 2} \\ \mathbf{0}_{2\times 5} & \bar{Q}_k \end{bmatrix}, \quad (5.6-34)$$

where

$$\bar{Q}_k = 4 \times 2 \times \begin{bmatrix} R_{k+1} + w_{4,k+1} - R_k - w_{4,k} & 0 \\ 0 & R^*_{k+1} + w_{5,k+1} - R^*_k - w_{5,k} \end{bmatrix},$$

and the design of the extended model noise covariance matrix Q_{2k} is

$$Q_{2k} = \begin{bmatrix} 0 & 0 & 0 & 0 & 0 & 0 & 0 \\ 0 & \dfrac{S_{w2,k-1}}{\Delta t} & 0 & 0 & 0 & 0 & 0 \\ 0 & 0 & 0 & 0 & 0 & 0 & 0 \\ 0 & 0 & 0 & \dfrac{S_{w4,k-1}}{\Delta t} & 0 & 0 & 0 \\ 0 & 0 & 0 & 0 & \dfrac{S_{w5,k-1}}{\Delta t} & 0 & 0 \\ 0 & 0 & 0 & 0 & 0 & 0 & 0 \\ 0 & 0 & 0 & 0 & 0 & 0 & 0 \end{bmatrix}. \quad (5.6-35)$$

Then, the design of the discrete system noise-driven covariance matrix of the extended model is

$$\boldsymbol{\Gamma}_k \approx \boldsymbol{m} \cdot \Delta t = \begin{bmatrix} \Delta t & 0 & 0 & 0 & 0 \\ 0 & \Delta t & 0 & 0 & 0 \\ 0 & 0 & \Delta t & 0 & 0 \\ 0 & 0 & 0 & \Delta t & 0 \\ 0 & 0 & 0 & 0 & \Delta t \end{bmatrix}. \quad (5.6-36)$$

Furthermore, the design measurement noise covariance matrix of the extended model is

$$R_k = \frac{S_v}{\Delta t} = (\sigma_v)^2. \quad (5.6-37)$$

Then, the online estimation and compensation method for the DRE parasitic loop using ESKF (Equation (5.6-38) and Equation (5.6-39)) is designed as follows

$$\boldsymbol{K}_k = -\bar{\boldsymbol{A}}_k \boldsymbol{P}_k \bar{\boldsymbol{C}}_k^{\mathrm{T}} \left(\bar{\boldsymbol{C}}_k \boldsymbol{P}_k \bar{\boldsymbol{C}}_k^{\mathrm{T}} + \frac{1}{\Delta t + \theta} \boldsymbol{R}_k \right)^{-1}, \quad (5.6-38)$$

$$\boldsymbol{P}_{k+1} = (\Delta t + \theta)(\bar{\boldsymbol{A}}_k + \boldsymbol{K}_k \bar{\boldsymbol{C}}_k) \boldsymbol{P}_k (\bar{\boldsymbol{A}}_k + \boldsymbol{K}_k \bar{\boldsymbol{C}}_k)^{\mathrm{T}} + \boldsymbol{K}_k \boldsymbol{R}_k \boldsymbol{K}_k^{\mathrm{T}} + \left(1 + \frac{1}{\theta}\right) \boldsymbol{Q}_{1,k} + \boldsymbol{\Gamma}_k \boldsymbol{Q}_{2,k} \boldsymbol{\Gamma}_k^{\mathrm{T}}. \quad (5.6-39)$$

5.6.4.3 Mathematical simulation and result analysis

In this section, the ESKF method is applied to achieve online estimation and compensation for the DRE parasitic loop by estimating the radome error slope R, beam pointing error slope R^*, and the LOS angular rate $\dot{q}_t$. The simulation parameters are designed in Table 5.6-1.

Table 5.6-1 The simulation parameters of the DRE model

Parameters	$\sigma_v/((°) \cdot s^{-1})$	$\sigma_{u1}/((°) \cdot s^{-1})$	$\sigma_{u2}/((°) \cdot s^{-1})$	S_{w2}	S_{w4}	S_{w5}
Values	0.2	0.1	0.1	0.00001	0.001	0.001

The strap-down phased array radar seeker forward loop gain was set to $K = 10$. The actual LOS angle was set as $q_t = 1t^*$. Consequently, the LOS angular rate is $\dot{q}_t = 1^*/s$. Also, it is assumed that the missile body motion varies sinusoidally with an amplitude of 3° and a frequency of 2Hz. The initial value of the state estimation was defined as $\hat{X}_0 = [0, 0, 0, 0, 0]$, $\hat{F}_0 = [0, 0]$, and the initial value of the extended model estimation error covariance matrix was set as $P_0 = I_{7 \times 7}$. For simplicity, "ESKF" will be used to denote the proposed method in the following simulations.

This method has the same inputs as the traditional EKF. To demonstrate the robustness and superior estimation accuracy of the proposed method, the EKF, which is widely used in engineering practice, and the STUKF will be used as benchmarks for comparison in the simulations.

The radome error slope R is a random value within the range of $(-0.03, 0.03)$ and the beam pointing error slope R^* is a random value within the range of $(-0.05, 0.05)$. This simulation focuses on comparing the performance differences in estimating two types of error slopes among the EKF, STUKF, and ESKF. The results of the simulation are presented in Fig. 5.6-11.

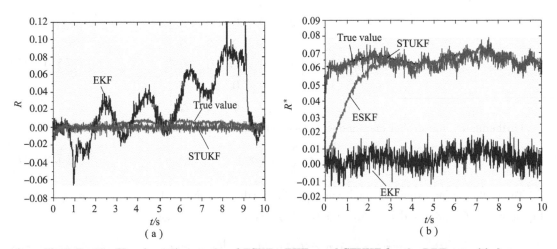

Fig. 5.6-11 The simulation results of ESKF, EKF, and STUKF for the DRE parasitic loop
(a) The estimation of radome error slope R; (b) The estimation of beam pointing error slope R^*

As illustrated in Fig. 5.6-11 (a), the EKF exhibits the poorest performance. The ESKF provides more accurate and stable estimates compared to the STUKF under conditions of small initial errors. The STUKF, however, shows some initial estimation errors. Fig. 5.6-11 indicates that the ESKF takes longer to achieve stable and precise estimation in low nonlinear conditions. While the STUKF provides faster estimates, the remaining estimation errors are significant and cannot be disregarded.

§ 5.7 Stabilization Loop and Tracking Loop Design of the Platform-based Seeker

5.7.1 Stabilization Loop Design

As discussed in previous sections, minimizing the impact of missile angular motion interference on seeker performance requires maximizing the open-loop gain of the stabilization loop. Given that the seeker must accommodate a certain load and the driving motor bandwidth is limited, the low-frequency gain of the stabilization loop can be enhanced by designing a PI compensator (or a lag compensator). This approach is effective because missile interference motion primarily affects the low-frequency range, thus improving the seeker's ability to counteract missile-seeker coupling.

The following example illustrates how to implement this approach. With a stabilization loop structure as shown in Fig. 5.7-1, the rate gyro has the undamped natural frequency of $\omega_{gn} = 80$ Hz, the damping coefficient of $\zeta_g = 0.7$, and the motor time constant of $T = 0.0004\ s$. The missile's angular motion interference frequency is about 2 Hz. It is required to select the optimal parameters of the PI compensator K_p and K_i so that under the design constraints of a gain margin greater than 6 dB and a phase margin greater than 40°, the stabilization loop is optimized to reject the disturbance of the missile body angular motion. The compensator parameter K_p and K_i should be chosen to maximize the loop gain at the interference frequency 2 Hz. This optimal design problem can be described as follows.

Objective function

$$\text{Max } M_{2\,\text{Hz}}(K_p, K_i), \qquad (5.7-1)$$

subject to

$$\text{Gain margin } L(K_p, K_i) \geq 6 \text{ dB}, \qquad (5.7-2)$$
$$\text{Phase margin } \Delta\Phi(K_p, K_i) \geq 40°. \qquad (5.7-3)$$

In Equation (5.7-1), $M_{2\,\text{Hz}}$ is the stabilization loop gain at 2 Hz.

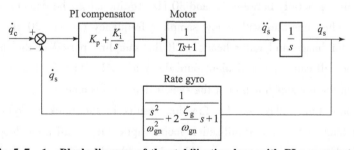

Fig. 5.7-1 Block diagram of the stabilization loop with PI compensator

With commercial non-linear programming software, the solution of this optimization problem can be obtained as $K_p = 181$, $K_i = 1.0 \times 10^4$ and the open loop gain at 2 Hz is 65.4. Fig. 5.7-2 illustrates the contour of the objective function. It is evident that the maximum gain at 2 Hz is constrained by the phase margin requirement.

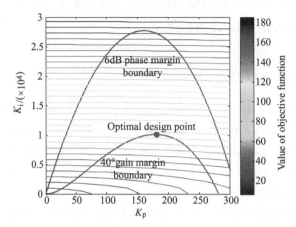

Fig. 5.7 – 2 Contour of the objective function with gain
margin and phase margin constraints

With this design, the stabilization open-loop gain at the 2 Hz is 65.4. When the stabilization loop is designed with the same gain and phase-margin constraints and with no PI compensator added, the open-loop gain at 2 Hz is only 22.2. So the open-loop gain at 2 Hz has been increased by a factor of 3 while maintaining the same phase and gain margin constraints, following the adoption of the PI compensation optimization design. This means that the decoupling level under the 2 Hz missile motion disturbance has also been improved by a factor of 3.

5.7.2 Tracking Loop Design

The design of the tracking loop in a platform-based seeker differs significantly from that of type II ground tracking radars. In type II radars, the inner loop is typically designed to achieve a bandwidth ratio of approximately 5 between the inner and outer loops to minimize phase shift at the open-loop crossover frequency of the tracking radar's outer loop. However, for missile seekers, the primary purpose of the inner loop is to decouple the missile motion, necessitating a much wider frequency bandwidth, generally between 15 and 30 Hz. Additionally, the detector used in the seeker angle tracking loop often operates with a low sampling frequency (e.g., 20 Hz for laser detectors and 50 to 100 Hz for image and radar heads). If the sampling period of the detector is τ_1, the sample holding time will cause an equivalent pure delay of $\tau_1/2$. Suppose that the pure delay of the detector is τ_2, then the transfer function of the seeker detector becomes $e^{-\tau s}$ ($\tau = \tau_1/2 + \tau_2$). The presence of this pure time delay transfer function results in the tracking loop bandwidth being significantly lower than that of the stabilization loop despite the stabilization loop's minimal phase shift at the tracking loop's crossover frequency. To enhance the dynamic response of the seeker, a PD compensation network can be incorporated into the tracking loop. The effectiveness of this design will be illustrated in the following example.

Assume the block diagram of the seeker tracking loop is depicted in Fig. 5.7 – 3.

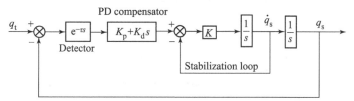

Fig. 5.7 – 3 Block diagram of the seeker tracking loop

It is known that the open loop gain of the stabilization loop $K = 188$ (bandwidth 30 Hz) and the seeker time delay $\tau = 40$ ms. Selecting the optimal parameters of the PD compensation network K_p and K_d is required to achieve the fastest step response in the tracking loop. Suppose $e(t)$ is the tracking error for a unit step input. Then, the design objective function can be taken as $J = \int_0^T t|e(t)|dt$, subject to the stability constraint of a gain margin of 6 dB and a phase margin of 40°.

The mathematical model for this design is an objective function

$$\min J = \int_0^T t|e(t)|dt,$$

subject to the tracking loop gain margin $L(K_p, K_d) \geqslant 6$ dB and the tracking loop phase margin $\Delta\Phi(K_p, K_d) \geqslant 40°$.

The integral time is designated as the tracking loop settling time $T = 0.6$ s. Fig. 5.7 – 4 illustrates the contour of the objective function, along with the gain and phase margins, obtained using nonlinear programming software to identify the optimal solution.

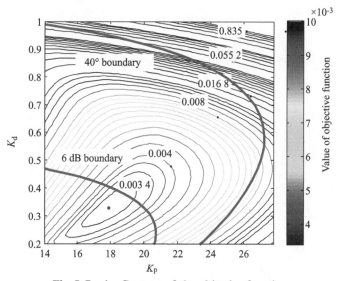

Fig. 5.7 – 4 Contour of the objective function

It can be seen that the result of this optimal design is $K_p = 17.8$ and $K_d = 0.33$ (that is, the open loop gain of the seeker is 17.8, and its corresponding bandwidth is 2.83 Hz). Fig. 5.7 – 5 displays the step response of the designed seeker and clearly demonstrates that incorporating PD compensator markedly enhances the seeker's response.

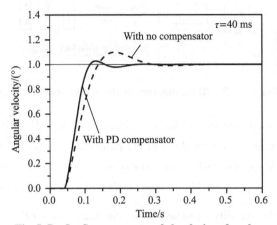

Fig. 5.7-5 Step response of the designed seeker

6

Autopilot Design

The previous chapters provided an overview of the missile guidance control system, including its fully nonlinear equations of motion, linearized simplified equations of motion, and basic mathematical models for missile control components. This chapter focuses on the design of the autopilot.

§ 6.1 Acceleration Autopilot

AUTOPILOT

The purpose of the guidance is to change the missile's flight path. Since $a = V\dot{\theta}$, controlling the missile's normal acceleration can lead to a change in the missile's velocity vector direction and flight path. An acceleration autopilot is designed to achieve precise tracking of the normal acceleration command and to improve the response speed of the autopilot through feedback mechanisms.

6.1.1 Two-loop Acceleration Autopilot

The structure of a typical two-loop acceleration autopilot is shown in Fig. 6.1 – 1.

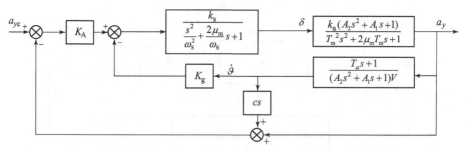

Fig. 6.1 – 1 Structure of a typical two-loop acceleration autopilot

The autopilot has a damping loop; its feedback coefficient is given as K_g. Assume that the position of the feedback accelerometer is placed at $c(m)$ in front of the center of gravity, which causes the missile angular acceleration $\ddot{\vartheta}$ introduced normal acceleration component $c\ddot{\vartheta}$ to be added to the output of the accelerometer.

Fig. 6.1 – 1 could also be given in the form of Fig. 6.1 – 2.

Omitting the small terms A_1 and A_2, the transfer function from the missile normal acceleration to the signal output at position A of Fig. 6.1 – 2 will be

$$K_g\left(\frac{T_\alpha}{K_A V}s + \frac{1}{K_A V}\right)a_y, \tag{6.1-1}$$

and the signal output at position B will be

$$\left(\frac{cT_\alpha}{V}s^2 + \frac{c}{V}s + 1\right)a_y. \tag{6.1-2}$$

Adding these two signal outputs will give the total feedback of the missile's normal acceleration as

$$\left[\frac{cT_\alpha}{V}s^2 + \left(\frac{K_t T_\alpha}{K_A V} + \frac{1}{V}c\right)s + \left(1 + \frac{K_g}{K_A V}\right)\right]a_y. \tag{6.1-3}$$

Considering all feedback, the feedback structure can be represented as a second-order lead compensation network (Fig. 6.1-3). In this structure, the angular velocity loop provides first-order compensation, while the accelerometer preposition contributes second-order compensation for the s^2 term. This approach allows for an increase in the autopilot's bandwidth through phase lead compensation, given a fixed actuator lag.

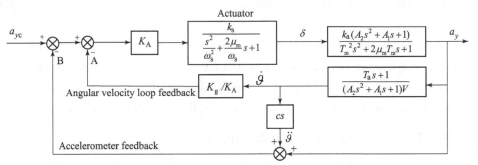

Fig. 6.1-2 Alternative two-loop acceleration autopilot block diagram

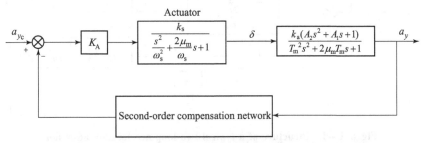

Fig. 6.1-3 Two-loop acceleration autopilot feedback structure

The phase lead compensation network is designed to provide a lead phase angle of approximately 70° at the autopilot crossover frequency. Given that the uncontrolled missile body exhibits a phase shift of around -180° at this frequency due to minimal aerodynamic damping and allowing for a phase lag of 20° to 25° from the actuator, this configuration can achieve a phase margin of 45° to 50° for the autopilot design.

The following example utilizes the two-loop autopilot structure with missile dynamic coefficients provided in Table 6.1-1 from the literature [5]. In addition, the first-order model of the actuator is given as $\dfrac{-0.017\,5}{0.013\,3s+1}$. It has a time constant of 0.013 3 s, a bandwidth of about 12 Hz, and the other autopilot parameters are $K_A = 0.000\,65$ and $K_g = 0.072\,8$. The corresponding control diagram of the autopilot is shown in Fig. 6.1 – 4.

Table 6.1 – 1 Missile dynamic coefficients

a_α/s^{-2}	a_δ/s^{-2}	a_ω/s^{-1}	b_α/s^{-1}	b_δ/s^{-1}	c/m	$V/(\mathrm{m\cdot s^{-1}})$
72.4	471	1.5	1.27	0.477	0.66	1 140

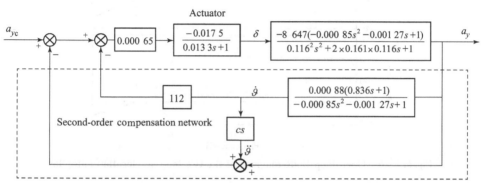

Fig. 6.1 – 4 Corresponding control diagram of the two-loop acceleration autopilot example

The second-order compensation network Bode diagram corresponding to Fig. 6.1 – 4 in this example is shown in Fig. 6.1 – 5. The advantages of this second-order lead compensation for the normal acceleration a_y output are clear from the graph. Without forward positioning of the accelerometer (as $c = 0$), the lead compensation becomes a first-order lead compensation network.

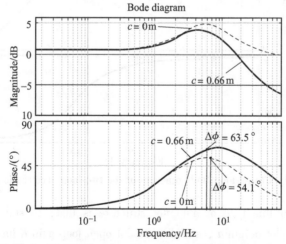

Fig. 6.1 – 5 First-order ($c = 0$) and second-order ($c = 0.66$ m) compensation network Bode diagrams

The open-loop Bode diagram for this example is presented in Fig. 6.1 – 6. The second-order lead compensation network delivers a leading phase angle of 63.5° at the crossover frequency of 5.96 Hz. At this frequency, the actuator introduces a lag angle of −26.5°, while the missile body contributes a lag phase angle of −176.7°. Consequently, the phase margin of this autopilot design is 40.3°.

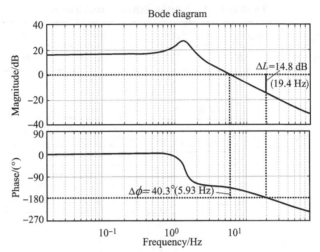

Fig. 6.1−6　Open-loop Bode diagram of the two-loop acceleration autopilot

The closed-loop Bode diagram for this example is shown in Fig. 6.1−7. The time-domain response to the unit step autopilot acceleration command is shown in Fig. 6.1−8.

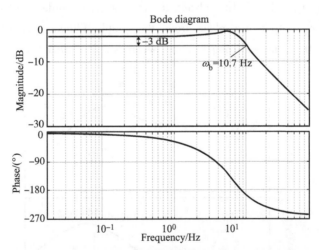

Fig. 6.1−7　Closed-loop Bode diagram of the two-loop acceleration autopilot

The disadvantage of this type of autopilot structure is that, to stabilize the gain from the command a_c to the autopilot output a, a high autopilot open loop gain K has to be taken. However, actuator bandwidth often constrains the autopilot's open-loop gain to a value lower than anticipated. For instance, in the current example, the open-loop gain is limited to 3.68, resulting in a closed-loop gain of only 0.787. Due to this limitation, which affects the robustness of the closed-loop gain, such autopilot configurations are seldom used in practical applications today.

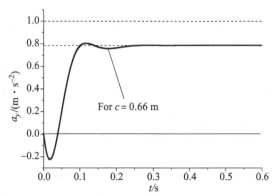

Fig. 6.1 – 8 The time-domain response to the unit step autopilot acceleration command

6.1.2 Two-loop Autopilot with PI Compensation

To enhance the robustness of the two-loop autopilot described previously, introducing a PI compensator or a lag compensator into the forward loop can be beneficial. This approach significantly increases the open-loop gain at low frequencies and improves the robustness of the autopilot's steady-state closed-loop gain.

Normally, the PI compensator transfer function is given as $1 + \dfrac{\omega^*}{s}$. Here, ω^* is the corner frequency of the PI compensator. A lower designed value of ω^* will lead to a slow, steady, stable error elimination transient, and a higher value of ω^* will lead to lower autopilot stability. A properly designed value of ω^* should eliminate the steady-state error roughly at the end of the autopilot transient. For this reason, $\omega^* = 0.3$ rad/s is chosen for the above example. The block diagram of the autopilot design is shown in Fig. 6.1 – 9.

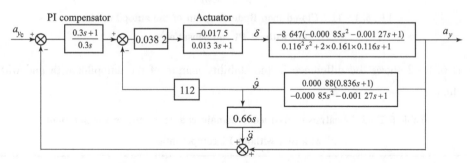

Fig. 6.1 – 9 Block diagram of the two-loop autopilot with PI compensator

The open-loop and closed-loop Bode diagrams of the autopilot with PI compensation, compared to the same diagrams without PI compensation, are illustrated in Fig. 6.1 – 10 and Fig. 6.1 – 11, respectively.

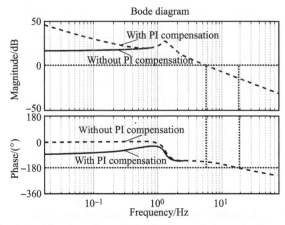

Fig. 6.1-10 Open-loop Bode diagram of the autopilot with
and without PI compensation

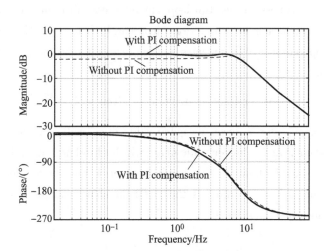

Fig. 6.1-11 Closed-loop Bode diagram of the autopilot with
and without PI compensation

Table 6.1-2 shows the difference in the stability margin of the autopilot with and without PI compensation.

Table 6.1-2 Analysis of frequency domain characteristics of the autopilot
with and without PI compensation

Items	Gain margin ΔL/dB	Phase margin $\Delta \phi$/(°)
Without PI compensation	14.8	40.3
With PI compensation	14.8	39.7

Fig. 6.1-12 shows the improvement in the steady state value of the autopilot time domain response with PI compensation.

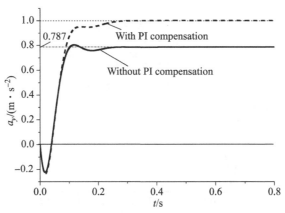

Fig. 6.1 – 12 Unit step response of the autopilot with and without PI compensation

The following example illustrates that it is possible to replace the PI compensator with a lag compensator and still get the same autopilot closed loop robustness. It is known that the standard structure of a PI compensator is $1 + \dfrac{\omega^*}{s}$, and that improvement of a lag compensator is $\beta\left(\dfrac{T_b s + 1}{\beta T_b s + 1}\right)$, where the value of β will increase the autopilot low-frequency gain by a factor of β. When the value of T_b is chosen as $\dfrac{1}{T_b} \approx \omega^*$ and β is large enough, the two compensator structures will have similar compensation effects for reducing autopilot steady static error. The Bode diagram of a PI compensator $1 + \dfrac{1}{0.3s}$ and lag compensator $10 \times \left(\dfrac{0.3s + 1}{10 \times 0.3s + 1}\right)$ is shown in Fig. 6.1 – 13. Both designs have a corner frequency of 3.33 rad/s (0.531 Hz). The low-frequency gain of the autopilot with PI compensation has increased tenfold compared to the system without compensation. Fig. 6.1 – 14 illustrates the unit step response of the autopilot to systems with PI compensation, lag compensation, and no compensation. Both PI and lag compensator effectively reduce the system's static error, with similar results observed for each compensation method.

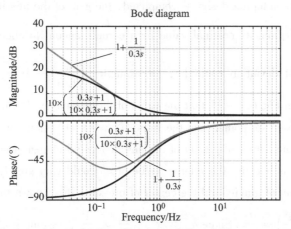

Fig. 6.1 – 13 Bode diagram of the designed PI compensator and lag compensator

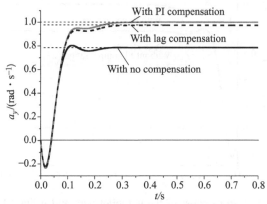

Fig. 6.1 – 14 Unit step response of the autopilot with lag compensation, PI compensation, and no compensation

6.1.3 Three-loop Autopilot with Pseudo Angle of Attack Feedback

Fig. 6.1 – 15 shows the structural diagram of the missile body transfer function $\frac{\alpha(s)}{\delta(s)}$ as the object being controlled for an acceleration autopilot.

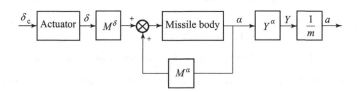

Fig. 6.1 – 15 Structural diagram of transfer function $\frac{\alpha(s)}{\delta(s)}$

Generally, after the autopilot design is completed, the gain of the missile body transfer function $\frac{\alpha(s)}{\delta(s)}$ has to be changed from its nominal value by a certain amount to check its effect on autopilot robustness. The range of deviation could be, for example, $\left(\frac{1}{1.6}\right) \sim 1.6$, that is from 0.625 to 1.6. It can be seen from Fig. 6.1 – 15 that the value of the parameters Y^α, M^δ and m generally do not change much at the chosen set points, but the value $M^\alpha = -x^* Y^\alpha$ (here, x^* is the distance between the center of pressure and center of gravity of the missile) could change a great deal when the static stability x^* is low. This is because the relative value of x^* could vary greatly with the uncertainty of the positions of the center of gravity and the center of pressure. It is known the missile body's transfer function gain $\frac{\alpha}{\delta} = \left(\frac{M^\delta}{M^\alpha}\right) \approx \frac{a_\delta}{a_\alpha}$. Since $a_\alpha \propto x^*$, $\frac{\alpha}{\delta} \propto \frac{1}{x^*}$, assuming ε represents the possible absolute variation of x^*, its effect on x^* is shown in Fig. 6.1 – 16.

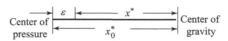

Fig. 6.1–16 The distance change between the center of gravity and the center of pressure

As an example, the dynamic coefficient values of a typical air-to-air missile given in the literature [7] are taken as the parameters of the missile body (Table 6.1–3), and a second-order actuator $\omega_n = 150$ rad/s, $\mu = 0.7$ is chosen for the autopilot. Suppose that the nominal distance between the missile's center of gravity and center of pressure is $x^* = 50$ mm, and the nominal control gain is set to $\dfrac{\alpha}{\delta} = \dfrac{a_\delta}{a_\alpha} = 1$. If the uncertainty in x^* for this example is $\varepsilon = \pm 5$ mm, the range of x^* variation will be 45 – 55 mm and the range of a_α will be 216 – 264 s^{-2}. With this x^* variation, the control gain $\dfrac{\alpha}{\delta}$ may vary by $-10\% - +11\%$.

Table 6.1–3 Model parameters of the missile body H_I

Height/m	a_α/s^{-2}	a_δ/s^{-2}	a_ω/s^{-1}	b_α/s^{-1}	b_δ/s^{-1}	$V/(\text{m}\cdot\text{s}^{-1})$
9.14×10^3	240	240	3	1.17	0.239	914

Let a damping feedback loop be added to the missile body H_I in Table 6.1–3 to form a new missile body H_II (Fig. 6.1–17). In this instance, the transfer function of the new missile body becomes $\dfrac{\alpha(s)}{\delta^*(s)}$ (here, δ^* is the input of the new missile body formed after the damping loop is added, and it can be called the pseudo actuator deflection angle). The angle of attack response to the unit step pseudo actuator input δ^* for different ε values and static stability is shown in Fig. 6.1–18. Obviously, the missile body with sufficient static stability demonstrates strong robustness.

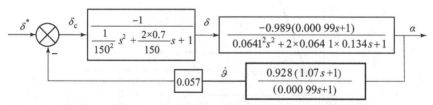

Fig. 6.1–17 Structural diagram of the new missile body H_II (missile body H_I with a damping loop) ($x^* = 50$ mm, $a_\alpha = 240$ s^{-2})

However, if a missile body L_I is of low static stability, for example, its nominal value of x^* is only 6 mm, then the nominal value of a_α will be reduced from 240 to $\left(\dfrac{6}{50}\right)\times 240 = 28.8$ s^{-2}, and the nominal control ratio will increase to approximately $\dfrac{\alpha}{\delta} = \dfrac{240}{28.8} = 8.3$. In other words, while a missile body with low static stability can significantly enhance its control ratio, it is crucial to be mindful of the potential reduction in robustness associated with such a configuration.

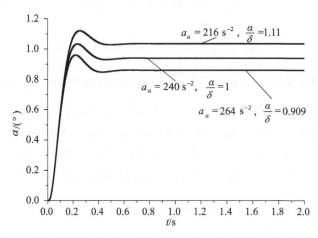

Fig. 6.1 – 18 Angle of attack response to the unit step pseudo actuator input for the new missile body H_II

If the uncertainty between the center of gravity and the center of pressure x^* is still taken as ± 5 mm, its variation will range from 1 mm to 11 mm. This makes the missile control ratio fluctuate from -44% to $+600\%$. The angle of attack response to the unit step actuator input δ with this different static stability change is shown in Fig. 6.1 – 19. Obviously, the uncertainty of x^* has a great influence on the performance of a weakly-stabilized missile body. Since the missile body is only a component of the autopilot loop, significant variations in its gain can render the autopilot design unacceptable during the final robustness evaluation.

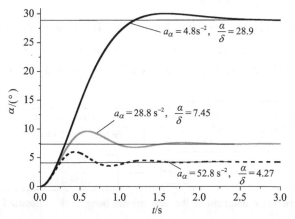

Fig. 6.1 – 19 Angle of attack response to the unit step actuator input for the missile body L_I

The solution to the above problems is to construct an angle of attack feedback loop around the missile body to form an artificial restoring moment and achieve stability augmentation for the missile body with low static stability, and to stabilize the control ratio $\dfrac{\alpha}{\delta^*}$ of the new missile body. Currently, the angle of attack feedback signal is typically derived from the angular rate gyro output, with the pseudo angle of attack signal generated using a transfer function that relates angular velocity to the angle of attack. The block diagram of the new missile body with the pseudo angle of attack loop is shown in Fig. 6.1 – 20, where K_g and K_S are respectively the design values of the autopilot damping loop and the stability augmentation loop for the pseudo angle of attack; k_s is the gain of the actuator; and T_s and μ_s are the time constant and damping of the second-order actuator. The definitions of the remaining parameters of the missile body are given in Chapter 2.

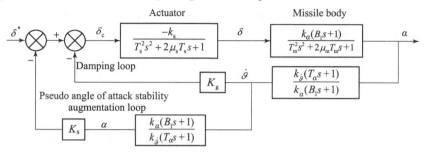

Fig. 6.1 – 20 Block diagram of the new missile body L_II with the pseudo angle of attack feedback loop

The block diagram and selected design parameters of the missile body with weak static stability are illustrated in Fig. 6.1 – 21.

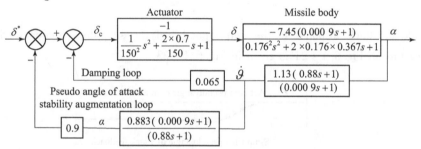

Fig. 6.1 – 21 Block diagram of the new missile body L_II with the pseudo angle of attack feedback loop for the given example

A new missile body L_II is formed after the pseudo angle of attack feedback is added. The angle of attack response of the new missile body L_II to the unit step pseudo actuator input δ^* is shown in Fig. 6.1 – 22.

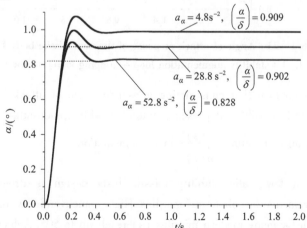

Fig. 6.1 – 22 Angle of attack response of the new missile body L_II to the unit step pseudo actuator input

It can be seen that the use of a pseudo angle of attack feedback loop allows a new missile with low static stability to have a stable gain from δ^* to α.

The possibility of simplification of the pseudo angle of attack feedback transfer function $\dfrac{\alpha(s)}{\dot{\vartheta}(s)}$ is discussed below. As established in Chapter 2, the value of B_1 in Equation (6.1-4) is very small and can be neglected. Therefore

$$\frac{\alpha(s)}{\dot{\vartheta}(s)} = \frac{k_\alpha(B_1 s + 1)}{k_{\dot{\vartheta}}(T_\alpha s + 1)} \approx \frac{k_\alpha}{k_{\dot{\vartheta}}}\left(\frac{1}{T_\alpha s + 1}\right) = \frac{k}{(T_\alpha s + 1)}, \qquad (6.1-4)$$

where

$$T_\alpha = \frac{a_\delta}{a_\delta b_\alpha - a_\alpha b_\delta} = \frac{1}{b_\alpha\left(1 - \dfrac{a_\alpha b_\delta}{a_\delta b_\alpha}\right)} = \frac{1}{b_\alpha(1 - x^*/\ell_\delta)} \approx \frac{1}{b_\alpha} \text{ (as } x^*/\ell_\delta \text{ is a small item)}.$$

So, the value of k could be simplified as

$$k = \frac{a_\delta}{a_\delta b_\alpha - a_\alpha b_\delta} \approx \frac{1}{b_\alpha}.$$

The angle of attack response to the unit input δ^* of the simplified pseudo angle of attack feedback structure in Fig. 6.1-21 is given in Fig. 6.1-23.

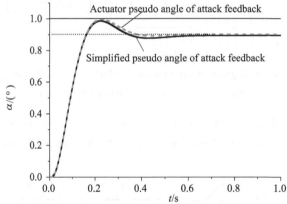

Fig. 6.1-23 Angle of attack response of the new missile body L_II
(missile body L_I + stability augmentation loop + damping loop) to the unit step input

Simulation results confirm that constructing a pseudo angle of attack loop with a simplified transfer function is entirely feasible. At present, autopilots with pseudo-attack angle feedback often adopt this simplified transfer function $\dfrac{\alpha(s)}{\dot{\vartheta}(s)}$ in implementation.

The adoption of a low static stability missile body design is driven by the increasing maneuverability requirements for air-to-air missiles. High maneuverability often necessitates high angles of attack, which for many modern missiles can reach up to 50°. Achieving such high angles requires significant actuator deflection, yet the allowable deflection is typically limited to no more than 30°. To meet these demands, reducing the missile body's static stability becomes essential. By incorporating a pseudo angle of attack feedback, it is possible to achieve a high control ratio while maintaining acceptable robustness in the missile's design. For example, if the missile of the above example is required to have an angle of attack output of 50°, the new missile design will require a pseudo actuator command $\delta^* = -55.5°$, but the actual required steady state actuator deflection angle is only $\delta_c(\infty) = -6.72°$. The response of an angle of attack with a steady state value of 50° and the related actual actuator angle δ_c and pseudo actuator angle δ^* inputs are given in Fig. 6.1 – 24.

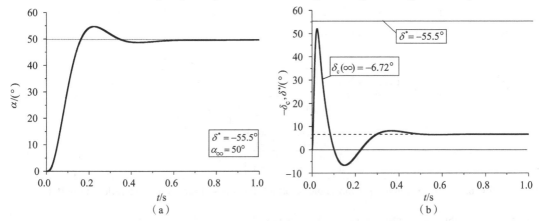

Fig. 6.1 – 24　Response of angle of attack and related δ_c and δ^* inputs

(a) The response of α; (b) δ_c and δ^* inputs

It should be noted that introducing a pseudo angle of attack feedback can only reduce the steady state actuator angle required. For the fast angle of attack response, a large transient actuator angle is still demanded. This problem can be solved by introducing a saturation limit for the actuator angle δ, with a slightly reduced angle of attack response. Fig. 6.1 – 25 shows the effect of an actuator angle limit of 25° on the angle of attack response and related δ_c input.

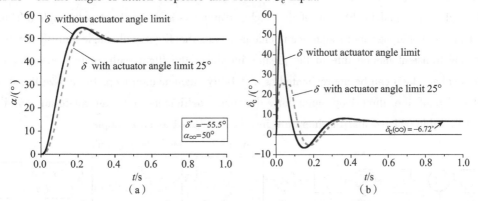

Fig. 6.1 – 25　Angle of attack response and actuator input δ_c with and without actuator angle limit

(a) The response of α with and without actuator angle limit; (b) δ_c input with and without actuator angle limit

In the simulation, the actual actuator is limited to an amplitude of 25°. This limitation is due to the fact that a missile with low static stability can only reduce the steady-state actuator deflection angle required to achieve the desired angle of attack. However, it cannot lessen the dynamic actuator deflection angle needed for the missile body's dynamic response.

Fig. 6.1 – 26 illustrates the structure of a standard pseudo angle of attack feedback three-loop acceleration autopilot. The design methods for the acceleration loop and PI or lag compensators are similar to those used in the two-loop autopilot with PI or lag compensation. This autopilot structure achieves a bandwidth close to the maximum allowed by the actuator bandwidth constraint. Additionally, the pseudo angle of attack feedback reduces the steady-state actuator angle required for a given angle of attack demand and provides robust missile body control. The inclusion of PI or lag compensators minimizes static error in the autopilot response. Due to these benefits, this autopilot configuration is widely used in modern systems.

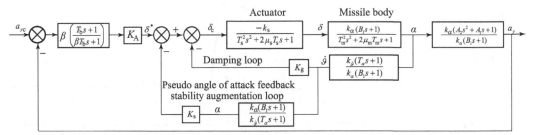

Fig. 6.1 – 26 Standard pseudo angle of attack feedback three-loop acceleration autopilot

6.1.4 Classic Three-loop Autopilot

The classic three-loop autopilot discussed here is the original design used by the US Sparrow Air-to-Air Missile, as shown in Fig. 6.1 – 27. This structure, frequently referred to as the three-loop autopilot in many documents, is commonly recognized by this term.

The inner loop of this structure is a damping loop, and the middle loop is a simplified stability augmentation loop. Here, instead of taking the angle of attack as the feedback, the integration of the angular velocity $\dot{\vartheta}$, that is ϑ, is used for stability augmentation feedback. It is known that the attitude angle ϑ is equal to the sum of the angle of attack α and the flight path angle θ Thus, the angle of attack undergoes more significant changes than the flight path angle during the missile's short-period transient motion due to the greater inertia of the flight path angle. Consequently, the attitude feedback loop can be approximated as a stability augmentation loop. To analyze the structural characteristics of the three-loop autopilot, dynamic coefficients from an air-to-air missile, as provided in Table 6.1 – 4 and the literature in [7], are used as an example.

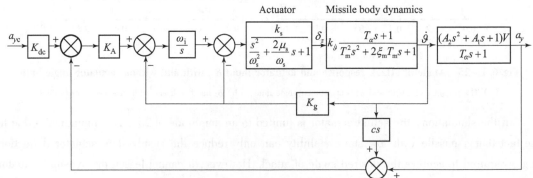

Fig. 6.1 – 27 Three-loop autopilot structure

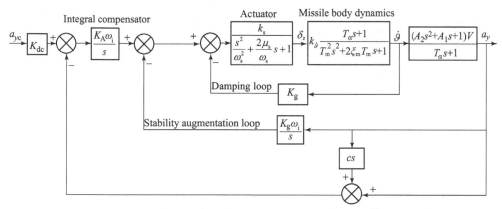

Fig. 6.1-27 Three-loop autopilot structure (Continued)

Table 6.1-4 Missile dynamic coefficients

Height/m	a_α/s^{-2}	a_δ/s^{-2}	a_ω/s^{-1}	b_α/s^{-1}	b_δ/s^{-1}	$V/(\text{m}\cdot\text{s}^{-1})$
9.14×10^3 m	240	204	0	1.17	0.239	914

Table 6.1-5 shows the design values K_g, ω_i, and K_A of the feedback gains of the three autopilot loops, and gain K_{dc} is introduced to adjust the autopilot closed-loop gain as 1. When the actuator dynamics are neglected, the block diagram of the autopilot is illustrated in Fig. 6.1-28. The unit step response, depicted in Fig. 6.1-29, closely resembles that of a first-order system.

Table 6.1-5 Design values of the autopilot

K_g	ω_i	K_A	K_{dc}
0.264	5.10	0.000 929	1.31

Conclusions regarding the characteristics of this autopilot structure can be drawn by analyzing the characteristic roots of each loop with the selected parameters (Table 6.1-6). The missile body itself has a natural frequency of 2.46 Hz and a damping ratio of 0.04. Introducing the damping loop raises the damping ratio to 1.62, with only a slight increase in the second-order root frequency to 2.70 Hz, which is typical for a general damping loop. The addition of the stability augmentation loop increases the frequency to 5.71 Hz and the damping ratio to 0.76, enhancing static stability and control ratio robustness. The outermost feedback loop employs integral compensator, introducing the primary first-order root. The final design results in a second-order root frequency of 5.44 Hz and a first-order root frequency of 0.53 Hz. Thus, it is concluded that the closed-loop characteristics of this autopilot structure are predominantly determined by the integral compensator of the outer loop. Given the high-frequency bandwidth of the stability augmentation loop, the autopilot's characteristics are minimally affected by missile body parameter variations, indicating that this autopilot structure is highly robust.

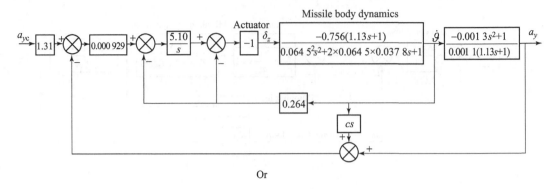

Or

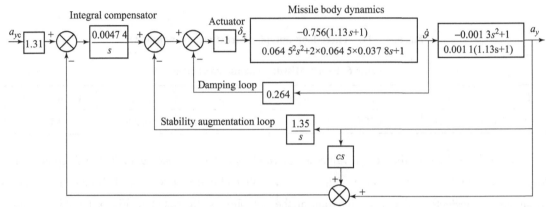

Fig. 6.1−28 Block diagram of the Sparrow Air-to-Air three-loop autopilot

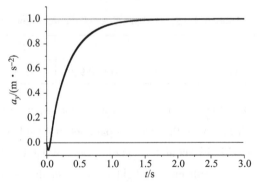

Fig. 6.1−29 a_y unit step response

Table 6.1−6 Selected parameters of the three-loop autopilot

Items	Closed loop second order frequency/Hz	Damping	First order time constant	First order root bandwidth frequency/Hz
Missile body itself	2.46	0.04	—	—
New missile body with damping loop	2.70	1.62	—	—
New missile body with stability augmentation loop	5.71	0.76	—	—
Three-loop autopilot	5.44	0.70	0.3	0.53

Fig. 6.1 – 30 illustrates the transient components for the first and second modes in the autopilot's step response. It is observed that the steady-state value of the first-order low-frequency mode accounts for approximately 114% of the closed-loop response steady-state value of the autopilot. In contrast, the second-order high-frequency mode contributes only about −13%. This indicates that the first-order low-frequency mode is the dominant component in this three-loop autopilot structure, while the second-order high-frequency mode has a minimal impact on the autopilot's time-domain transient response. Consequently, variations in missile parameters will have a negligible effect on the characteristics of this autopilot structure. However, this advantage comes at the cost of reduced autopilot response speed. Given the current emphasis on fast autopilot response, designers must carefully balance the benefits and drawbacks of this traditional autopilot design approach.

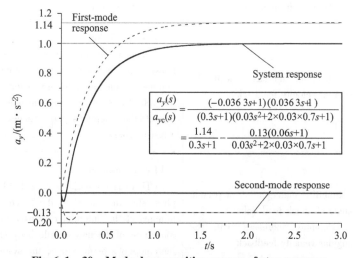

Fig. 6.1 – 30　Mode decomposition curves of step response

One of the disadvantages of this autopilot structure is that its closed-loop gain deviates from the standard integral compensation control loop. As illustrated in Fig. 6.1 – 17, the closed-loop gain is not equal to 1 but rather

$$\frac{a_y(\infty)}{a_{yc}(\infty)} = \frac{k_A \cdot V}{k_g + k_A \cdot V}. \tag{6.1-5}$$

The gain of this autopilot structure will change with the change in missile flight speed. Therefore, there is a need to introduce a parameter K_{dc} to adjust the autopilot gain as 1.

$$K_{dc} = \frac{k_g + k_A \cdot V}{k_A \cdot V} \tag{6.1-6}$$

For the given example, the unadjusted autopilot gain is $\dfrac{a_y(\infty)}{a_{yc}(\infty)} = \dfrac{k_A \cdot V}{k_g + k_A \cdot V} = \dfrac{0.000\,929 \times 914}{0.264 + 0.000\,929 \times 914} = 0.763$. Therefore, the parameter K_{dc} value for a nominal missile speed of 914 m/s should be taken as 1.31.

6.1.5 Discussion of Variable Acceleration Autopilot Structures

The various acceleration autopilot structures discussed — namely the two-loop autopilot, the two-loop autopilot with PI compensation, the three-loop autopilot with pseudo angle of attack feedback, and the classic three-loop autopilot—can be unified into a basic framework as depicted in Fig. 6.1 – 31. Table 6.1 – 7 summarizes the characteristics of each structure, including the features of the compensated missile body and its corresponding compensation network.

Table 6.1 – 7 Comparison of the structural characteristics of different acceleration autopilots

Items	Feedback compensation formed new missile body	Compensation network and its characteristics
Two-loop autopilot	Damping increases with the angular velocity feedback	No compensator in forward loop. With static error
Two-loop autopilot with PI compensation	Damping increases with the angular velocity feedback	PI compensation (There is no static error with PI compensation. The phase shift introduced by the PI compensator should be minimal at the autopilot crossover frequency. Although the robustness of the new missile body significantly impacts the autopilot's performance, the system ensures a fast response.)
Three-loop autopilot with pseudo angle of attack feedback	Stability augmentation of pseudo angle of attack feedback. Damping increases with the angular velocity feedback	PI compensation (There is no static error with PI compensation. The phase shift introduced by the PI compensator should be minimal at the autopilot crossover frequency. While the robustness of the new missile body significantly impacts the autopilot's performance, the system maintains a fast response.)
Classic three-loop autopilot	Stability augmentation of missile attitude feedback. Damping increases with the angular velocity feedback	Integral compensation (The integral compensator introduces a 90° phase shift at the autopilot crossover frequency. The robustness of the new missile body has a minimal impact on the autopilot's performance, which remains high. The static error is influenced by the missile's flight speed.)

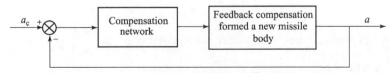

Fig. 6.1 – 31 Basic structure of the acceleration autopilot

6.1.6 Hinge Moment Autopilot

One of the simplest ways to generate the normal acceleration a_y required for guidance is to use the following actual angle δ and steady-state missile normal acceleration relation.

$$a_y = \frac{\frac{1}{2}\rho V^2 S c_y^\alpha}{m}\alpha = \frac{\frac{1}{2}\rho V^2 S c_y^\alpha}{m}\left(\frac{a_\delta}{a_\alpha}\right)\delta = K\delta. \qquad (6.1-7)$$

It is possible to use the actuator command $\delta_c = \frac{a_{yc}}{K}$ to replace the acceleration command a_{yc} to form a very simple aerodynamic feedback autopilot. When the dynamic pressure of the missile $q = \frac{1}{2}\rho V^2$ does not change much, and its static stability is relatively high (i.e., the value of a_δ/a_α changes little with the change of the positions of the center of gravity and the center of pressure). Even without artificial acceleration feedback, the missile body itself constructs a very simple aerodynamic autopilot (Fig. 6.1 – 32).

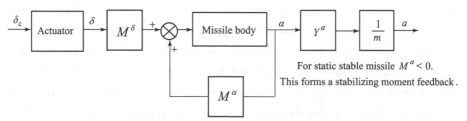

Fig. 6.1 – 32 Schematic diagram of an aerodynamic autopilot

Typically, anti-tank missiles maintain a constant speed close to the ground with the support of a sustainer rocket. Because their dynamic pressure remains relatively stable, a high static stability aerodynamic design can provide sufficient stabilizing moments to maintain control effectiveness α/δ. Consequently, many anti-tank missiles forego artificial acceleration autopilot systems to simplify the design and reduce costs.

However, the flight speed and flight height of ground-to-air missiles and air-to-air missiles could vary greatly (i.e., their dynamic pressure q changes greatly). A straightforward approach to address the impact of dynamic pressure variations is to make the actuator deflection angle inversely proportional to changes in missile dynamic pressure. This ensures that the missile's acceleration output remains consistent despite fluctuations in dynamic pressure. Early air-to-air missiles, such as the American "Sidewinder", Israel's "Python 3", and several others, employed this simple autopilot structure.

To implement this concept, these autopilots use the actuator driving moment command as the control command (Fig. 6.1 – 33). Since the actuator's hinge moment produced by the actuator incident angle of attack $\delta + \alpha$ is proportional to the missile flight dynamic pressure, the actuator hinge moment is balanced by the actuator driving moment. The generated corresponding normal acceleration will not be sensitive to the dynamic pressure change.

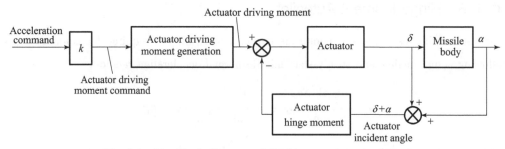

Fig. 6.1-33 Block diagram of the hinge moment autopilot

The actuator hinge moment H consists of two parts. One part is related to the actuator deflection angle δ and the other part is related to the angle of attack α of the missile. The hinge moment transfer function expression

$$H(s) = S \cdot q \cdot d(C_{H_\delta}(s) \cdot \delta(s) + C_{H_\alpha}(s) \cdot \alpha(s)), \qquad (6.1-8)$$

$$\frac{H(s)}{\delta(s)} = S \cdot q \cdot d\left(C_{H_\delta}(s) + C_{H_\alpha}(s) \frac{\alpha(s)}{\delta(s)}\right), \qquad (6.1-9)$$

where S and d are, respectively, the reference area and reference length of the actuator center of pressure position to the actuator axis, and q is the dynamic pressure.

The characteristics of this type of autopilot can be illustrated using an example from the literature [3], *Automatic Control of Aircraft and Missiles*. This autopilot design typically has two configurations. One of the more complex structures is depicted in Fig. 6.1-34. This configuration utilizes rate gyro feedback for the damping loop and incorporates an actuator deflection angle feedback loop to enhance the actuator response speed. To further stabilize the actuator mechanism, a phase lead compensator is employed. It is important to note that in this design, the primary feedback is the hinge moment feedback; without it, the fundamental nature of the actuator would be significantly altered.

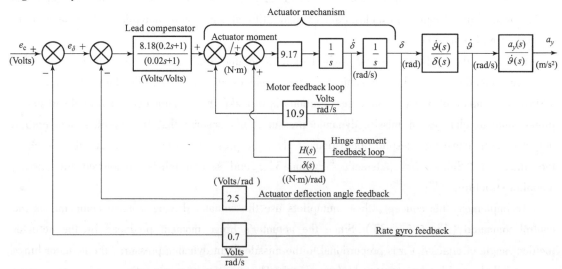

Fig. 6.1-34 Structure of the hinge moment autopilot (I)

The early hinge moment autopilot featured a straightforward design that did not include a rate gyro or an actuator deflection angle feedback loop. Its structure is illustrated in Fig. 6.1 – 35.

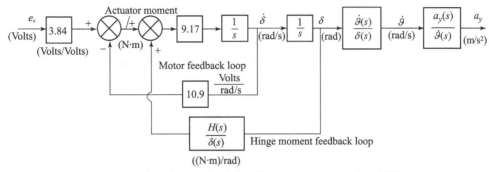

Fig. 6.1 – 35 Structure of the hinge moment autopilot (II)

For the above two examples, the related transfer functions are

$$\frac{H(s)}{\delta(s)} = \frac{-12.73(s^2 + 1.56s + 811.72)}{s^2 + 1.652s + 424.63}, \left(\frac{N \cdot m}{rad}\right), \quad (6.1-10)$$

$$\frac{\dot{\vartheta}(s)}{\delta(s)} = \frac{303.6(s + 1.75)}{s^2 + 1.652s + 424.63}, \left(\frac{rad/s}{rad}\right), \quad (6.1-11)$$

and

$$\frac{a_y(s)}{\dot{\vartheta}(s)} = \frac{-0.55(s^2 + 8.7s + 8050)}{s + 1.75}, \left(\frac{m/s^2}{rad/s}\right). \quad (6.1-12)$$

The normal acceleration time responses to the unit step actuator driving moment input e_c (V) for these two autopilots are shown in Fig. 6.1 – 36. As shown in the figure, the hinge moment autopilot maintains a steady-state normal acceleration that remains unaffected by variations in dynamic pressure. Additionally, the hinge moment autopilot equipped with an angular rate gyro and actuator deflection angle feedback offers a faster and smoother response.

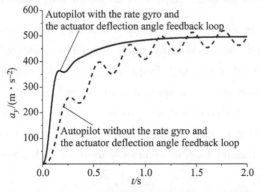

Fig. 6.1 – 36 Normal acceleration time responses of hinge moment autopilots to the unit step actuator driving moment input

6.1.7 Several Considerations in Acceleration Autopilot Design

1) Impact of missile elastic vibration modes on autopilot design

When accounting for missile body elasticity, actuator deflection not only induces rigid body rotation but also generates elastic angular motion in the missile. This elastic motion is theoretically comprised of multiple vibration modes; however, in autopilot design, typically only the first-order mode, and occasionally the second-order mode, need to be considered. Fig. 6.1 – 37 illustrates the positions of the autopilot rate gyro and accelerometer that are least sensitive to elastic vibrations.

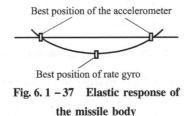

Fig. 6.1 – 37 Elastic response of the missile body

When the missile body elastic motion is taken into account, the block diagram of the autopilot angular velocity loop will be changed to that shown in Fig. 6.1 – 38. Here, ω_R is the missile's elastic variation angular velocity at the rate of gyro position.

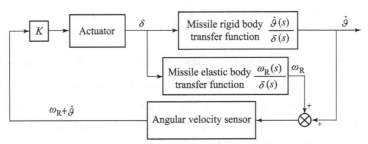

Fig. 6.1 – 38 Block diagram of the autopilot angular velocity loop with the elastic response of the missile body included

Due to the very low damping of missile elastic vibrations, the transfer function's Bode diagram can exhibit a high magnitude at its dominant mode frequency. This could result in the autopilot's open-loop gain exceeding 0 dB at this high frequency, potentially causing instability. Additionally, actuator friction and saturation at these frequencies can be problematic.

To address this issue, autopilot designs typically include a notch filter before the actuator input. This filter helps prevent control system instability and mitigates negative effects on actuator performance. The standard transfer function commonly used for such a notch filter is

$$G_F(s) = \frac{\frac{1}{\omega_i^2}s^2 + \frac{2\xi_1}{\omega_i}s + 1}{\frac{1}{\omega_i^2}s^2 + \frac{2\xi_2}{\omega_i}s + 1}, \quad (6.1-13)$$

where ω_i is the center frequency of the notch filter and is usually taken as the estimated value of the main elastic frequency of the missile body, and the denominator damping coefficient ξ_2 generally takes the value of 1. The notch depth and width are dependent on the value of the numerator damping coefficient ξ_1.

Fig. 6.1 – 39 shows the Bode diagram of a notch filter when $\omega_i = 40$ Hz, $\xi_2 = 1$, $\xi_1 = 0.1$, 0.15, and 0.2. The corresponding notch depths are – 20 db, – 16.5 db, and – 14 db, respectively.

In practical engineering design, it is crucial to ensure that the notch filter provides adequate signal attenuation even if the actual elastic frequency deviates slightly from its nominal value.

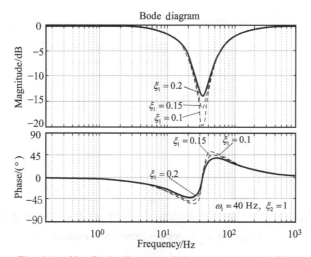

Fig. 6.1 – 39 Bode diagram of the example notch filter

2) Canard control and rear control

Based on Section 6.1.1, the transfer function of the missile body, which is controlled by the autopilot, is

$$\frac{a_y(s)}{\delta_z(s)} = \frac{k_\alpha(A_2 s^2 + A_1 s + 1)}{T_m^2 s^2 + 2\mu_m T_m s + 1}. \tag{6.1-14}$$

For rear-controlled missiles, $A_1 < 0$, $A_2 < 0$. Therefore, its transfer function can also be given as

$$\frac{a_y(s)}{\delta_z(s)} = \frac{k_\alpha(-|A_2| s^2 - |A_1| s + 1)}{T_m^2 s^2 + 2\mu_m T_m s + 1}. \tag{6.1-15}$$

Since this transfer function has a zero in the right half plane, a rear-controlled missile is a non-minimum phase system.

For canard-controlled missiles, $A_1 > 0$, $A_2 > 0$. So a canard-controlled missile has a transfer function as

$$\frac{a_y(s)}{\delta_z(s)} = \frac{k_\alpha(A_2 s^2 + A_1 s + 1)}{T_m^2 s^2 + 2\mu_m T_m s + 1}. \tag{6.1-16}$$

The zeros and the poles of this transfer function are all in the left half plane, so a canard-controlled missile is a minimum-phase system.

Fig. 6.1 – 40 shows the typical acceleration autopilot response of a rear-controlled missile to a

$a_{yc} = 50$ m/s² step acceleration command. In the figure, the related actuator deflection angle δ and angle of attack α are also given. It should be noted that initially, the output of the acceleration autopilot in rear control may oppose the expected acceleration response. The desired acceleration direction will only be achieved once the lift generated by the angle of attack surpasses the negative actuator force after the angle of attack reaches a sufficient magnitude. This issue is characteristic of non-minimum phase systems. The positive actuator deflection during the intermediate process helps to mitigate response overshoot. The ratio of the angle of attack to the actuator deflection in a steady state reflects the missile's control ratio.

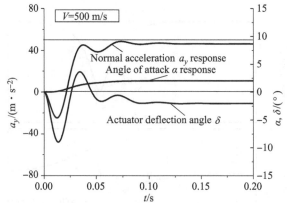

Fig. 6.1–40 The rear-controlled missile autopilot response to a $a_{yc} = 50$ m/s² step acceleration command

Fig. 6.1–41 shows the response of a canard-controlled acceleration autopilot to a $a_{yc} = 50$ m/s² step acceleration command. Also shown are the actuator deflection angle δ and the angle of attack α response. In this scenario, the forces initially generated by the actuator align with the lift produced by the angle of attack, leading to more efficient control. During the intermediate phase of control, the actuator deflection angle is reduced to suppress response overshoot. The ratio of the angle of attack to the actuator deflection angle in the steady state reflects the missile's control ratio. The autopilot controlled by canards demonstrates the typical transient response characteristics of a minimum-phase control system.

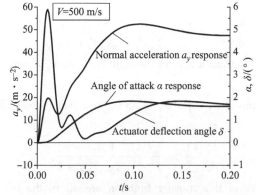

Fig. 6.1–41 The canard-controlled autopilot response to a $a_{yc} = 50$ m/s² step acceleration command

From the above analysis, it is evident that for a canard-controlled missile, the steady-state acceleration response is the result of the combined lift and actuator forces, both acting in the same direction. In contrast, for a rear-controlled missile, the steady-state acceleration response arises from the difference between the lift force and the actuator force. Suppose that the actuator angle generated force is $b_\delta \delta$. Then, the corresponding steady-state angle of attack α produced force will be $b_\alpha \alpha = b_\alpha \left(\dfrac{a_\delta}{a_\alpha} \right) \delta$. Therefore, the ratio of the two forces will be $K = b_\alpha \cdot \dfrac{a_\delta}{a_\alpha} \cdot \delta \cdot \dfrac{1}{b_\delta \delta} = \dfrac{b_\alpha}{b_\delta} \cdot \dfrac{a_\delta}{a_\alpha}$. That is to say, if the lift value is 1, the actuator force will be $\dfrac{1}{K}$. So, the gain ratio of the canard-controlled missile and the rear-controlled missile will be

$$\frac{\text{Canard controlled missile gain}}{\text{Rear controlled missile gain}} = \frac{1 + \dfrac{1}{K}}{1 - \dfrac{1}{K}}.$$

Suppose that $K = \dfrac{a_\delta b_\alpha}{a_\alpha b_\delta} = 6$, then the gain ratio of the canard control and the rear control will be $\dfrac{\left(1 + \dfrac{1}{6}\right)}{\left(1 - \dfrac{1}{6}\right)} = \dfrac{1.17}{0.83} = 1.4$.

Although canard control offers several advantages, it also has a drawback: the actuator's incident angle is the sum of the angle of attack and the actuator's deflection angle. In contrast, for rear control, the actuator's incident angle is the difference between the angle of attack and the actuator's deflection angle. Consequently, the usable actuator deflection angle in canard-controlled missiles is reduced compared to rear-controlled missiles due to saturation limits on the actuator's incident angle. As a result, most modern tactical missile designs prefer the rear-controlled aerodynamic configuration.

3) Static unstable missile control

The problem of static unstable missile control may occur for a short period of time in the initial phase of the powered flight for some tactical missiles. A static, unstable missile body itself is a divergent system without an autopilot. Fig. 6.1 – 42 shows the response of the actuator deflection angle, angle of attack, and acceleration to a $a_{yc} = 50$ m/s² step acceleration command for an acceleration autopilot of a rear-controlled static unstable missile.

Similarly, an initial negative actuator deflection angle is required to rotate the missile body and achieve a positive angle of attack. However, unlike a statically stable missile body, the actuator's static deflection angle must be positive in this case to counterbalance the destabilizing moment generated by the acceleration output needed to achieve the desired angle of attack (Fig. 6.1 – 43).

Additionally, the pseudo angle of attack feedback previously described can also be employed to stabilize the newly compensated missile body.

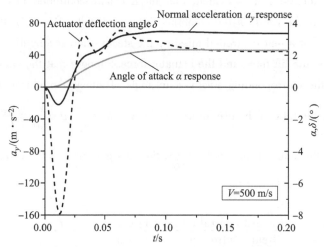

Fig. 6.1 – 42 Response of the rear-controlled static unstable missile body acceleration autopilot to a $a_{yc} = 50$ m/s² step acceleration command

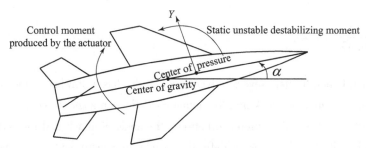

Fig. 6.1 – 43 Moment equilibrium relationship when a static unstable missile body produces a steady angle of attack

4) Discussion on angle-of-attack autopilot

Some researchers have explored the application of advanced robust control theory in designing acceleration autopilots. However, a rear-controlled missile body $\dfrac{a_y(s)}{a_{yc}(s)}$ is a non-minimum phase system. But the transfer function from the actuator angle to the angle of attack is a minimum phase system, since here B_1 is greater than 0.

$$\frac{\alpha(s)}{\delta(s)} = k_\alpha \frac{B_1 s + 1}{T_m^2 s^2 + 2\mu_m T_m s + 1}. \tag{6.1-17}$$

Theoretically, an angle-of-attack autopilot can be used indirectly to achieve acceleration control (Fig. 6.1 – 44).

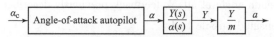

Fig. 6.1 – 44 Acceleration control with angle-of-attack autopilot

However, without an acceleration feedback loop, the variation of the aerodynamic parameters in the transfer function $\dfrac{Y(s)}{\alpha(s)}$ will result in an acceleration control error. The pros and cons of this approach in real engineering applications are to be carefully weighed by the designers.

6.1.8 Acceleration Autopilot Design Methods

6.1.8.1 Pole-placement method

The state feedback pole-placement method involves designing a state feedback control system for LTI system by assigning a set of desired poles (eigenvalues) to achieve specific performance criteria. As detailed in the literature [13], for autopilots with various structures, ignoring actuator dynamics, the state space model can be considered controllable and observable when state variables and output variables are appropriately chosen, as outlined in Table 6.1-8. Consequently, the pole-placement method can streamline autopilot design. This section will apply the pole-placement concept and analytical techniques to examine the design methods for classic two-loop and three-loop autopilots. For additional autopilot design methods and the derivation of pole-placement algorithms for specific state feedback and output feedback, please refer to the literature [13].

Table 6.1-8 Pole placement of several typical structural autopilots

Autopilot structures	State variables	Output variables	State/output controllability	Observability	Pole placement
Classic two-loop autopilot	$[\alpha \;\; \dot{\vartheta}]^T$	$[\alpha_y \;\; \dot{\vartheta}]^T$	√	√	√ (2 poles)
Two-loop autopilot with PI compensation	$[\alpha \;\; \vartheta \;\; \dot{\vartheta}]^T$	$[\alpha_y \;\; \int \alpha_y \;\; \dot{\vartheta}]^T$	√	√	√ (3 poles)
Classic three-loop autopilot	$[\alpha \;\; \vartheta \;\; \dot{\vartheta}]^T$	$[\int \alpha_y \;\; \vartheta \;\; \dot{\vartheta}]^T$	√	√	√ (3 poles)
Three-loop autopilot with pseudo angle of attack feedback	$[\alpha \;\; \vartheta \;\; \dot{\vartheta}]^T$	$[\int \alpha_y \;\; \alpha \;\; \dot{\vartheta}]^T$	√	√	√ (3 poles)
Attitude autopilot	$[\alpha \;\; \vartheta \;\; \dot{\vartheta}]^T$	$[\vartheta \;\; \dot{\vartheta}]^T$	√	×	× (3 poles)
Velocity vector autopilot	$[\alpha \;\; \vartheta \;\; \dot{\vartheta}]^T$	$[\theta \;\; \dot{\vartheta}]^T$	√	×	× (3 poles)

1) Design a two-loop acceleration autopilot with the pole-placement method

As a preliminary theoretical design, to simplify the design process, the hardware dynamics such as accelerometer, angular rate gyro, and actuator are not considered temporarily. The accelerometer gain is introduced, and the structure of the two-loop acceleration autopilot is changed into the form of Fig. 6.1-45 based on Fig. 6.1-1. Where k_{ac}, K_g, and k_s are accelerometer, angular rate gyro, and actuator gains respectively, and K_A and K_g are design parameters – the closed-loop gain adjustment coefficient.

All the feedback in Fig. 6.1-45 is equivalent to the actuator as

$$\delta_e = -k_s k_{ac} K_A (a_y + c\ddot{\vartheta}) - k_s K_g \dot{\vartheta}. \qquad (6.1-18)$$

Substituting it into the missile's equation of motion yields

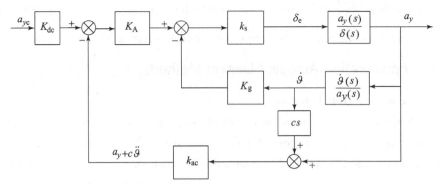

Fig. 6.1-45 Block diagram of the two-loop acceleration autopilot

$$\delta_e = -\frac{k_s k_{ac} K_A (Vb_\alpha - ca_\alpha)}{1 + k_s k_{ac} K_A (Vb_\delta - ca_\delta)}\alpha - \frac{k_s K_g - k_s k_{ac} K_A ca_\omega}{1 + k_s k_{ac} K_A (Vb_\delta - ca_\delta)}\dot{\vartheta}, \quad (6.1-19)$$

the gain matrix of state feedback K is

$$K = \begin{bmatrix} k_1 \\ k_2 \end{bmatrix}^T = \begin{bmatrix} \dfrac{k_s k_{ac} K_A (Vb_\alpha - ca_\alpha)}{1 + k_s k_{ac} K_A (Vb_\delta - ca_\delta)} \\ \dfrac{k_s K_g - k_s k_{ac} K_A ca_\omega}{1 + k_s k_{ac} K_A (Vb_\delta - ca_\delta)} \end{bmatrix}^T. \quad (6.1-20)$$

If output feedback is adopted, the equation $F = (I - KC^{-1}D)^{-1}KC^{-1}$ can be used to get

$$F = \begin{bmatrix} f_1 \\ f_2 \end{bmatrix}^T = \begin{bmatrix} k_s k_{ac} K_A \\ k_s K_g \end{bmatrix}^T = \begin{bmatrix} \dfrac{k_1}{(Vb_\alpha - ca_\alpha) - k_1(Vb_\delta - ca_\delta)} \\ \dfrac{k_1 ca_\omega + k_2(Vb_\alpha - ca_\alpha)}{(Vb_\alpha - ca_\alpha) - k_1(Vb_\delta - ca_\delta)} \end{bmatrix}^T. \quad (6.1-21)$$

Thus, the relationship between the design parameters of the two-loop autopilot and the state feedback and output feedback gains is given by

$$\begin{cases} K_A = \dfrac{f_1}{k_s k_{ac}} = \dfrac{k_1/(k_{ac} k_s)}{(Vb_\alpha - ca_\alpha) - k_1(Vb_\delta - ca_\delta)}, \\ K_g = \dfrac{f_2}{k_s} = \dfrac{k_1 ca_\omega/k_s + k_2(Vb_\alpha - ca_\alpha)/k_s}{(Vb_\alpha - ca_\alpha) - k_1(Vb_\delta - ca_\delta)} \end{cases} \quad (6.1-22)$$

According to the closed-loop transmission function of the two-loop autopilot, the expression of the closed-loop gain of the autopilot K can be obtained as

$$K = \frac{k_s K_A k_{\dot\vartheta} V}{1 + k_s k_{\dot\vartheta}(K_g + k_{ac} K_A V)} = \frac{-k_s K_A V(a_\delta b_\alpha - a_\alpha b_\delta)}{(a_\alpha + a_\omega b_\alpha) - k_s(K_g + k_{ac} K_A V)(a_\delta b_\alpha - a_\alpha b_\delta)}. \quad (6.1-23)$$

To ensure that the steady-state error in tracking acceleration commands is zero, the system must satisfy the final value theorem

$$\lim_{s \to 0} \frac{a_y}{a_{yc}} = 1. \quad (6.1-24)$$

Therefore, the closed-loop gain adjustment coefficient is

$$K_{dc} = \frac{(a_\alpha + a_\omega b_\alpha) - k_s(K_g + k_{ac}K_A V)(a_\delta b_\alpha - a_\alpha b_\delta)}{-k_s K_A V(a_\delta b_\alpha - a_\alpha b_\delta)}. \tag{6.1-25}$$

2) Design classic three-loop autopilot with pole-placement method

Similarly, the structure adjustment of the three-loop autopilot in Section 6.1.4 is shown in Fig. 6.1-46. Note that the design parameter variables K_g, ω_i, and K_A are different from those in Fig. 6.1-27.

All the feedback in Fig. 6.1-45 is equivalent to the actuator as

$$\delta_e = k_s k_{ac} K_A V \alpha - k_s(k_{ac}K_A V + \omega_i)\vartheta - k_s(K_g + c k_{ac} K_A)\dot{\vartheta}. \tag{6.1-26}$$

The gain matrix of state feedback $\boldsymbol{K}$ is

$$\boldsymbol{K} = \begin{bmatrix} k_1 \\ k_2 \\ k_3 \end{bmatrix}^T = \begin{bmatrix} -k_s k_{ac} K_A V \\ k_s(\omega_i + k_{ac} K_A V) \\ k_s(K_g + c k_{ac} K_A) \end{bmatrix}^T. \tag{6.1-27}$$

The gain matrix of output feedback $\boldsymbol{F}$ is

$$\boldsymbol{F} = \begin{bmatrix} f_1 \\ f_2 \\ f_3 \end{bmatrix}^T = \begin{bmatrix} k_s k_{ac} K_A \\ k_s \omega_i \\ k_s K_g \end{bmatrix}^T = \begin{bmatrix} -k_1/V \\ k_1 + k_2 \\ ck_1/V + k_3 \end{bmatrix}^T. \tag{6.1-28}$$

Therefore, the design parameters of the three-loop autopilot are

$$\begin{cases} K_A = f_1/(k_s k_{ac}) = -k_1/(k_s k_{ac} V), \\ \omega_i = f_2/k_s = (k_1 + k_2)/k_s, \\ K_g = f_3/k_s = (ck_1 + k_3 V)/(k_s V). \end{cases} \tag{6.1-29}$$

The closed-loop gain is

$$K = K_A V/(\omega_i + k_{ac} K_A V). \tag{6.1-30}$$

Equation (6.1-30) shows that the closed-loop gain K is related to the velocity and design parameters. It can be seen that, as described in Section 6.1.5, although the forward channel of the three-loop autopilot contains integral terms, it is not a structure without static error. To realize the accurate tracking of acceleration instruction (to ensure zero steady-state error), $K_{dc}K = 1$ is needed to obtain the expression of the pilot's closed-loop gain adjustment coefficient K_{dc}.

$$K_{dc} = (\omega_i + k_{ac} K_A V)/(K_A V). \tag{6.1-31}$$

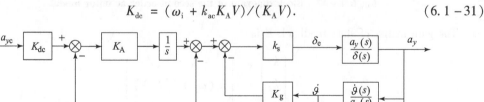

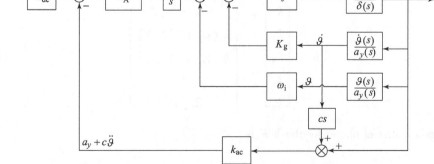

Fig. 6.1-46 Block diagram of the modified three-loop autopilot

In the previous autopilot pole-placement design, actuator dynamics were neglected, allowing for the arbitrary placement of the closed-loop poles without considering state feedback or output feedback. However, actual actuators cannot be infinitely fast. When the actuator frequency approaches the open-loop cutoff frequency of the autopilot, its impact on the closed-loop poles becomes significant. Therefore, actuator dynamics cannot be disregarded. Even if actuator dynamics are not included in the initial design, they should be examined once the autopilot design is complete.

In this context, the actuator is treated as an independent component within the autopilot design. Although actuator dynamics are incorporated in this section, they are used solely as constraints for pole placement and are not redesigned alongside the missile body for output or state feedback.

For illustration, consider a second-order actuator model, as depicted in Fig. 6.1-47. The transfer function of this actuator is

$$\frac{\delta_e}{\delta_{ec}} = \frac{k_s}{s^2/\omega_s^2 + 2\zeta_s s/\omega_s + 1} \tag{6.1-32}$$

In Fig. 6.1-46 and Fig. 6.1-47, the system state equation considering the steering gear dynamics is constructed, and the feedback is equivalent to u, then the following equation is obtained.

$$u = -k_s[-k_{ac}K_A V\alpha + (\omega_i + k_{ac}K_A V)\vartheta + (K_g + ck_{ac}K_A)\dot{\vartheta}] - \delta_e - (2\zeta_s/\omega_s)\dot{\delta}_e \tag{6.1-33}$$

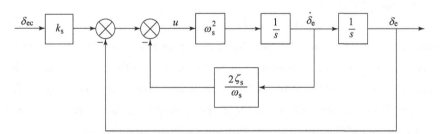

Fig. 6.1-47 Block diagram of the second-order actuator model

The gain matrix of state feedback K is

$$K = \begin{bmatrix} k_1 \\ k_2 \\ k_3 \\ k_4 \\ k_5 \end{bmatrix}^T = \begin{bmatrix} -k_s k_{ac} K_A V \\ k_s(\omega_i + k_{ac}K_A V) \\ k_s(K_g + ck_{ac}K_A) \\ 1 \\ 2\zeta_s/\omega_s \end{bmatrix}^T. \tag{6.1-34}$$

The gain matrix of output feedback F is

$$\boldsymbol{F} = \begin{bmatrix} f_1 \\ f_2 \\ f_3 \\ f_4 \\ f_5 \end{bmatrix}^{\mathrm{T}} = \begin{bmatrix} -k_1/V \\ k_1 + k_2 \\ ck_1/V + k_3 \\ k_4 \\ k_5 \end{bmatrix}^{\mathrm{T}}. \qquad (6.1-35)$$

The design parameters of the autopilot are

$$\begin{cases} K_A = f_1/(k_s k_{ac}) = -k_1/(k_s k_{ac} V), \\ \omega_i = f_2/k_s = (k_1 + k_2)/k_s, \\ K_g = f_3/k_s = (ck_1 + k_3 V)/(k_s V). \end{cases} \qquad (6.1-36)$$

Similarly, in the design process, the constraint $f_4 \approx 1$, $f_5 \approx 2\zeta_s/\omega_s$ needs to be satisfied.

6.1.8.2 Analytical method

This section explores the analytical design method for classical two-loop and three-loop autopilots, which is fundamentally similar to the pole-placement method. Thus, autopilots designed using the pole-placement method can also be designed using the analytical method described here.

1) Design of classic three-loop autopilot using the analytical method

To simplify the design process, actuator dynamics are initially excluded. The open-loop structure of the autopilot, with the actuator disconnected, is illustrated in Fig. 6.1-48.

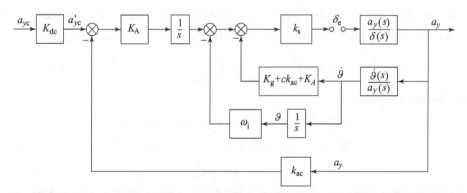

Fig. 6.1-48 Block diagram of the open-loop structure of the classic three-loop autopilot

In Fig. 6.1-48, the open-loop transmission function of the autopilot system is

$$\mathrm{HG}(s) = k_s \frac{a_y(s)}{\delta(s)} \left[(K_g + ck_{ac} K_A) \frac{\dot{\vartheta}(s)}{a_y(s)} + \omega_i \frac{\vartheta(s)}{a_y(s)} + \frac{k_{ac} K_A}{s} \right]. \qquad (6.1-37)$$

The closed-loop transmission function is

$$\frac{a_y(s)}{a'_{yc}(s)} = \frac{\delta(s)}{a'_{yc}(s)} \cdot \frac{a_y(s)}{\delta(s)} = \frac{k_s K_A/s}{1 + \mathrm{HG}(s)} \cdot \frac{a_y(s)}{\delta(s)}. \qquad (6.1-38)$$

By introducing intermediate variables M_2, M_1, M_0 and connecting Equation (6.1-37) and Equation (6.1-38), the open-loop transmission and closed-loop transmission of the autopilot can be expressed as

$$\mathrm{HG}(s) = \frac{M_2 s^2 + M_1 s + M_0}{s(s^2/\omega_m^2 + 2\zeta_m s/\omega_m + 1)}, \qquad (6.1-39)$$

$$\frac{a_y(s)}{a'_{yc}(s)} = \frac{k_s k_\vartheta^* K_A V(A_2 s^2 + A_1 s + 1)}{s(s^2/\omega_m^2 + 2\zeta_m s/\omega_m + 1) + M_2 s^2 + M_1 s + M_0}, \quad (6.1-40)$$

$$\begin{cases} M_2 = k_s k_\vartheta^* (K_g T_\alpha + c k_{ac} K_A T_\alpha + k_{ac} K_A V A_2), \\ M_1 = k_s k_\vartheta^* (K_g + c k_{ac} K_A + \omega_i T_\alpha + k_{ac} K_A V A_1), \\ M_0 = k_s k_\vartheta^* (\omega_i + k_{ac} K_A V). \end{cases} \quad (6.1-41)$$

Then, introduce intermediate variables $K_{\vartheta A}$, $K_{\vartheta I}$, and $K_{\vartheta G}$, and set

$$\begin{cases} K_{\vartheta A} = k_s k_\vartheta^* K_A, \\ K_{\vartheta I} = k_s k_\vartheta^* \omega_i, \\ K_{\vartheta G} = k_s k_\vartheta^* K_g, \end{cases} \quad (6.1-42)$$

M_2, M_1, and M_0 can be re-expressed as

$$\begin{cases} M_2 = T_\alpha K_{\vartheta G} + (cT_\alpha + VA_2) k_{ac} K_{\vartheta A}, \\ M_1 = K_{\vartheta G} + T_\alpha K_{\vartheta I} + (c + VA_1) k_{ac} K_{\vartheta A}, \\ M_0 = K_{\vartheta I} + k_{ac} V K_{\vartheta A}. \end{cases} \quad (6.1-43)$$

The matrix representation is

$$\begin{bmatrix} M_2 \\ M_1 \\ M_0 \end{bmatrix} = \begin{bmatrix} k_{ac}(cT_\alpha + VA_2) & 0 & T_\alpha \\ k_{ac}(c + VA_1) & T_\alpha & 1 \\ k_{ac}V & 1 & 0 \end{bmatrix} \begin{bmatrix} K_{\vartheta A} \\ K_{\vartheta I} \\ K_{\vartheta G} \end{bmatrix}. \quad (6.1-44)$$

At the same time, the autopilot closed-loop pole parameters τ, ω, and ζ are taken as constraint indexes, and the desired characteristic equation of the autopilot is assumed to satisfy

$$(1 + \tau s)\left(\frac{s^2}{\omega^2} + \frac{2\zeta}{\omega}s + 1\right) = \frac{s^3}{M_0 \omega_m^2} + \left(\frac{2\zeta_m}{M_0 \omega_m} + \frac{M_2}{M_0}\right)s^2 + \frac{1 + M_1}{M_0}s + 1. \quad (6.1-45)$$

Thus, the resulting equation is

$$\begin{bmatrix} M_2 \\ M_1 \\ M_0 \end{bmatrix} = \begin{bmatrix} 1/\tau\omega_m^2 + 2\tau(\zeta\omega - \zeta_m \omega_m)/\tau\omega_m^2 \\ 2\zeta\omega/\tau\omega_m^2 + \tau(\omega^2 - \omega_m^2)/\tau\omega_m^2 \\ \omega^2/\tau\omega_m^2 \end{bmatrix}. \quad (6.1-46)$$

In Equation (6.1-46), τ, ζ, ω, ω_m, and ζ_m are known. So M_2, M_1, and M_0 can also be calculated. In this way, the equation can be expressed as

$$\begin{bmatrix} K_{\vartheta A} \\ K_{\vartheta I} \\ K_{\vartheta G} \end{bmatrix} = \begin{bmatrix} \dfrac{1}{k_{ac}V(A_2 + T_\alpha^2 - A_1 T_\alpha)} & \dfrac{-T_\alpha}{k_{ac}V(A_2 + T_\alpha^2 - A_1 T_\alpha)} & \dfrac{T_\alpha^2}{k_{ac}V(A_2 + T_\alpha^2 - A_1 T_\alpha)} \\ \dfrac{-1}{A_2 + T_\alpha^2 - A_1 T_\alpha} & \dfrac{T_\alpha}{A_2 + T_\alpha^2 - A_1 T_\alpha} & \dfrac{A_2 - A_1 T_\alpha}{A_2 + T_\alpha^2 - A_1 T_\alpha} \\ \dfrac{VT_\alpha - c - VA_1}{V(A_2 + T_\alpha^2 - A_1 T_\alpha)} & \dfrac{cT_\alpha + VA_2}{V(A_2 + T_\alpha^2 - A_1 T_\alpha)} & \dfrac{-(cT_\alpha + VA_2)T_\alpha}{V(A_2 + T_\alpha^2 - A_1 T_\alpha)} \end{bmatrix} \begin{bmatrix} M_2 \\ M_1 \\ M_0 \end{bmatrix}.$$

$$(6.1-47)$$

By combining Equation (6.1-42) and Equation (6.1-47), the final formula for calculating the autopilot design parameters is derived as follows

$$\begin{cases} K_A = \dfrac{M_2 - T_\alpha M_1 + T_\alpha^2 M_0}{k_s k_\vartheta^* k_{ac} V(A_2 + T_\alpha^2 - A_1 T_\alpha)}, \\ \omega_i = \dfrac{-M_2 + T_\alpha M_1 + M_0}{k_s k_\vartheta^* (A_2 + T_\alpha^2 - A_1 T_\alpha)}, \\ K_g = \dfrac{(VT_\alpha - c - VA_1)M_2 + (cT_\alpha + VA_2)M_1 - (cT_\alpha + VA_2)T_\alpha M_0}{k_s k_\vartheta^* V(A_2 + T_\alpha^2 - A_1 T_\alpha)}. \end{cases} \quad (6.1-48)$$

It should be pointed out that although the open-loop crossing frequency index ω_{CR} is not introduced obviously in the design process, it is equivalent to the indirect introduction of ω_{CR} due to the approximate correspondence between ω and ω_{CR} as shown in Equation (6.1-49) (see the literature [1])

$$\omega = [\tau(\omega_{CR} + 2\zeta_m \omega_m) - 1]/2\zeta\tau. \quad (6.1-49)$$

At the same time, because the actuator dynamics are neglected in the autopilot design, there is a discrepancy between the open-loop crossover frequency of the autopilot designed using the above approach and the actual value. When considering actuator dynamics in the design, if the desired open-loop crossover frequency is to be achieved, adjustments to the set value of ω may be necessary, and the design process should be iterated until the target crossover frequency is attained. In this case, the autopilot parameters represent the final design outcome.

2) Design classic two-loop autopilot using the analytical method

The classic two-loop autopilot is designed using the aforementioned analytical method.

The open-loop transfer function of the classic two-loop autopilot is given by

$$HG(s) = \dfrac{M_2' s^2 + M_1' s + M_0'}{s^2/\omega_m^2 + 2\zeta_m s/\omega_m + 1}. \quad (6.1-50)$$

The closed-loop transmission function is

$$\begin{cases} M_2' = k_s k_{ac} k_\vartheta^* K_A (VA_2 + cT_\alpha), \\ M_1' = k_s k_\vartheta^* (k_{ac} K_A VA_1 + ck_{ac} K_A + K_g T_\alpha), \\ M_0' = k_s k_\vartheta^* (K_g + k_{ac} K_A V), \end{cases} \quad (6.1-51)$$

$$\dfrac{a_y(s)}{a_{yc}'(s)} = \dfrac{k_s K_A V k_\vartheta^* (A_2 s^2 + A_1 s + 1)}{(s^2/\omega_m^2 + 2\zeta_m s/\omega_m + 1) + (M_2' s^2 + M_1' s + M_0')}. \quad (6.1-52)$$

Taking ζ and ω as the design indices for the two-loop autopilot, it is assumed that the desired characteristic equation of the pilot satisfies

$$\left(\dfrac{s^2}{\omega^2} + \dfrac{2\zeta}{\omega}s + 1\right) = \left[\dfrac{1}{\omega_m^2(1+M_0')} + \dfrac{M_2'}{1+M_0'}\right]s^2 + \left[\dfrac{2\zeta_m}{\omega_m(1+M_0')} + \dfrac{M_1'}{1+M_0'}\right]s + 1. \quad (6.1-53)$$

By equating the coefficients of the corresponding terms, the following results are obtained

$$\begin{aligned} \dfrac{1}{\omega^2} &= \dfrac{1}{\omega_m^2(1+M_0')} + \dfrac{M_2'}{1+M_0'}, \\ \dfrac{2\zeta}{\omega} &= \dfrac{2\zeta_m}{\omega_m(1+M_0')} + \dfrac{M_1'}{1+M_0'}. \end{aligned} \quad (6.1-54)$$

Then introduce intermediate variables $K'_{\vartheta A}$ and $K'_{\vartheta G}$, and set

$$\begin{cases} K'_{\vartheta A} = k_s k_{\vartheta}^* K_A, \\ K'_{\vartheta G} = k_s k_{\vartheta}^* K_g. \end{cases} \quad (6.1-55)$$

After further derivation, the following results are obtained

$$\frac{\omega^2}{\omega_m^2} - 1 = K'_{\vartheta A} k_{ac} (V - VA_2 \omega^2 - cT_\alpha \omega^2) + K'_{\vartheta G},$$

$$\frac{2\omega \zeta_m}{\omega_m} - 2\zeta = K'_{\vartheta A} k_{ac} (2\zeta V - VA_1 \omega - c\omega) + K'_{\vartheta G} (2\zeta - T_\alpha \omega). \quad (6.1-56)$$

The system in matrix form is

$$\begin{bmatrix} (\omega^2/\omega_m^2 - 1) \\ 2\omega \zeta_m/\omega_m - 2\zeta \end{bmatrix} = \begin{bmatrix} k_{ac}(V - VA_2 \omega^2 - cT_\alpha \omega^2) & 1 \\ k_{ac}(2\zeta V - VA_1 \omega - c\omega) & 2\zeta - T_\alpha \omega \end{bmatrix} \begin{bmatrix} K'_{\vartheta A} \\ K'_{\vartheta G} \end{bmatrix}. \quad (6.1-57)$$

The design equation of K_A and K_g can be obtained by simultaneous Equation (6.1-55) and Equation (6.1-57)

$$\begin{bmatrix} K_A \\ K_g \end{bmatrix} = \begin{bmatrix} k_{ac}(V - VA_2 \omega^2 - cT_\alpha \omega^2) & 1 \\ k_{ac}(2\zeta V - VA_1 \omega - c\omega) & 2\zeta - T_\alpha \omega \end{bmatrix}^{-1} \begin{bmatrix} (\omega^2/\omega_m^2 - 1)/k_s k_{\vartheta}^* \\ 2(\omega \zeta_m/\omega_m - \zeta)/k_s k_{\vartheta}^* \end{bmatrix}.$$

$$(6.1-58)$$

6.1.8.3 Autopilot design examples

As shown in Table 6.1-9, the dynamic coefficients of a specific air-to-air missile, as provided in the literature [12], are selected for the autopilot-controlled object during mid-air flight. In this context, the notation $a_\alpha = -250$ indicates that the missile body is statically unstable. Table 6.1-10 presents a set of onboard hardware parameters and phase lags at various open-loop crossover frequencies of the autopilot. The data in Table 6.1-9 and Table 6.1-10 serve as the basis for the design and analysis in this section.

Table 6.1-9 Aerodynamic data of the missile

$V(\text{m} \cdot \text{s}^{-1})$	a_α/s^{-2}	a_δ/s^{-2}	a_ω/s^{-1}	b_α/s^{-1}	b_δ/s^{-1}	c/m
914.4	±250	280	1.5	1.6	0.23	0.68

Table 6.1-10 Hardware parameters and corresponding phase lags

Hardware	Damping ζ	Frequency $\omega/(\text{rad} \cdot \text{s}^{-1})$	Phase lag/(°)			
			35 rad/s	40 rad/s	45 rad/s	50 rad/s
Actuator	0.65	220	12.0	13.7	15.5	17.3
Rate gyro	0.65	300	8.7	10.0	11.3	12.6
Accelerometer	0.65	300	8.7	10.0	11.3	12.6
Filter	0.50	314	6.4	7.4	8.3	9.3
The total phase lag			27.1	31.1	35.1	39.2

It should be noted that since the rate gyro and the accelerometer are arranged in parallel and their dynamics are identical, it is sufficient to calculate the phase lag for only one of these components when determining the total phase lag in Table 6.1-10.

Based on the index selection principle outlined in the literature [1], the expected performance index for the autopilot is set as follows.

For two-loop acceleration autopilot: $\zeta = 0.9$, $\omega_{CR} = 45$ rad/s;

For three-loop autopilot: $\tau = 0.2s$, $\zeta = 0.9$, $\omega_{CR} = 45$ rad/s.

Table 6.1-11 shows the related design results of the two-loop and three-loop acceleration autopilot for statically stable missiles ($a_\alpha = 250$ s^{-2}), statically neutral stable missiles ($a_\alpha = 0$ s^{-2}), and statically unstable missiles ($a_\alpha = -250$ s^{-2}). ① means that the autopilot design only considers the actuator dynamics, and ② means that the autopilot design considers all hardware dynamics listed in Table 6.1-10. $\Delta\phi_K$ represents the phase lead angle that can be provided by the controller transmission function from output a_y to actuator at ω_{CR}; $\Delta\phi_{Total}$ represents the sum of the phase lag angles of the on-board hardware dynamics at ω_{CR} (as shown in the last line of Table 6.1-10); and $\Delta\phi_M$ represents the phase lag angle of the missile acceleration transfer function at ω_{CR}.

Since the phase lag of the missile $a_y(s)/\delta(s)$ at the autopilot's open-loop crossover frequency is close to $-180°$, achieving a phase margin of over $30°$ requires that the feedback correction loop of the autopilot provides an advance phase of more than $75°$ for a statically stable missile, and more than $60°$ for a statically unstable missile. Additionally, the dynamics of the angular rate gyro, accelerometer, and structural filters significantly affect both the magnitude and phase margin of the autopilot, and these effects must be carefully considered in the design.

Observing the magnitude and phase margins in Table 6.1-11 reveals that the magnitude margin of the classical three-loop autopilot is superior to that of the classical two-loop autopilot for statically unstable missiles.

Table 6.1-11 Design results of two kinds of classical acceleration autopilot

(a) Classical two-loop acceleration autopilots										
		K_A/ (rad·ms^{-2})	K_g/s	K_{dc}	ω_{CR}/ (rad·s^{-1})	ΔL/ dB	$\Delta\phi$/ (°)	$\Delta\phi_K$/ (°)	$\Delta\phi_{Total}$/ (°)	$\Delta\phi_M$/ (°)
Statically stable missiles	①	0.000 67	0.14	2.38	45	16.3	68.6	80.6	-15.5	-176.5
	②	0.000 63	0.14	2.35	45	7.8	49.3	80.9	-35.1	-176.5
Statically neutral stable missiles	①	0.001 4	0.15	1.12	45	14.5	58.1	70.6	-15.5	-177.0
	②	0.001 4	0.15	1.12	45	6.6	38.7	70.8	-35.1	-177.0
Statically unstable missiles	①	0.001 9	0.16	0.81	45	13.3	50.2	63.1	-15.5	-177.4
	②	0.001 9	0.16	0.81	45	5.6	30.8	63.3	-35.1	-177.4

(b) Classical three-loop acceleration autopilots											
		K_A/ (rad·ms^{-2})	ω_i/ s	K_g/s	K_{dc}	ω_{CR}/ (rad·s^{-1})	ΔL/ dB	$\Delta\phi$/ (°)	$\Delta\phi_K$/ (°)	$\Delta\phi_{Total}$/ (°)	$\Delta\phi_M$/ (°)
Statically stable missiles	①	0.004 9	1.13	0.14	1.25	45	16.8	66.8	78.8	−15.5	−176.5
	②	0.004 8	1.08	0.14	1.25	45	7.8	47.6	79.2	−35.1	−176.5
Statically neutral stable missiles	①	0.003 8	2.35	0.15	1.67	45	15.7	57.3	69.8	−15.5	−177.0
	②	0.003 7	2.30	0.15	1.68	45	6.8	37.9	70.0	−35.1	−177.0
Statically unstable missiles	①	0.002 6	3.51	0.16	2.46	45	14.9	50.0	62.9	−15.5	−177.4
	②	0.002 5	3.44	0.16	2.49	45	6.0	30.5	63.0	−35.1	−177.4

Fig. 6.1−49 shows the unit step response curves of the two-loop and three-loop acceleration autopilot when a_α is 250 s^{-2}, 0 s^{-2}, and −250 s^{-2}, respectively. The results indicate that, as discussed in Section 6.1.5, although the typical damping loop offers limited control over a statically

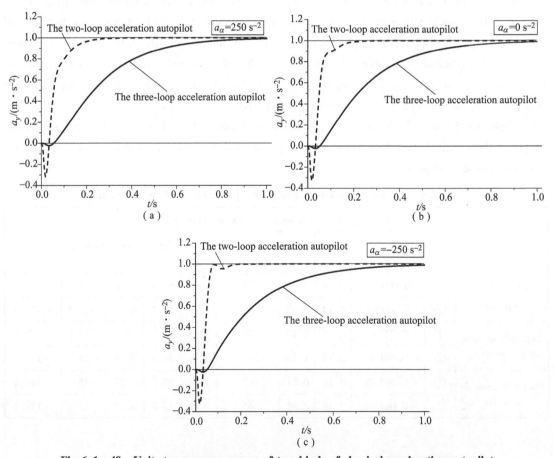

Fig. 6.1−49　Unit step response curves of two kinds of classical acceleration autopilot
(a) The statically stable missile; (b) The statically neutral stable missile; (c) The statically unstable missile

unstable missile, incorporating attitude angle feedback or acceleration feedback (i.e., angular velocity feedback) significantly enhances the autopilot's control capability for such missiles. For all three autopilot configurations, the addition of angular velocity feedback slightly reduces the autopilot's control ability over the statically unstable missile. However, this reduction is minimal, suggesting that angular velocity feedback effectively maintains the autopilot's control performance for statically unstable missiles.

Fig. 6.1 −50, Fig. 6.1 −51, and Fig. 6.1 −52 show the comparison of unit step responses for acceleration, rudder angle, and angle of attack, respectively, of the classical three-loop autopilot for missiles with different stability characteristics. Fig. 6.1 − 52 illustrates that a statically unstable missile requires a slightly smaller steady-state rudder angle compared to a statically stable missile, which needs a larger negative rudder angle at a steady state. During the initial phase, both the statically unstable missile and the statically neutral missile initially require a negative rudder angle. In the steady state, the statically unstable missile stabilizes with a positive rudder angle, while the statically neutral missile requires a rudder angle that is approximately zero.

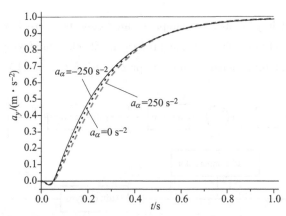

Fig. 6.1 −50 Unit step response curve of the acceleration for the classical three-loop autopilot (missiles with different stabilities)

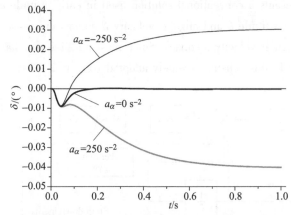

Fig. 6.1 −51 Unit step response curve of the rudder angle for the classical three-loop autopilot (missiles with different stabilities)

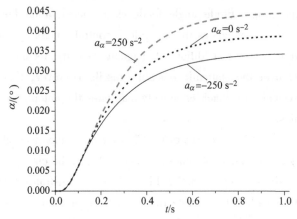

Fig. 6.1-52 Unit step response curve of the angle of attack for the classical three-loop autopilot (missiles with different stabilities)

§ 6.2　Pitch/Yaw Attitude Autopilot

The design of a pitch/yaw attitude autopilot can follow two possible structures. Structure I utilizes the output of an attitude gyro as the primary feedback, employing a lead compensator to generate the necessary damping signal for the damping loop (Fig. 6.2-1).

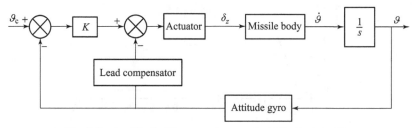

Fig. 6.2-1　Block diagram of an early attitude autopilot

This approach represents a conventional solution used in early attitude autopilots.

In Structure II, the attitude ϑ and angular velocity $\dot{\vartheta}$ signals are respectively used to form the attitude feedback and angular velocity feedback. Since most missiles now have a strap-down inertial navigation system onboard, this scheme is widely adopted (Fig. 6.2-2).

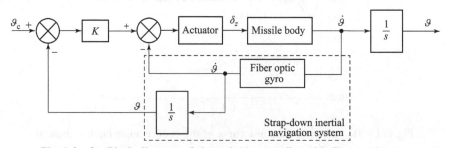

Fig. 6.2-2　Block diagram of the attitude autopilot with fiber optic gyro

The attitude autopilot was employed in the boosting phase of early ballistic missile programs for trajectory control. By utilizing the missile's angle of attack and the normal acceleration produced by the attitude control, indirect control of the flight path angle θ can be achieved (Fig. 6.2-3). Since no flight path angle feedback is used in this approach, all parameter variations in the transfer function from ϑ to θ will affect the trajectory control accuracy.

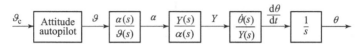

Fig. 6.2-3 Indirect control of θ with the help of an attitude autopilot

At present, as the flight path angle θ can be given by the strap-down integrated navigation system on a missile, the control of the θ angle can be accurately completed by the flight path angle autopilot without the help of attitude autopilots. Therefore, the use of attitude autopilots has become less common in recent times.

Currently, several potential applications for attitude autopilots include,

(1) Air-to-air missile applications. When a missile is launched from an aircraft, an attitude autopilot is often employed to stabilize the missile's attitude. This stabilization is crucial to prevent unwanted changes in missile orientation, as variations in the missile's attitude can alter its thrust direction. Such changes in trajectory could directly impact the safety of the launching aircraft.

(2) To achieve effective penetration, an air-to-ground missile must minimize its inertial angle of attack upon impact. The inertial angle of attack is defined as the angle between the missile's inertial velocity direction and its body x-axis, as provided by the onboard inertial navigation system. Therefore, an attitude autopilot is necessary to ensure that, during the final phase of flight, the missile's attitude aligns with its inertial velocity direction, thus minimizing the inertial angle of attack before penetration. It is important to note that, for simplification, some penetration missiles use an acceleration autopilot to achieve the desired effect. This autopilot can ensure a zero missile wind angle of attack (the angle between the missile's wind-affected velocity and its body axis) when the missile's normal acceleration is zero. However, this method may not be effective in the presence of wind.

(3) To maximize the electromagnetic damage when the missile approaches its target, it is crucial to stabilize the missile's direction so that it aligns with the direction that optimizes the electromagnetic warhead's damage effect.

An example illustrating the design features of this type of autopilot is provided.

Table 6.2-1 presents the parameters of the missile body and the actuator specifications for the required attitude autopilot. The design block diagram is shown in Fig. 6.2-4.

Table 6.2-1 Parameters of the example missile and actuator

Velocity	Parameters of missile dynamics					Actuator dynamics	
$V/(\text{m} \cdot \text{s}^{-1})$	a_α/s^{-2}	a_δ/s^{-2}	a_ω/s^{-1}	b_α/s^{-1}	b_δ/s^{-1}	T_s	k_s
306	62.3	71.6	1.17	1.02	0.17	0.025	-0.0703

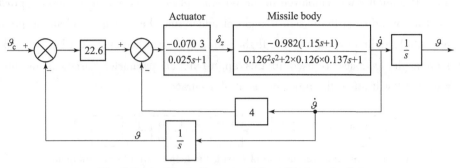

Fig. 6.2-4 Block diagram of the attitude autopilot

Fig. 6.2-5 shows the missile attitude ϑ response and flight path angle θ response for this attitude autopilot when a unit attitude angle command ϑ_c is given. When ϑ changes, an angle of attack α will be generated, and the related lift will change the missile flight path angle θ. According to the above sequence, θ will always be lagging behind ϑ response.

It is important to note that the angular velocity loop in an attitude autopilot is designed to enhance the damping of the attitude loop. Therefore, its phase angle compensation frequency is set close to the attitude loop crossover frequency. Given that the bandwidth difference between the attitude loop and the damping loop is relatively small, designing these two loops independently is not practical. The application of effective nonlinear programming optimization methods now allows for the simultaneous design of both loops, as discussed in the section on roll attitude autopilot design.

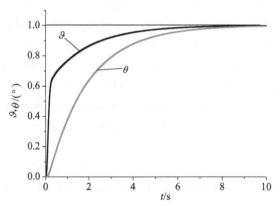

Fig. 6.2-5 Attitude autopilot step response

It should be noted that the relationship between the frequency bandwidths of the angular velocity loop and the angle loop can be categorized into three structures.

(1) When the angular velocity loop is used to enhance the damping of the angle loop, as is the case with the attitude autopilot described in this section, the frequency bandwidths of the two loops are similar. In such cases, independent design of the loops is not practical.

(2) In systems with a Type II angle loop structure, such as those used in ground guidance radars (see Chapter 4), the presence of two integrators and a $-180°$ phase lag requires that the angular velocity loop's frequency bandwidth be significantly higher than that of the angle

loop. Typically, the angular velocity loop bandwidth is set to be approximately five times greater than the angle loop bandwidth. Under these conditions, the two loops can be designed independently.

(3) When the angular velocity loop is designed to counteract the disturbance moments caused by the missile's angular motion affecting the seeker's angle tracking loop (see Chapter 5), a strong suppression of such disturbances necessitates that the seeker's angular velocity loop bandwidth be about ten to fifteen times greater than the frequency bandwidth of the seeker's angular tracking loop. Consequently, the two loops can also be designed independently in this scenario.

§ 6.3 Flight Path Angle Autopilot

Currently, flight path angle autopilots are commonly employed for controlling the initial programmed turn of ballistic missiles or rockets, as well as for velocity pursuit guidance. Two possible structures for flight path angle autopilots are illustrated in Fig. 6.3 – 1 and Fig. 6.3 – 2. The structure of Fig. 6.3 – 2 is superior to that of Fig. 6.3 – 1 as the acceleration autopilot inner loop of Fig. 6.3 – 2 has a better normal acceleration control than the aerodynamic acceleration control loop of Fig. 6.3 – 1.

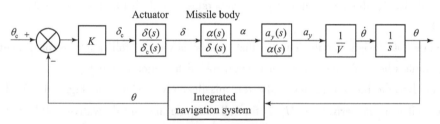

Fig. 6.3 – 1 Flight path angle autopilot, structure I

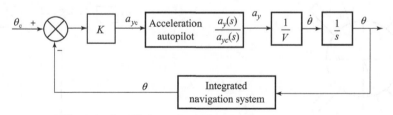

Fig. 6.3 – 2 Flight path angle autopilot, structure II

Fig. 6.3 – 3 illustrates the step responses of the velocity vector autopilot and the attitude autopilot for controlling the flight path angle. With the attitude autopilot, the function of suppressing attitude overshoot significantly limits the angle of attack required for changing the flight path angle. In contrast, the flight path angle autopilot allows for a larger attitude overshoot, which can result in a much quicker response in the flight path angle. Additionally, the flight path angle feedback improves the accuracy of flight path angle control.

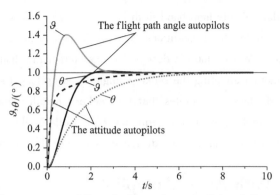

Fig. 6.3-3 Step responses of the velocity vector autopilot and the attitude autopilot

§ 6.4 Roll Attitude Autopilot

The primary distinction between the roll attitude autopilot and the pitch/yaw attitude autopilot lies in the need to account for potentially larger disturbance moments in the roll channel during the design process. This occurs when the missile's total angle of attack α_T plane (the missile maneuvering plane) is not aligned with the symmetrical planes of the missile body, the missile body aerodynamic asymmetry relative to the maneuvering plane generates a roll disturbance moment, as shown in Fig. 6.4-1.

The autopilot design should ensure that the steady-state roll angle error remains within acceptable limits when subjected to the maximum roll disturbance moment.

The aerodynamic block diagram of the roll missile body is shown in Fig. 6.4-2. Here, $M_x^{\delta_x}$ is the actuator roll moment derivative, $M_x^{\omega_x}$ is the roll damping moment derivative ($M_x^{\omega_x} < 0$), and J_x is the moment of inertia of the missile around x-axis.

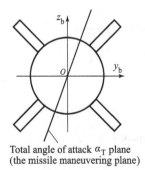

Fig. 6.4-1 Sketch of the maneuvering plane of an axial symmetrical missile

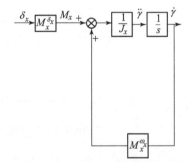

Fig. 6.4-2 Aerodynamic block diagram of the roll missile body

Therefore, the roll aerodynamic transfer function of the missile body will be

$$\frac{\dot{\gamma}(s)}{M_x(s)} = \frac{1}{M_x^{\omega_x}\left(\dfrac{1}{-M_x^{\omega_x}/J_x}s + 1\right)} = \frac{1}{M_x^{\omega_x}(T_r s + 1)}, \qquad (6.4-1)$$

where $T_r = \dfrac{J_x}{-M_x^{\omega_x}}$.

The transfer function from the actuator angle δ_x to roll angular velocity $\dot{\gamma}$ is

$$\frac{\dot{\gamma}(s)}{\delta_x(s)} = \frac{k_r}{T_r s + 1}, \qquad (6.4-2)$$

where $k_r = \dfrac{-M_x^{\delta_x}}{M_x^{\omega_x}}$.

Fig. 6.4-3 shows the typical structure of a roll attitude autopilot. Here, K_A and $K_{\omega x}$ are the autopilot design parameters, ω_s and μ_s are the actuator's natural frequency and damping. The value of K_A and $K_{\omega x}$ can be determined by selecting an appropriate objective function and design constraints and using an available nonlinear programming optimization software to search for their optimum solutions.

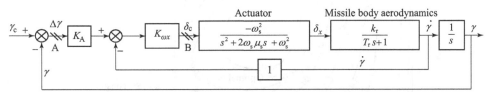

Fig. 6.4-3 Typical structure of a roll attitude autopilot

An example of the design process will now be explained in detail.

Suppose that, for example, missile $J_x = 0.96$ kg·m², $M_x^{\delta_x} = -13\,500$ (N·m/rad), $M_x^{\omega_x} = -37.3$ (N·m/(rad·s^{-1})), $T_r = 0.025\,7$s. Therefore

$$\frac{\dot{\gamma}(s)}{M_x(s)} = \frac{1}{37.3(0.025\,7s + 1)}. \qquad (6.4-3)$$

If the actuator second order model has $\omega_s = 180$ rad/s, $\mu_s = 0.7$, Fig. 6.4-4 shows the block diagram of the roll channel autopilot to be designed. The values of K_A and $K_{\omega x}$ for the roll autopilot can then be determined using the following optimization algorithm.

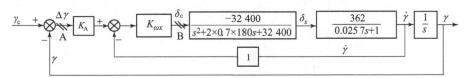

Fig. 6.4-4 Block diagram of the example roll channel autopilot

Assuming the design constraints include a phase margin of at least 55° and a gain margin of at least 8 dB, the design process will be as follows. An optimum transient to a step command γ_c will be obtained by minimizing the following objective function. Here, T is the autopilot transient settling time. The mathematical description of this optimization problem is

$$\begin{cases} \text{Constraint}: \Delta\phi(K_A, K_{\omega x}) > 55° \text{ and } \Delta L(K_A, K_{\omega x}) > 8 \text{ dB}, \\ \min J(K_A, K_{\omega x}) = \min \int_0^T |\gamma(t) - \gamma_{\text{steady state}}| \, dt (\text{take } \gamma_c = 1, \gamma_{\text{steady state}} = 1, T = 0.4 \text{ s}). \end{cases}$$

Fig. 6.4-5 shows the constraint boundaries of gain margin ΔL and phase margin $\Delta\phi$, as well as the contour line of the objective function for design parameters K_A and $K_{\omega x}$. It can be seen that the minimum value of the objective function is 0.034, and the corresponding optimal design results are $K_A = 50.0$ and $K_{\omega x} = 0.00399$ with the satisfaction of the above-mentioned gain and phase margin constraints. It is evident that, for this design, the system response speed is constrained by the phase margin requirement.

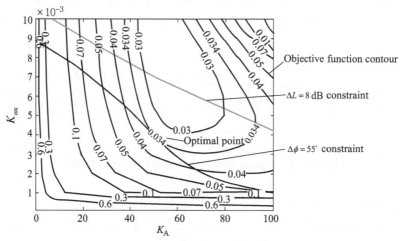

Fig. 6.4-5 Contour line of the objective function and frequency domain constraints

Fig. 6.4-6 displays the time-domain step response of the roll autopilot.

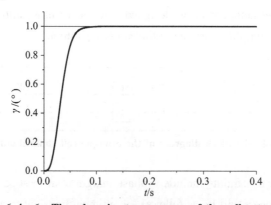

Fig. 6.4-6 Time-domain step response of the roll autopilot

A roll autopilot is usually affected by a disturbance moment L as the missile is maneuvering in the missile asymmetric plane and is required to have sufficient anti-disturbance ability. Given that the maximum allowable roll angle error under the action of the disturbance moment L is $\Delta\gamma$, then the autopilot gain from the allowable steady-state error to the counter-balance control moment must be no less than the following value K. As the autopilot gain value is constrained by stability considerations, a lag compensator in the form $\beta\left(\dfrac{T_b s + 1}{\beta T_b s + 1}\right)$ could be added to increase the autopilot low-frequency gain by a factor of β. Here, the value of β should be chosen to satisfy the constraint K and the value of T_b should be taken to reduce the roll error to the allowed value at the end of the autopilot transient. The block diagrams of the roll autopilot with phase lag compensation are shown in Fig. 6.4 – 7 and Fig. 6.4 – 8.

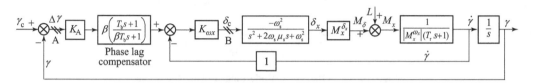

Fig. 6.4 – 7 Block diagram of the roll attitude autopilot with phase lag compensation

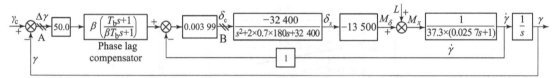

Fig. 6.4 – 8 Block diagram of the roll attitude autopilot with phase lag compensation for the given example

Assuming that the maximum disturbance moment acting on the missile body is $L = 1\ 000$ NM, and the maximum allowable roll angle error is $\Delta\gamma = 0.05$ rad $= 2.9°$, then the transfer ratio of the allowed roll angle error to the autopilot countering balancing control moment should be $K = \dfrac{1\ 000}{0.05} = 20\ 000$. However, the gain of the roll angle to the control moment of Fig. 6.4 – 4 is only 2 692.4. To make the previous system have a sufficient anti-disturbance gain in the low-frequency range, the value of β for the lag compensator should be no less than the following value:

$$\beta = \frac{20\ 000}{2\ 692.4} = 7.43. \qquad (6.4-4)$$

To ensure that the stability margin, transient overshoot, and roll error elimination speed of the designed autopilot meet their respective design requirements, the time constant of the phase lag network is selected as $T_1 = 0.12$ s. Therefore, the parameters of the lag network are determined to be:

$$\beta\left(\frac{T_b s + 1}{\beta T_b s + 1}\right) = 7.43 \times \left(\frac{0.12s + 1}{0.892s + 1}\right). \qquad (6.4-5)$$

The open-loop Bode diagrams for both the inner and outer loops of the roll autopilot, with and without lag compensation, are shown in Fig. 6.4 – 9 and Fig. 6.4 – 10.

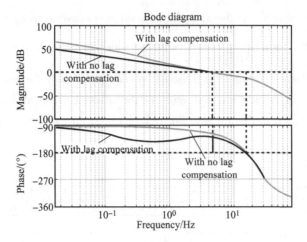

Fig. 6.4 – 9　Open-loop Bode diagram for the outer loop of the roll autopilot

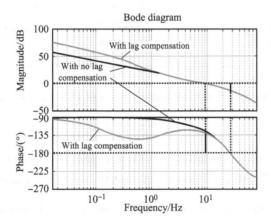

Fig. 6.4 – 10　Open-loop Bode diagram for the inner loop of the roll autopilot

Table 6.4 – 1 presents the crossover frequencies of the inner and outer loops, both with and without lag compensation, along with the corresponding phase and gain margins.

Table 6.4 – 1　Characteristics of system magnitude frequency

Items	Systems	ΔL/dB	Magnitude cross over frequency/Hz	$\Delta \varphi$/(°)	Phase cross over frequency/Hz
Outer loop	With no lag compensation	11.3	16.4	66.6	4.71
	With lag compensation	10.8	15.7	52.6	4.87
Inner loop	With no lag compensation	12.2	27.5	55	9.7
	With lag compensation	12.2	27.4	53.1	9.2

Fig. 6.4 – 11 shows the closed-loop Bode diagram of the system with lag compensation.

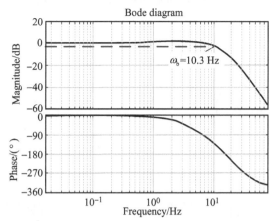

Fig. 6.4 – 11 Closed-loop Bode diagram with lag compensation

Fig. 6.4 – 12 and Fig. 6.4 – 13 illustrate the step response of the autopilot with and without lag compensation, respectively, as well as the transition process of the autopilot in the presence of disturbances. These figures demonstrate that, with the aid of lag compensation, the roll error caused by the disturbance moment L is reduced to the permissible value of 2.9° by the end of the autopilot's transient response.

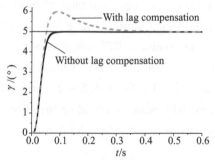

Fig. 6.4 – 12 Step response to a command $\gamma_c = 5°$

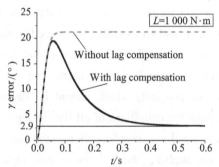

Fig. 6.4 – 13 System response under the disturbance moment L with command $\gamma_c = 0°$

§ 6.5 BTT Autopilot

The aerodynamic layout characteristic of a BTT missile is typically surface-symmetrical. In general, its lift surface is large in one direction and small in the perpendicular direction. The missile's maneuvering is achieved by simultaneously controlling its roll and pitch, thereby generating lift in the desired maneuvering plane through the angle of attack. Possible aerodynamic configurations for a BTT missile are illustrated in Fig. 6.5 – 1.

Fig. 6.5 – 1 (d) represents the aerodynamic layout of the air-to-air missile "Meteor". This design incorporates a ramjet rocket, which maintains high missile speed even towards the end of the engagement. However, many tactical missiles use rear control surfaces for roll control instead of ailerons.

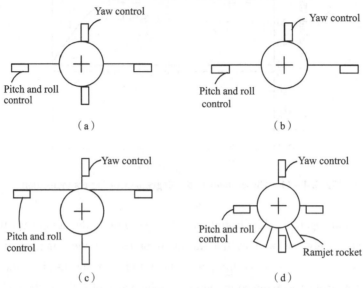

Fig. 6.5 – 1   Possible schemes of the BTT missile aerodynamic layout
(a) Scheme 1; (b) Scheme 2; (c) Scheme 3; (d) Scheme 4

For axis-symmetric missiles, aerodynamic couplings between yaw, roll, and pitch channels are typically small and can usually be ignored during the autopilot design phase. Their effects are often assessed only in the final six-degree-of-freedom simulations. In contrast, BTT missiles, with their unique non-axisymmetric aerodynamic configurations, experience significant aerodynamic coupling between yaw and roll, which cannot be disregarded. Table 6.5 – 1, Table 6.5 – 2, and Table 6.5 – 3 show the definitions of all the dynamic coefficients for BTT missiles including the effects of yaw-roll and roll-yaw coupling. Here, they are defined respectively as: δ_e (the elevator angle), δ_r (the rudder angle), and δ_a (the aileron angle).

Table 6.5 – 1   Dynamic coefficients of the uncoupled model

Items	Symbols	Expressions	Significances
Pitch	$a_\alpha = -M_z^\alpha/J_z$	$-m_z^\alpha qSL/J_z$	The effect of pitch moment caused by the angle of attack m_z^α, for static stable missile $m_z^\alpha < 0$
	$a_{\omega_z} = -M_z^{\omega_z}/J_z$	$-m_z^{\omega_z}qSL^2/(J_zV)$	The effect of pitch damping moment caused by pitch angular velocity $m_z^{\omega_z} < 0$
	$a_{\delta_e} = -M_z^{\delta_e}/J_z$	$-m_z^{\delta_e}qSL/J_z$	The effect of pitch moment caused by the elevator $m_z^{\delta_e} < 0$
	$b_\alpha = (P+Y^\alpha)/(mV)$	$(P+c_y^\alpha qS)/(mV)$	The effect of pitch force caused by the angle of attack $c_y^\alpha > 0$
	$b_{\delta_e} = Y^{\delta_e}/(mV)$	$c_y^{\delta_e}qS/(mV)$	The effect of pitch force caused by the elevator $c_y^{\delta_e} > 0$

Continued

Items	Symbols	Expressions	Significances
Yaw	$a_\beta = -M_y^\beta/J_y$	$-m_y^\beta qSl/J_y$	The effect of yaw moment caused by the sideslip angle $m_y^\beta < 0$
	$a_{\omega_y} = -M_y^{\omega_y}/J_y$	$-m_y^{\omega_y} qSl^2/(J_y V)$	The effect of yaw damping moment caused by the yaw anglar velocity $m_y^{\omega_y} < 0$
	$a_{\delta_r} = -M_y^{\delta_r}/J_y$	$-m_y^{\delta_r} qSl/J_y$	The effect of yaw moment caused by the rudder $m_y^{\delta_r} < 0$
	$b_\beta = -Z^\beta/(mV)$	$-c_z^{\delta_r} qS/(mV)$	The effect of yaw force caused by the sideslip angle $c_z^{\delta_r} < 0$
	$b_{\delta_r} = -Z^{\delta_r}/(mV)$	$-c_z^{\delta_r} qS/(mV)$	The effect of yaw force caused by the rudder $c_z^{\delta_r} < 0$
Roll	$c_{\omega_x} = -M_x^{\omega_x}/J_x$	$-m_x^{\omega_x} qSl^2/(2J_x V)$	The effect of roll moment caused by the roll angular velocity $m_x^{\omega_x} < 0$
	$c_{\delta_a} = -M_x^{\delta_a}/J_x$	$-m_x^{\delta_a} qSl/J_x$	The effect of roll moment caused by the aileron $m_x^{\delta_a} < 0$

Table 6.5 – 2 Dynamic coefficients of the yaw-roll coupling

Symbols	Expressions	Significances
$c_\beta = -M_x^\beta/J_x$	$-m_x^\beta qSl/J_x$	The effect of roll moment caused by the yaw sideslip angle $m_x^\beta < 0$
$c_{\omega_y} = -M_x^{\omega_y}/J_x$	$-m_x^{\omega_y} qSl^2/(J_x V)$	The effect of roll moment caused by the yaw angular velocity $m_x^{\omega_y} < 0$
$c_{\delta_r} = -M_x^{\delta_r}/J_x$	$-m_x^{\delta_r} qSl/J_x$	The effect of roll moment caused by the yaw rudder $m_x^{\delta_r} < 0$

Table 6.5 – 3 Dynamic coefficients of the roll-yaw coupling

Symbols	Expressions	Significances
$a_{\omega_x} = M_y^{\omega_x}/J_y$	$m_y^{\omega_x} qSL^2/(2J_y V)$	The effect of yaw moment caused by the roll angular velocity $m_y^{\omega_x} > 0$
$a_{\delta_a} = -M_y^{\delta_a}/J_y$	$-m_y^{\delta_a} qSl/J_y$	The effect of yaw moment caused by the roll aileron $m_y^{\delta_a}$

The following illustrates the physical mechanisms behind several important yaw and roll coupling moments.

(1) The roll moment M_x^β produced by the coupling of the sideslip angle.

The aerodynamic coefficient c_β related to the roll moment generated by the coupling of the sideslip angle is defined as

$$c_\beta = -m_x^\beta qSl/J_x = -M_x^\beta/J_x. \qquad (6.5-1)$$

As shown in Fig. 6.5 – 2, when the missile has a sideslip angle and an angle of attack α for a swept wing configuration missile, the velocity components V_R, V_L perpendicular to the leading edges of the two wings are different ($V_R > V_L$), and the lift of the right-wing will be greater than that of the left-wing, thus creating a coupling roll moment.

(2) The roll moment $M_x^{\omega_y}$ is caused by the yaw angular velocity.

The aerodynamic coefficient c_{ω_y} associated with the coupling roll moment due to yaw angular velocity is defined as

$$c_{\omega_y} = -m_x^{\omega_y} qSl^2/(J_x V). \qquad (6.5-2)$$

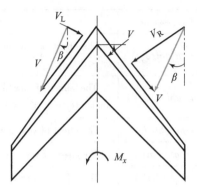

Fig. 6.5-2 Effect of wing sweep angle on the roll/yaw coupling

As shown in Fig. 6.5-3, in the case of the angle of attack $\alpha > 0°$, a yaw angular velocity exists. Here, when the right wing moves forward, its relative airspeed and lift force increase. Conversely, as the left wing moves backward, its relative airspeed and lift force decrease, resulting in a negative roll moment.

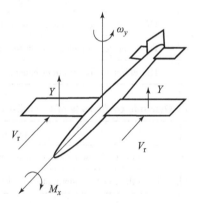

Fig. 6.5-3 Sketch of the roll moment produced by yaw angular velocity

(3) The yaw moment $M_y^{\omega_x}$ is produced by the roll angular velocity.

As shown in Fig. 6.5-4, if a roll angular velocity exists and $\alpha = 0°$, the roll angular velocity will make the right wing have a positive angle of attack and the left wing a negative angle of attack. The difference in the horizontal projection of the lift forces between the two wings generates a coupling yaw moment.

The LTI mathematical model for a BTT missile, incorporating these couplings, is described by the following equations.

Pitch:

$$\begin{aligned} \dot{\alpha} &= -b_\alpha \alpha + \omega_z - b_{\delta_e}\delta_e - \omega_x \beta \\ \dot{\omega}_z &= -a_\alpha \alpha - a_{\omega_z}\omega_z - a_{\delta_e}\delta_e \end{aligned}, \quad (6.5-3)$$

Yaw:

$$\dot{\beta} = -b_\beta \beta + \omega_y - b_{\delta_\gamma}\delta_\gamma + \omega_x \alpha - b_{\delta_a}\delta_a$$
$$\dot{\omega}_y = -a_\beta \beta - a_{\omega_y}\omega_y - a_{\delta_\gamma}\delta_\gamma + a_{\omega_x}\omega_x - a_{\delta_a}\delta_a, \tag{6.5-4}$$

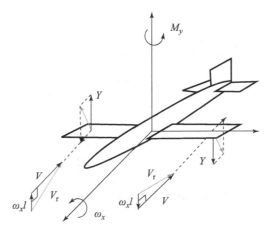

Fig. 6.5-4 Sketch of the yaw moment generated by roll angular velocity

Roll:
$$\dot{\gamma} = \omega_x$$
$$\dot{\omega}_x = -c_{\omega_x}\omega_x - c_{\delta_a}\delta_a - c_\beta \beta - c_{\omega_y}\omega_y - c_{\delta_\gamma}\delta_\gamma. \tag{6.5-5}$$

In Equation (6.5-3), Equation (6.5-4), and Equation (6.5-5), the $\omega_x\beta$ term in pitch and the $\omega_x\alpha$ terms in yaw are kinematic coupling-related terms. Since in BTT control, angular velocity ω_x is often large, it is not possible to consider these two items as second-order small quantities and to ignore them. However, incorporating these couplings into the autopilot design would transform the model into a nonlinear system, thereby rendering linear control theory inapplicable. Consequently, the standard approach is to initially exclude the kinematic couplings from the autopilot design and apply linear control theories. The nonlinear effects of these couplings are then thoroughly assessed in a six-degree-of-freedom simulation at the final stage.

The following is a brief description of the kinematic coupling mechanism. Assuming the missile experiences a sudden roll $\Delta\gamma$ while maintaining a constant angle of attack α_0, the resulting polar graph is illustrated in Fig. 6.5-5.

The point x_b on the graph is the missile-axis direction, the plane $x_b y_{b0}$ is the pitch symmetrical plane of the missile before rolling, the point V is the direction of the velocity axis, and the angle between x_b axis and V axis is the angle of attack α_0. Since the inertia of the missile velocity vector is much larger than that of the missile attitude, the velocity axis can be supposed to be in its original direction when the missile is rolling by an angle $\Delta\gamma$. It can be seen that a sideslip angle $\beta = \Delta\gamma \cdot \alpha_0$ and an angle of attack increment $\Delta\alpha = -\Delta\gamma \cdot \beta$ are formed relative to the pitch symmetrical plane of the missile $x_b y_b$ after a roll angle $\Delta\gamma$. As $\omega_x = \dfrac{\Delta\gamma}{\Delta t}$, the two terms related to the $\dot{\alpha}$ and $\dot{\beta}$ should be $\omega_x\beta$ and $\omega_x\alpha$ in the system equations respectively.

The following analysis uses a model of a plane-symmetric air-to-air missile equipped with a ramjet rocket to examine the characteristics of the BTT missile coupling model. The dynamic equations for this missile are given below.

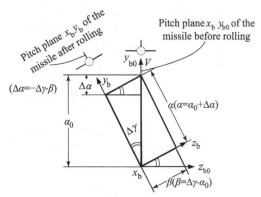

Fig. 6.5 – 5 Polar graph of the BTT missile kinematic coupling

Pitch channel:
$$\dot{\alpha} = \omega_z - 3.0286\alpha - 0.7785\delta_e - \omega_x\beta,$$
$$\dot{\omega}_z = -0.028\omega_z + 149\alpha - 701\delta_e. \quad (6.5-6)$$

Yaw channel:
$$\dot{\beta} = \omega_y - 0.853\beta - 0.8\delta_r + \alpha\omega_x,$$
$$\dot{\omega}_y = 0.0028\omega_x - 0.028\omega_y - 295\beta - 85\delta_a - 695\delta_r. \quad (6.5-7)$$

Roll channel:
$$\dot{\gamma} = \omega_x,$$
$$\dot{\omega}_x = -2.61\omega_x - 0.014\omega_y - 6438\beta - 11113\delta_a. \quad (6.5-8)$$

Firstly, the impact of roll coupling on yaw is analyzed as follows.

In the yaw angular acceleration equation, the dynamic coefficient of the roll aileron coupling to yaw a_{δ_a} is 85, but the yaw rudder dynamic coefficient a_{δ_r} is 695. Therefore, the roll aileron coupling effect is very small compared to the rudder. In addition, the dynamic coefficient of the roll angular velocity coupling a_{ω_x} to yaw is 0.0028, while the dynamic coefficient of the yaw angular velocity effect a_{ω_y} is 0.028. Therefore, compared to the effect of yaw angular velocity, the roll angular velocity coupling is relatively weak. This indicates that the roll coupling to yaw is considered weak.

Next, the coupling effect of yaw on the roll will be discussed. The coupling dynamic coefficient of the yaw angular velocity to the roll c_{ω_y} is 0.014, but the roll angular velocity coefficient c_{ω_x} is 2.61. Therefore, the influence of the yaw angular velocity on the roll is also very weak. Then, let us analyze the influence of the sideslip angle β on the roll. The dynamic coefficient of the sideslip angle to the roll coupling c_β is 6438, and the dynamic coefficient of the roll aileron c_{δ_a} is 11113. This suggests that to counteract a 1° sideslip angle-induced disturbance roll moment, a roll aileron deflection angle of 0.6° is needed. Clearly, the roll coupling moment due to the sideslip angle is a significant term in the roll equation. Thus, the coupling of yaw sideslip to roll is considered strong.

The response of the sideslip angle β to a step yaw disturbance moment is given in Fig. 6.5 – 6.

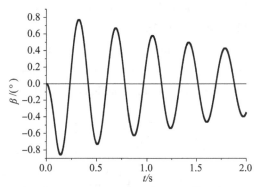

Fig. 6.5-6 The β response to a step yaw disturbance moment

This yaw sideslip response will be coupled to the roll channel owing to the coupling moment m_x^β effect, and its related coupling transfer function is given as follows

$$\frac{\gamma(s)}{\beta(s)} = \frac{-2\,466.8(0.001\,2s + 1)}{s(0.383s + 2.61)(0.001\,2s + 1)}. \tag{6.5-9}$$

At the yaw oscillation frequency of 2.73 Hz (also the coupling roll oscillation frequency), the transfer function $\dfrac{\gamma(s)}{\beta(s)}$ has a gain of 21.6. It means that a very strong coupling effect exists from yaw to roll. This is because the missile roll moment of inertia is small. Fig. 6.5-7 shows the roll response to the coupling moment m_x^β.

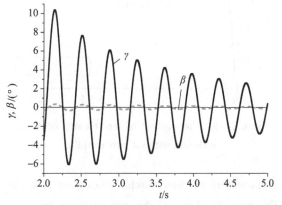

Fig. 6.5-7 Relation curves of β and γ for the Dutch roll mode

Typically, the coupling between yaw and roll motions is referred to as the Dutch roll mode.

Based on the analysis, it is evident that for a BTT missile, where yaw and roll represent a strong coupling system, the Multiple-Input-Multiple-Output (MIMO) control theory is generally recommended for such designs. However, BTT missile control is unique in that only roll and pitch channels are needed for maneuvering in a specific direction, while the yaw channel is not essential. The roll moment caused by the yaw sideslip is considered an interference in roll channel control. Therefore, an independent yaw autopilot can be designed to minimize the sideslip angle β to nearly zero during roll and pitch control, thereby reducing the yaw-to-roll interference to a weak coupling. This approach allows for the separate design of roll and pitch controls for the BTT missile.

The maneuver described, involving a sideslip-suppressing yaw autopilot for a BTT aircraft, is commonly referred to as a BTT coordinated turn.

The structure and parameters of the roll autopilot designed for this BTT missile example are illustrated in Fig. 6.5 – 8. Here, the actuator has a second-order model with an undamped natural frequency $\omega_n = 150$ rad/s $= 24$ Hz and damping coefficient $\xi = 0.7$.

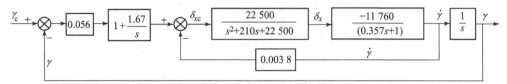

Fig. 6.5 – 8 Block diagram of the roll autopilot for the BTT missile

Fig. 6.5 – 9 shows the missile roll autopilot response to a 20° step roll command for an initial angle of attack $\alpha_0 = 3°$, with and without the yaw coupling effect. In Fig. 6.5 – 10, the yaw sideslip angle β response introduced by the roll control is given.

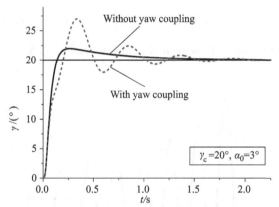

Fig. 6.5 – 9 Comparison of the roll step responses
with and without yaw coupling

The structure and parameters of the yaw autopilot designed to achieve a coordinated turn are shown in Fig. 6.5 – 11. The main feedback of the autopilot was taken as the sideslip angle β. To improve stability, a lead compensation network was added to the main loop of the autopilot. More specifically, the angular velocity ω_y damping loop also has a high-pass filter $\dfrac{\tau s}{\tau s + 1}$ ($\tau = 0.15$ s) added. Its function is to allow high-frequency yaw oscillation signals to pass through while blocking low-frequency coordinated turn signals.

Fig. 6.5 – 12 gives the transient of the yaw autopilot to an initial 1° sideslip angle β disturbance. It shows the effectiveness of the yaw autopilot for depressing the yaw sideslip angle disturbance.

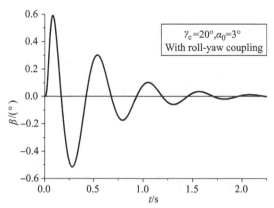

Fig. 6.5 – 10 Effect of roll control on yaw sideslip angle response

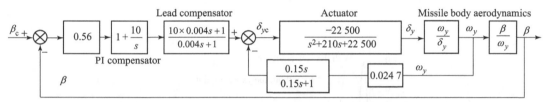

Fig. 6.5 – 11 Structure and parameters of the yaw autopilot designed to achieve a coordinated turn

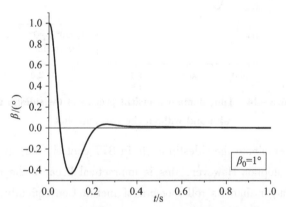

Fig. 6.5 – 12 Suppression curve of initial perturbation by the yaw autopilot

Fig. 6.5 – 13 shows the roll angle response of the roll autopilot (Fig. 6.5 – 8) under the same conditions as Fig. 6.5 – 9, with and without yaw autopilot coupling suppression.

Fig. 6.5 – 14 demonstrates the suppression effect of the above yaw autopilot on the sideslip angle.

Another potential structure for a yaw autopilot designed to achieve coordinated turns involves using yaw acceleration as feedback. Since the missile's lateral acceleration is primarily influenced by the yaw sideslip angle, this approach can effectively approximate the suppression of the sideslip angle.

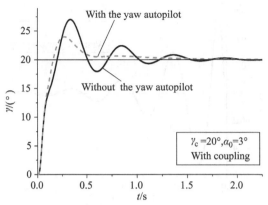

Fig. 6.5 – 13   Time-domain transient process of the roll autopilot with and without a yaw autopilot

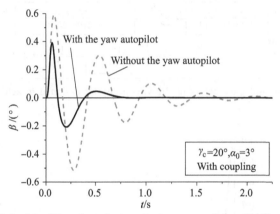

Fig. 6.5 – 14   Time-domain transient process of the sideslip angle with and without the yaw autopilot

In theory, to fully eliminate the sideslip angle in BTT control, the yaw autopilot would need to be faster than the roll autopilot. However, this is impractical because the missile's yaw moment of inertia is significantly larger than its roll moment of inertia. Consequently, BTT control is mainly utilized for midcourse guidance, where lower maneuverability is acceptable. For terminal guidance, which demands rapid missile response, STT control is typically employed.

§ 6.6   Thrust Vector Control and Thruster Control

To obtain a fast change of flight path direction (high value of $\dot{\theta}$) at low missile speed and high altitude (low air density), thrust vector control and thruster control are often used for tactical missiles. The classification of thrust vector control and thruster control for tactical missiles is given in Table 6.6 – 1.

6 Autopilot Design

Table 6.6 – 1 The types and applications of thrust vector control and thruster control

Items	Types	Applications
Endoatmosphere application	Jet vane (thruster vector control)	To rapidly change missile flight direction to incoming target for surface-to-air missile at low initial missile speed. Example: Patriot surface-to-air missile; S – 300. To rapidly change air-to-air dogfight missile flight direction to incoming enemy target. The required missile $\dot{\theta}$ value could be as high as $100°/s$. Example: AIM – 9X; Archer AA – 11; Mica; IRIS – T
	Pulse thruster (thruster control)	From Chapter 8, it is known that fast missile autopilot response will lead to smaller interception miss distance. For this reason, some modern surface-to-air missiles at high attitude end game phase use pulse thrusters to increase missile autopilot response speed. Example: PAC – 3 surface-to-air missile
	Trajectory control thruster (thruster control)	In this case, the control thrusters are positioned near the missile's center of gravity to directly generate the trajectory control force. This arrangement significantly enhances lateral control force and accelerates the autopilot response speed. Example: Aster surface-to-air missile
Exoatmosphere application	Attitude thruster control combined with trajectory thruster control (thruster control)	Exoatmosphere ground-to-air missile application. Example: SM-3 anti-ballistic missile

1) Jet vane

Jet vanes are allocated at the missile rocket engine exhaust (Fig. 6.6 – 1). Its deflection δ_V will change the rocket thrust direction, leading to the generation of lateral control force F and control moment M, their mathematical models are often simply given as $F = k\delta_V$ and $M = kL_V\delta_V$.

Here, L_V is the distance between the jet vane and the missile's center of gravity, and the parameter k in the expression for the control force can be measured by a rocket engine firing test on the 6-degrees-of-freedom test bench.

The dynamic coefficients related to missile rotational control a_δ and flight path control b_δ for aerodynamic control and jet vane control are given in Table 6.6 – 2.

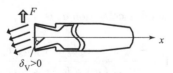

Fig. 6.6 – 1 Jet vane

Table 6.6-2 The dynamic coefficients of aerodynamic control and jet vane control

Items	Missile rotational control efficiency a_δ	Flight path control efficiency b_δ
Aerodynamic control	$a_{\delta_{ae}} = \dfrac{-m_z^{\delta_{ae}} \rho V^2 SL}{2J_z}$	$b_{\delta_{ae}} = \dfrac{c_y^{\delta_{ae}} \rho VS}{2m}$
Jet vane control	$a_{\delta_V} = \dfrac{-kL_V}{J_z}$	$b_{\delta_V} = \dfrac{k}{mV}$

It can be seen from Table 6.6-2 that aerodynamic control $a_{\delta_{ae}}$ is proportional to ρV^2, and $b_{\delta_{ae}}$ is proportional to ρV. Therefore, it is clear that the values of its rotational control efficiency $a_{\delta_{ae}}$ and flight path control efficiency $b_{\delta_{ae}}$ are low at low missile speed and high altitude (low air density).

But for jet vane control, its a_{δ_V} is not related to V and ρ, and b_{δ_V} is proportional to $1/V$. As a result, jet vane rotational control efficiency remains unaffected by low missile speeds and high altitudes, and flight path control efficiency is even greater at lower speeds. This is why jet vane control is widely used in modern surface-to-air and air-to-air missile systems.

The block diagram of a typical aerodynamic and jet vane hybrid control acceleration autopilot is shown in Fig. 6.6-2.

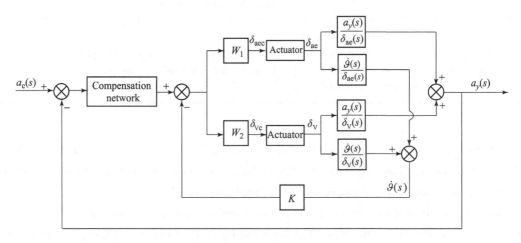

Fig. 6.6-2 Block diagram of a aerodynamic and jet vane hybrid control acceleration autopilot

For aerodynamic control, the transfer function $a_y(s)/\delta_{ae}(s)$ from the aerodynamic actuator deflection angle δ_{ae} to the missile lateral acceleration a_y is

$$\frac{a_y(s)}{\delta_{ae}(s)} = -V \cdot \frac{-b_{\delta_{ae}} s^2 - a_\omega b_{\delta_{ae}} s + (a_{\delta_{ae}} b_\alpha - a_\alpha b_{\delta_{ae}})}{s^2 + (a_\omega + b_\alpha)s + (a_\alpha + a_\omega b_\alpha)}. \qquad (6.6-1)$$

The transfer function $\dot{\vartheta}(s)/\delta_{ae}(s)$ from the aerodynamic actuator deflection angle δ_{ae} to the missile angular velocity $\dot{\vartheta}$ is

$$\frac{\dot{\vartheta}(s)}{\delta_{ae}(s)} = -\frac{a_{\delta_{ae}}s + (a_{\delta_{ae}}b_\alpha - a_\alpha b_{\delta_{ae}})}{s^2 + (a_\omega + b_\alpha)s + (a_\alpha + a_\omega b_\alpha)}. \tag{6.6-2}$$

For jet vane control, the transfer function $a_y(s)/\delta_V(s)$ from the jet vane deflection angle δ_V to the missile lateral acceleration a_y is

$$\frac{a_y(s)}{\delta_V(s)} = -V \cdot \frac{-b_{\delta_V}s^2 - a_\omega b_{\delta_V}s + (a_{\delta_V}b_\alpha - a_\alpha b_{\delta_V})}{s^2 + (a_\omega + b_\alpha)s + (a_\alpha + a_\omega b_\alpha)}. \tag{6.6-3}$$

The transfer function $\dot{\vartheta}(s)/\delta_V(s)$ from jet vane deflection angle δ_V to missile angular velocity $\dot{\vartheta}$ is

$$\frac{\dot{\vartheta}(s)}{\delta_V(s)} = -\frac{a_{\delta_V}s + (a_{\delta_V}b_\alpha - a_\alpha b_{\delta_V})}{s^2 + (a_\omega + b_\alpha)s + (a_\alpha + a_\omega b_\alpha)}. \tag{6.6-4}$$

2) Pulse thruster control

To enhance autopilot response at high altitudes, some modern surface-to-air missiles use a pulse thruster control scheme. A notable example is the American PAC – 3 surface-to-air missile, which features 180 solid propellant pulse thrusters arranged perpendicular to the missile's centerline for pitch and yaw control. These thrusters are distributed evenly around the missile in rings, each containing 18 motors, with a total of 10 rings along the missile's length (Fig. 6.6 – 3).

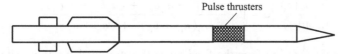

Fig. 6.6 – 3   American PAC – 3 surface-to-air missile

In this case, the control variable n is the number of pulse thrusters fired in the sampling interval. Its average force generated in the time interval Δt is nf (Fig. 6.6 – 4).

Fig. 6.6 – 4   The control force
generated by the pulse thrusters

Here, f is the control force generated by one pulse thruster firing.

If the distance between the pulse thruster and the missile center of gravity is L_P, the pulse thruster control-related dynamic coefficients will be given as

$$a_n = \frac{-fL_P}{J_z}, \tag{6.6-5}$$

$$b_n = \frac{f}{mV}. \tag{6.6-6}$$

The block diagram of a typical aerodynamic and pulse thruster hybrid control autopilot is shown in Fig. 6.6 – 5.

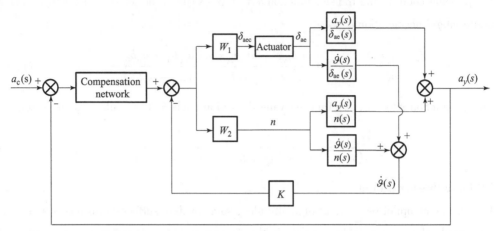

Fig. 6.6 – 5 Block diagram of a aerodynamic and pulse thruster hybrid control autopilot

For pulse thruster control, the transfer function $a_y(s)/n(s)$ from the fired pulse thruster number n to the missile lateral acceleration a_y is

$$\frac{a_y(s)}{n(s)} = -V \cdot \frac{-b_n s^2 - a_\omega b_n s + (a_n b_\alpha - a_\alpha b_n)}{s^2 + (a_\omega + b_\alpha)s + (a_\alpha + a_\omega b_\alpha)}. \qquad (6.6-7)$$

The transfer function $\dot{\vartheta}(s)/n(s)$ from the fired pulse thruster number n to missile angular velocity $\dot{\vartheta}$ is

$$\frac{\dot{\vartheta}(s)}{n(s)} = -\frac{a_n s + (a_n b_\alpha - a_\alpha b_n)}{s^2 + (a_\omega + b_\alpha)s + (a_\alpha + a_\omega b_\alpha)}. \qquad (6.6-8)$$

The Russian literature [4], examined the performance of the American PAC – 3 pulse thruster-controlled autopilot using the following parameters.

Pulse thruster operation time $\Delta t = 16$ ms, pulse force $f = 2\,500$ N, missile roll angle velocity $\dot{\gamma} = 3$ Hz $= 1\,080°/s$, total pulse thruster number 180, and the distance between the pulse thruster and missile center of gravity $L_P = 1$ m. The pulse force direction change in 16 ms is $\Delta r = 1\,080 \times 0.016 = 17.3°$, which is acceptable. The result of the analysis showed that at high altitudes, the acceleration autopilot could have a response time of $t_{63} = 0.05$ s.

3) Trajectory control thruster

In the pulse thruster control scheme, the distance between the pulse thruster and the missile center of gravity is relatively large. This setup primarily aims to generate a significant control moment to enhance autopilot response. If a control thruster with a higher power is used and positioned closer to the missile's center of gravity, the primary design goal shifts to generating a substantial trajectory control force, with control moment benefits becoming secondary.

Fig. 6.6−6 illustrates the trajectory control thruster arrangement for the French Aster surface-to-air missile.

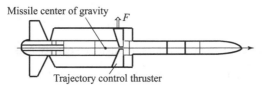

Fig. 6.6−6 French Aster surface-to-air missile

Suppose the trajectory control thruster control variable is taken as u. Then, the control force will be $F = k_1 u$, and the related dynamic coefficients will be

$$a_u = \frac{-k_1 L_u}{J_z}, \qquad (6.6-9)$$

$$b_u = \frac{k_1}{mV}. \qquad (6.6-10)$$

Here, L_u is the distance between the trajectory thruster and the missile's center of gravity.

This design significantly improves the missile's lateral acceleration and provides a quicker autopilot response. However, this enhancement comes with the trade-offs of increased missile weight, size, and cost.

Fig. 6.6−7 illustrates the block diagram of a typical aerodynamic and trajectory thruster hybrid control autopilot.

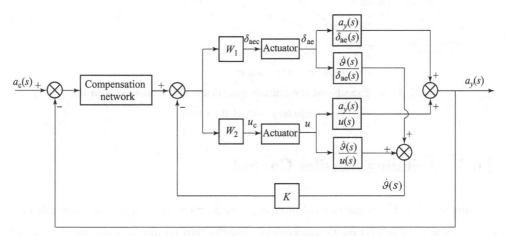

Fig. 6.6−7 Block diagram of a aerodynamic and trajectory thruster hybrid control autopilot

For trajectory thruster control, the transfer function $a_y(s)/u(s)$ from u to the missile lateral acceleration a_y is

$$\frac{a_y(s)}{u(s)} = -V \cdot \frac{-b_u s^2 - a_\omega b_u s + (a_u b_\alpha - a_\alpha b_u)}{s^2 + (a_\omega + b_\alpha)s + (a_\alpha + a_\omega b_\alpha)}. \qquad (6.6-11)$$

The transfer function $\dot{\vartheta}(s)/u(s)$ from u to missile angular velocity $\dot{\vartheta}$ is

$$\frac{\dot{\vartheta}(s)}{u(s)} = -\frac{a_u s + (a_u b_\alpha - a_\alpha b_u)}{s^2 + (a_\omega + b_\alpha)s + (a_\alpha + a_\omega b_\alpha)}. \qquad (6.6-12)$$

4) Attitude thruster control combined with trajectory thruster control

During exoatmospheric flight, aerodynamic forces are not available for missile control. Consequently, both attitude and trajectory control must depend on thruster control. Fig. 6.6 – 8 illustrates the engagement trajectories during the final phase.

In this application, the attitude control thruster could keep the missile centerline always pointing to the target (by way of the strap-down seeker angular error ε feedback). The guidance required for the LOS rate $\dot{q}$ signal could be obtained in the following relation $\dot{q} = \dot{\vartheta} + \dot{\varepsilon}$. Since the guidance acceleration produced by the trajectory control thruster is perpendicular to the missile's centerline, it must also be perpendicular to the LOS between the missile and the target. This orientation aligns with the control direction required for the proportional navigation guidance law.

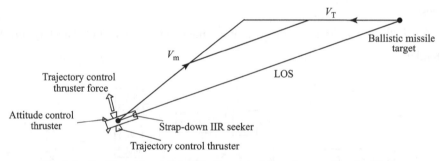

Fig. 6.6 – 8 Example of the attitude control thruster combined with the trajectory control thruster

§ 6.7 Spinning Missiles Control

Spinning missiles offer significant advantages, such as increased robustness against disturbances like thrust asymmetries and fin misalignments. As a result, they require less advanced manufacturing technology compared to non-spinning missiles and are, therefore, widely used in engineering applications, including low-cost gun-launched terminal-guided projectiles and mortars. However, spinning missiles introduce cross-coupling effects between pitch and yaw due to the spinning airframe, which can lead to undesirable coning motion. In this subsection, spinning missiles are modeled, and the unique phenomenon of coning motion is described.

To simplify the analysis, the derivation is performed in a non-spinning reference frame, which is defined as follows:

The nonspinning body coordinate system $Ox_{nb}y_{nb}z_{nb}$, Ox_{nb} is coincident with the velocity vector, Oy_{nb} is perpendicular to Ox_{nb} in the vertical plane, and Oz_{nb} is determined by the right-hand rule.

Note that the new frame can be obtained by rotating the body-fixed coordinate system $Ox_b y_b z_b$. The time-varying spinning angle γ about the Ox_b axis is shown in Fig. 6.7-1.

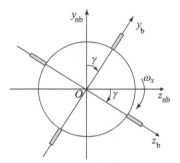

Fig. 6.7-1 The relationship between nonspinning frame and spinning frame

There exist relations between two frames

$$\sigma_b = L(\gamma)\sigma_{nb}, \sigma_{nb} = L^{-1}(\gamma)\sigma_b, \quad (6.7-1)$$

where the coordinate transformation matrix $L(\gamma)$ is

$$L(\gamma) = \begin{bmatrix} \cos\gamma & \sin\gamma \\ -\sin\gamma & \cos\gamma \end{bmatrix}. \quad (6.7-2)$$

The governing equations of dynamics are derived using the angular momentum theorem. However, the moments and forces acting on spinning missiles differ slightly from those on non-spinning missiles due to coupling effects. The model for spinning missiles must account for additional moments and forces arising from these coupling effects.

The aerodynamic cross-coupling effects include the gyroscopic effect, the Magnus effect, and the control coupling effect. Each of these aspects will be discussed in detail.

6.7.1 Aerodynamic Coupling

The Magnus effect, identified by Magnus in 1852, results from asymmetric vortices shed from the spinning missile and is often considered a significant source of cross-coupling. It has been observed that the Magnus effect can cause a spinning missile to deviate from its intended trajectory and potentially become unstable. Additional terms in the dynamical equations account for Magnus moments and forces, as well as gyroscopic moments.

The Magnus moments in a non-spinning frame can be expressed as

$$M_{y\_Mag} = a_M \alpha$$
$$M_{z\_Mag} = -a_M \dot{\beta} \tag{6.7-3}$$

Nevertheless, as the spinning rate ω_x is high, the terms $J_x \omega_x \omega_z$ and $J_x \omega_x \omega_y$ are significant enough to be ignored.

$$M_{y\_Gyro} = -a_G \dot{\vartheta}$$
$$M_{z\_Gyro} = a_G \dot{\psi} \tag{6.7-4}$$

where

$$a_G = \frac{J_x}{J_y} \omega_x. \tag{6.7-5}$$

Fig. 6.7-2 is the block diagram of the aerodynamics model for the spinning missiles.

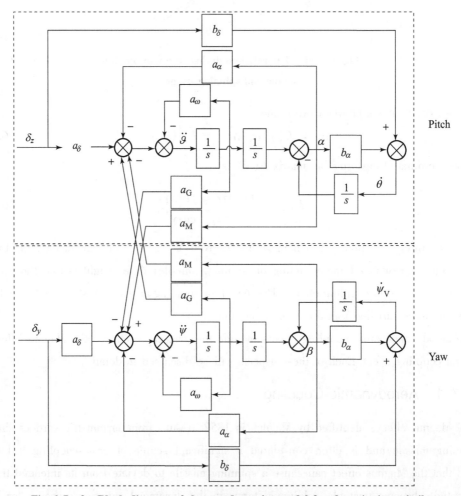

Fig. 6.7-2 Block diagram of the aerodynamics model for the spinning missiles

6.7.2 Control Coupling

For spinning missiles, the dynamics of actuators in a non-spinning coordinate system introduce control coupling effects. To achieve high control precision, it is essential to analyze the control coupling resulting from the actuator dynamics.

The original dynamic of actuator $G_b(s)$ in the body-fixed frame can be expressed as a typical second-order element.

$$G_b(s) = \frac{\delta_{y_b}}{\delta_{ybc}} = \frac{\delta_{z_b}}{\delta_{zbc}} = \frac{1}{T_a^2 s^2 + 2T_a \mu_a s + 1}, \quad (6.7-6)$$

where T_a and μ_a stand for the time constant and damping ratio of the actuator, respectively.

The relation between the command vector and response vector of control surface deflections in the body-fixed frame can be rewritten as

$$\begin{bmatrix} \delta_{z_b} \\ \delta_{y_b} \end{bmatrix} = \begin{bmatrix} G_b(s) & 0 \\ 0 & G_b(s) \end{bmatrix} \begin{bmatrix} \delta_{zbc} \\ \delta_{ybc} \end{bmatrix}. \quad (6.7-7)$$

Likewise

$$\begin{bmatrix} \dot{\delta}_{z_b}(t) \\ \dot{\delta}_{y_b}(t) \end{bmatrix} = \begin{bmatrix} \cos\gamma & -\sin\gamma \\ \sin\gamma & \cos\gamma \end{bmatrix} \begin{bmatrix} \dot{\delta}_z(t) \\ \dot{\delta}_y(t) \end{bmatrix} + \omega_x \begin{bmatrix} -\sin\gamma & -\cos\gamma \\ \cos\gamma & -\sin\gamma \end{bmatrix} \begin{bmatrix} \delta_z(t) \\ \delta_y(t) \end{bmatrix}, \quad (6.7-8)$$

$$\begin{bmatrix} \ddot{\delta}_{z_b}(t) \\ \ddot{\delta}_{y_b}(t) \end{bmatrix} = \begin{bmatrix} \cos\gamma & -\sin\gamma \\ \sin\gamma & \cos\gamma \end{bmatrix} \begin{bmatrix} \ddot{\delta}_z(t) \\ \ddot{\delta}_y(t) \end{bmatrix} + 2\omega_x \begin{bmatrix} -\sin\gamma & -\cos\gamma \\ \cos\gamma & -\sin\gamma \end{bmatrix} \begin{bmatrix} \dot{\delta}_z(t) \\ \dot{\delta}_y(t) \end{bmatrix} + \\ \omega_x^2 \begin{bmatrix} -\cos\gamma & \sin\gamma \\ -\sin\gamma & -\cos\gamma \end{bmatrix} \begin{bmatrix} \delta_z(t) \\ \delta_y(t) \end{bmatrix}. \quad (6.7-9)$$

The transfer function matrix of the equivalent dynamic of Equation (6.7-6) is

$$\boldsymbol{G}_a(s) \triangleq \begin{bmatrix} G_f(s) & G_{co}(s) \\ -G_{co}(s) & G_f(s) \end{bmatrix}, \quad (6.7-10)$$

where

$$G_f(s) = \frac{T_a^2 s^2 + 2T_a \mu_a s + 1 - T_a^2 \omega_x^2}{(T_a^2 s^2 + 2T_a \mu_a s + 1 - T_a^2 \omega_x^2)^2 + 4\omega_x^2 (T_a^2 s + T_a \mu_a)^2}$$

$$G_{co}(s) = \frac{2\omega_x (T_a^2 s + T_a \mu_a)}{(T_a^2 s^2 + 2T_a \mu_a s + 1 - T_a^2 \omega_x^2)^2 + 4\omega_x^2 (T_a^2 s + T_a \mu_a)^2} \quad (6.7-11)$$

$G_f(s)$ and $G_{co}(s)$ are the transfer functions of the forward channel and the coupling channel, respectively. When ω_x equals zero, $G_{co}(s) = 0$, making the transfer function matrix diagonal with no cross-coupling, which can refer to the non-spinning case.

For spinning missiles, the diagonal elements of the dynamics differ from those in the original system, and the non-diagonal elements representing cross-coupling are non-zero.

The poles of the equivalent dynamics for first-order elements will be doubled, and according to Theorem 1, all will lie in the left half-plane. However, the positions of these additional poles vary with the autopilot gains, potentially affecting system stability, which will be examined further below.

$$\begin{bmatrix} \dddot{\vartheta}(t) \\ \ddot{\varphi}(t) \\ \dot{\alpha}(t) \\ \dot{\beta}(t) \end{bmatrix} = \begin{bmatrix} -a_\omega & a_G & -a_\alpha & -a_M \\ -a_G & -a_\omega & a_M & -a_\alpha \\ 1 & 0 & -b_\alpha & 0 \\ 0 & 1 & 0 & -b_\alpha \end{bmatrix} \begin{bmatrix} \dot{\vartheta}(t) \\ \dot{\varphi}(t) \\ \alpha(t) \\ \beta(t) \end{bmatrix} + \begin{bmatrix} -a_s & 0 \\ 0 & -a_s \\ -b_s & 0 \\ 0 & -b_s \end{bmatrix} \begin{bmatrix} \delta_z(t) \\ \delta_y(t) \end{bmatrix} \quad (6.7-12)$$

$$\begin{bmatrix} \dot{\vartheta}(t) \\ \ddot{\phi}(t) \\ \dot{\alpha}(t) \\ \dot{\beta}(t) \end{bmatrix} = \begin{bmatrix} -a_\omega & a_G & -a_\alpha & -a_M \\ -a_G & -a_\omega & a_M & -a_\alpha \\ 1 & 0 & -b_\alpha & 0 \\ 0 & 1 & 0 & -b_\alpha \end{bmatrix} \begin{bmatrix} \dot{\vartheta}(t) \\ \dot{\phi}(t) \\ \alpha(t) \\ \beta(t) \end{bmatrix} + \begin{bmatrix} -a_s & 0 \\ 0 & -a_s \\ -b_s & 0 \\ 0 & -b_s \end{bmatrix} \begin{bmatrix} \delta_z(t) \\ \delta_y(t) \end{bmatrix} \quad (6.7-13)$$

$$D(s) = a_\alpha^2 + a_M^2 + 2[a_\alpha(s + a_\omega) - a_M a_G](s + b_\alpha) + [a_G^2 + (s + a_\omega)^2](s + b_\alpha)^2 \quad (6.7-14)$$

$$D(s) = a_\alpha^2 + a_M^2 + 2[a_\alpha(s + a_\omega) - a_M a_G](s + b_\alpha) + [a_G^2 + (s + a_\omega)^2](s + b_\alpha)^2 \quad (6.7-15)$$

$$D(s) = a_\alpha^2 + a_M^2 + 2[a_\alpha(s + a_\omega) - a_M a_G](s + b_\alpha) + [a_G^2 + (s + a_\omega)^2](s + b_\alpha)^2 \quad (6.7-16)$$

$$D(s) = a_\alpha^2 + a_M^2 + 2[a_\alpha(s + a_\omega) - a_M a_G](s + b_\alpha) + [a_G^2 + (s + a_\omega)^2](s + b_\alpha)^2 \quad (6.7-17)$$

$$D(s) = a_\alpha^2 + a_M^2 + 2[a_\alpha(s + a_\omega) - a_M a_G](s + b_\alpha) + [a_G^2 + (s + a_\omega)^2](s + b_\alpha)^2 \quad (6.7-18)$$

With Equation (6.7-12), the integrated model of spinning missiles is given by

$$D(s) = a_\alpha^2 + a_M^2 + 2[a_\alpha(s + a_\omega) - a_M a_G](s + b_\alpha) + [a_G^2 + (s + a_\omega)^2](s + b_\alpha)^2$$

(6.7 – 19)

Due to the couplings between pitch and yaw dynamics, the model of spinning missiles becomes a MIMO system, making it necessary to consider both dynamics together in autopilot design. Additionally, the cross-coupling effects in rolling missiles lead to coning motion, which can degrade system stability.

7

LOS Guidance

§ 7.1 LOS Guidance System

There are various types of LOS guidance systems, but they all operate on the same principle: guiding the missile along the LOS to the target via guidance radar. This method is often called three-point guidance. In this approach, deviations of the missile's trajectory from the LOS are measured by the guidance radar, communicated to the missile as commands, and corrected by its control system, making it a form of command guidance.

LOS guidance is typically categorized into semi-automatic command guidance and automatic command guidance. The key difference is that in semi-automatic command guidance, an operator tracks the target, while in automatic command guidance, the tracking is carried out automatically by guidance radar or optoelectronic tracking devices.

Most current anti-tank guided missiles that employ command guidance use semi-automatic command guidance. In this system, the operator tracks the target using an optical or infrared sight. The deviation of the missile's infrared beacon from the goniometer axis is measured by the goniometer. Guidance commands, based on these measurements, are then transmitted to the missile via a wire or radio link to complete the guidance loop. In this scheme, the misalignment $\Delta\theta$ of the optic axis of the operator sight and the goniometer axis will influence the performance of the guidance (Fig. 7.1 – 1).

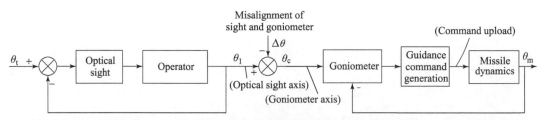

Fig. 7.1 – 1 Block diagram I of the semi-automatic anti-tank missile guidance system with the help of the operator sight and goniometer loop

When the operator uses an infrared imaging system, the sight can simultaneously observe the target and measure the error angle of the missile's infrared beacon. In this case, since the sight and the goniometer are integrated into the same device, there is no misalignment error (Fig. 7.1 – 2).

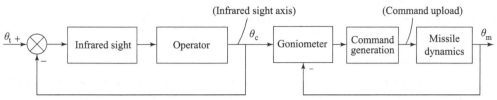

Fig. 7.1 – 2 Block diagram Ⅱ of the infrared semi-automatic anti-tank missile guidance loop

For an anti-tank missile equipped with a laser beam riding system, the missile's deviation from the LOS is automatically measured by detecting the coded laser signal field emitted by the laser guidance device controlled by the operator. The missile generates the guidance commands internally, eliminating the need for external command uploads. However, since the sight and the laser guidance device are separate components, any misalignment between them can still impact guidance accuracy. The block diagram for this type of system is shown in Fig. 7.1 – 3.

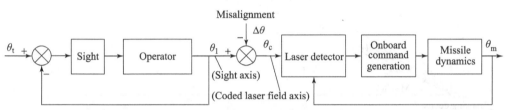

Fig. 7.1 – 3 Block diagram of the anti-tank missile equipped with the laser beam riding system

Current ground-to-air missiles with LOS guidance typically use an automatic command guidance scheme. In this approach, target tracking is performed automatically by tracking radar or optoelectronic tracking devices. Since the radar can simultaneously track the target and measure angles using both the target's echo and the missile's radio beacon signals, this system avoids issues of misalignment (Fig. 7.1 – 4). A detailed analysis and design of the guidance radar loop can be found in Chapter 4.

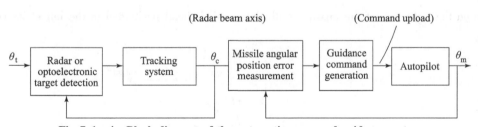

Fig. 7.1 – 4 Block diagram of the automatic command guidance system

§ 7.2 Analysis of the Required Acceleration for the Missile with LOS Guidance

Fig. 7.2 – 1 shows the trajectory diagram of the target and missile in a LOS guidance.

In Fig. 7.2 – 1, suppose that the target T makes a constant velocity straight-line flight, and the missile is launched at a speed higher than that of the target at the initial time T_0. The plane containing the missiles speed vector and the target speed vector is called the missile flight plane. The LOS at the moments 1, 2, and 3 after launch are respectively given as OT_1, OT_2, OT_3, and so on. Since the three-point guidance law aims to keep the missile in the LOS, the actual trajectory will be curved, with the curvature increasing significantly during the terminal phase. Additionally, the direction of the missile's velocity at any point will align with the tangent to the instantaneous trajectory, which may differ from the direction of the LOS, especially in the terminal phase.

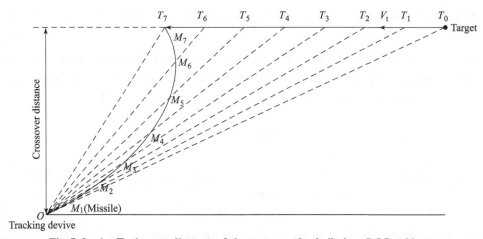

Fig. 7.2 – 1   Trajectory diagram of the target and missile in a LOS guidance

The LOS usually refers to the direction of the tracking system's beam. The angle between the missile's velocity direction and the LOS is known as the "trajectory-beam angle". The angle between the missile's body x-axis and the beam direction, which is often more relevant in guidance system design, is called the "missile body-beam angle". This angle differs from the trajectory-beam angle by the angle of attack (Fig. 7.2 – 2). As illustrated in the engagement scenario shown, the missile's lateral acceleration must be directed to the left of the LOS, meaning the velocity direction will be on the right side of the missile, with the missile's head positioned to the left of its velocity vector.

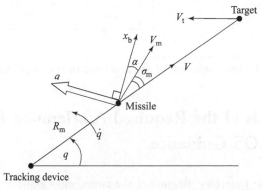

Fig. 7.2 – 2   Definitions of the symbols related to the LOS guidance

Typically, ground-to-air LOS guidance missiles are equipped with a radio beacon that transmits the deviation of the missile from the tracking radar axis. However, the beacon has a limited beam width. If the "missile body-beam angle" exceeds this width, the tracking device may lose the missile's return signal. Generally, the "missile body-beam angle" is restricted to about 40°. Furthermore, during curved flight, a necessary lateral acceleration, which is perpendicular to the LOS, must be applied to prevent the missile from deviating from the LOS. A larger "missile body-beam angle" results in less effective lateral acceleration or requires a higher normal acceleration command to maintain LOS guidance. Thus, the actual value of the "missile body-beam angle" during flight is a crucial parameter that requires close attention.

To simplify the analysis, let's assume that the missile's angle of attack during, that is the guidance is not significant and may be omitted. Additionally, the missile has a lateral acceleration a, that is perpendicular to the missile's velocity V_m direction (Fig. 7.2-3). Taking the trajectory-beam angle as σ_m, the LOS rotation angular velocity and acceleration as $\dot{q}$ and $\ddot{q}$, the velocity of the missile along the LOS should be $V = V_m \cos\sigma_m$. In addition, due to the presence of thrust and aerodynamic drag, the missile will have an acceleration component $\dot{V}_m$ in the direction of the velocity vector.

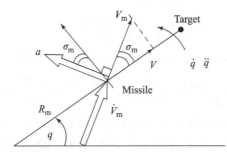

Fig. 7.2-3 Definitions of related parameters on the flight plane

To realize the flight of the missile along the rotating LOS, the acceleration projection perpendicular to the LOS direction should be

$$a\cos\sigma_m + \dot{V}_m \sin\sigma_m = 2V_m \cos\sigma_m \dot{q} + R_m \ddot{q}. \tag{7.2-1}$$

In Equation (7.2-1), the right-hand side of the equation is the acceleration of the missile perpendicular to the LOS required to fly along the rotating LOS, and the left-hand side is the source of this acceleration (the combined action of the missile's normal acceleration a and the missile axis acceleration $\dot{V}_m$).

That is, to achieve the motion of the missile along the rotation LOS, the required missile normal acceleration will be:

$$a = 2V_m \dot{q} + \frac{R_m}{\cos\sigma_m} \ddot{q} - \dot{V}_m \tan\sigma_m. \tag{7.2-2}$$

The values of $\dot{q}$ and $\ddot{q}$ depend on the velocity and position of the target. According to the geometric relationship shown in Fig. 7.2-4,

$$\dot{q} = V_m \frac{\sin\sigma_m}{R_m} = V_t \frac{\sin\sigma_t}{R_t}. \tag{7.2-3}$$

That is

$$\sin\sigma_m = \sin\sigma_t \frac{R_m}{R_t} \cdot \frac{V_t}{V_m}. \qquad (7.2-4)$$

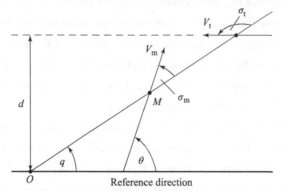

Fig. 7.2-4 Geometric relationship of the flight

If the target flies parallel to the reference direction, its cross distance d will be a constant. Therefore, $\dot{q}$ can also be expressed as

$$\dot{q} = \frac{V_t \sin^2 \sigma_t}{d}. \qquad (7.2-5)$$

Note that $\ddot{q}$ can be obtained from the equations $\dot{q} = -\dot{\sigma}_t$ and $dq = -d\sigma_t$

$$\ddot{q} = \frac{d^2 q}{dt^2} = \left[\frac{d}{dq}\left(\frac{dq}{dt}\right)\right] \cdot \left(\frac{dq}{dt}\right) = -\frac{d\dot{q}}{d\sigma_t} \cdot \dot{q} = -\frac{2V_t^2}{d^2} \sin^3 \sigma_t \cos\sigma_t. \qquad (7.2-6)$$

Inserting the expressions of $\dot{q}$ and $\ddot{q}$ into the expression of the normal acceleration a, and assuming that the missile is flying at a constant speed ($\dot{V}_m = 0$). The following expression can be written as

$$a = \frac{2V_m V_t}{d}\left(\sin^2\sigma_t - \frac{\sin\sigma_t \cos\sigma_t \sin\sigma_m}{\cos\sigma_m}\right)$$

$$= \frac{2V_m V_t}{d}(\beta_1 + \beta_2). \qquad (7.2-7)$$

In Equation (7.2-7), $\beta_1 = \sin^2\sigma_t$, $\beta_2 = -\dfrac{\sin\sigma_t \cos\sigma_t \sin\sigma_m}{\cos\sigma_m}$, and β_1 is the Coriolis acceleration factor corresponding to the LOS rotation angular velocity $\dot{q}$. β_2 is the translational acceleration factor corresponding to the LOS rotation angular acceleration $\ddot{q}$.

It is seen from the expression of the normal acceleration a that its value is proportional to $V_m V_t$ and inversely proportional to d. That is, the larger the target velocity V_t and the missile velocity V_m are, the greater the normal acceleration is required. When the target is flying over the tracking device, the smaller the cross distance d is, the greater the missile's required normal acceleration is. Fig. 7.2-5 shows the relative magnitudes of the Coriolis acceleration factor β_1 caused by $\dot{q}$ and the translational acceleration factor β_2 caused by $\ddot{q}$ at different target positions σ_t, when $V_m/V_t = 2.5$, and at the target impact time (with $R_m/R_t = 1$). It is clear that the Coriolis acceleration caused by $\dot{q}$ is the primary factor of the normal acceleration required when hitting the target in the LOS guidance.

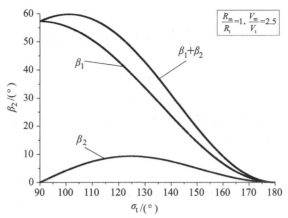

Fig. 7.2-5 Variation of the two normal acceleration components with the change of σ_t, when $R_m/R_t = 1$ and $V_m/V_t = 2.5$

From the expression $\beta_1 = \sin^2\sigma_t$, it is known that when $\dot{q}$ does not change, Coriolis acceleration has nothing to do with the relative position R_m/R_t of the missile on the LOS, but when the aircraft is an incoming target, the LOS angle q will gradually increase from zero to 90°. Since $\dot{q} = \dfrac{V_t \sin^2 q}{d}$, $\dot{q}$ increases as the target approaches. When the missile hits the target, $\dot{q}$ will be the largest in comparison with its other positions on the LOS; that is, the lateral acceleration of the missile required in the LOS guidance is the maximum as the missile meets the target, which is the biggest drawback of this guidance law.

Fig. 7.2-6 shows the variation of the trajectory-beam angle σ_m of the missile as the missile moves to different LOS positions R_m/R_t and the target approaches at different σ_t positions, with $V_m/V_t = 2.5$. It is known that as the target approaches, the value of σ_m is the greatest when the missile encounters the target ($R_m/R_t = 1$) (if the angle of attack is included, the actual missile body-beam angle will be greater than the trajectory-beam angle σ_m). In other words, the missile's normal acceleration efficiency is the lowest at the end of the attack, and the signal direction from the missile beacon deviates the most from the LOS. These are key disadvantages of LOS guidance.

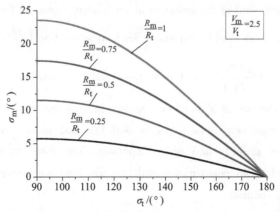

Fig. 7.2-6 Variation of the trajectory-beam angle σ_m with the change of σ_t and R_m/R_t

§ 7.3　Analysis of the LOS Guidance Loop

Fig. 7.3 – 1 shows the block diagram of the LOS guidance loop, in which $\Delta\theta$ is the angular deviation of the missile from the LOS, h_c is the linear deviation of the missile, R is the distance between the target tracking device and the missile, Δy_m is the missile linear displacement response, and θ_m is the angular response of the missile.

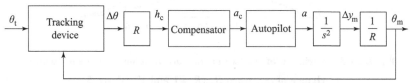

Fig. 7.3 – 1　Block diagram of the LOS guidance loop

The loop of Fig. 7.3 – 1 is an angular tracking loop, but R and $1/R$ in the loop can cancel each other. Therefore, the effect of R can be neglected in the control system loop design, and only the control loop of the linear deviation h needs to be considered. That is, Fig. 7.3 – 1 can be simplified as Fig. 7.3 – 2.

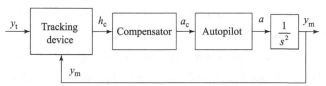

Fig. 7.3 – 2　Control loop of the LOS guidance with only linear deviation considered

Since there are two integrators in this loop, which brings a $-180°$ phase shift, the system is unstable without a compensation network after the autopilot lag is added. After the lead compensation design is introduced, the phase margin of the system should be the compensation lead phase $\Delta\phi_{\text{compensator}}$ minus the phase shift of the autopilot at the system crossover frequency.

The compensation network could be a single lead compensation $\left(\dfrac{\alpha Ts + 1}{Ts + 1}\right)$ or a cascaded lead compensation $\left(\dfrac{\alpha^* Ts + 1}{Ts + 1}\right)\left(\dfrac{\alpha^* Ts + 1}{Ts + 1}\right)$. When the lead compensator provides positive angle compensation, its gain increases at higher frequencies, causing the crossover frequency to shift to a higher range. This can negatively impact system stability and reduce noise attenuation. This is detrimental to the system stability and noise depression. Therefore, to maintain the same gain increase effect, the single lead compensator and cascade lead compensator parameter α and α^* should have the following relation

$$\alpha^* = \sqrt{\alpha}. \tag{7.3 – 1}$$

Fig. 7.3 – 3 shows the difference in gain and phase characteristics between the transfer function of the single lead compensator and the cascaded lead compensator when $\alpha = 10$ and $\alpha^* = \sqrt{10}$. Note that the maximum phase compensation of the cascaded lead compensator is 62.6°, which exceeds the 55° maximum phase compensation of a single lead compensator. However, the cascaded compensator has a slightly narrower compensation bandwidth. Despite this, both types of compensators have the same gain values: 10 dB at maximum phase compensation and 20 dB in the high-frequency band.

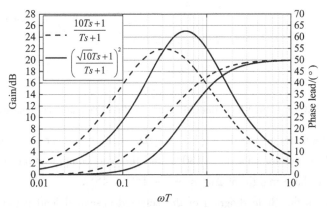

Fig. 7.3 – 3 Gain and phase characteristics of the two lead compensator transfer functions

The following gives an example of the design of a LOS guidance control loop. The missile's acceleration autopilot transfer function selected in this system is of second order, in which $\omega_m = 12$ rad/s $= 1.91$ Hz and damping coefficient $\mu = 0.6$ (it should be noted that the transient speed of the LOS guidance loop is almost completely determined by the acceleration autopilot bandwidth). If a cascaded lead compensation network is adopted in this design with $\alpha^* = \sqrt{10}$, it is required to select the proper system open loop gain K and the lead compensation network time constant T so that the loop gain margin is greater than 6 dB and the phase margin is greater than 45°. Fig. 7.3 – 4 shows the block diagram of this loop.

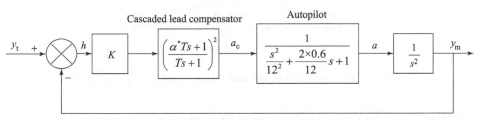

Fig. 7.3 – 4 Block diagram of the LOS guidance control loop

Taking $K = 3$ and $T = 0.15$, the open-loop Bode diagram of the system, both with the compensation network and without a compensation network, is shown in Fig. 7.3 – 5. The design result is a phase margin of 45° and a gain margin of 9.68 dB.

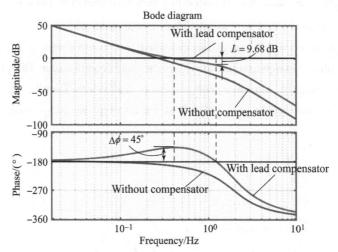

Fig. 7.3 − 5 Open-loop Bode diagram of the LOS guidance loop above

The crossover frequency in the design is 0.41 Hz. At the crossover frequency, the autopilot lag is −14°, the double-integrator's lag is −180°, and the lead compensation phase lead is 59°. So, the final design has a phase margin of $\Delta\phi = 59° - 14° = 45°$.

Fig. 7.3 − 6 shows the Bode diagram of the selected cascaded lead compensator. The diagram indicates that this compensation network can provide a phase lead at the system crossover frequency, approaching its maximum possible value.

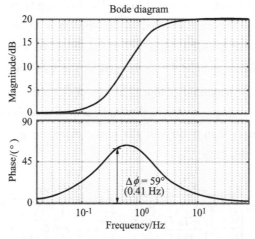

Fig. 7.3 − 6 Bode diagram of the lead compensator $\left(\dfrac{\sqrt{10} \times 0.15s + 1}{0.15s + 1} \right)^2$

Fig. 7.3 − 7 shows the time response of the system with a unit step command h_c input. Since the LOS guidance control loop is a type II system, achieving a fast response is challenging, and the overshoot during the transition process is often significant.

The LOS control loop of the above design is used to intercept a flyover aircraft target. Take the target speed as $V_t = 250$ m/s and the flyover distance $d = 4\,000$ m.

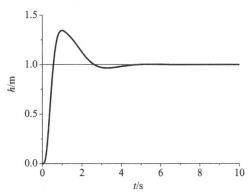

Fig. 7.3 – 7 Time response of the example LOS guidance loop

Suppose the system accomplishes accurate target tracking and launches the missile in the direction of the target when $q = 45°$, with the missile velocity $V_m = 600$ m/s. When the tracking error of the tracking radar is neglected, the flight trajectories of the target and the missile for this given scenario are shown in Fig. 7.3 – 8.

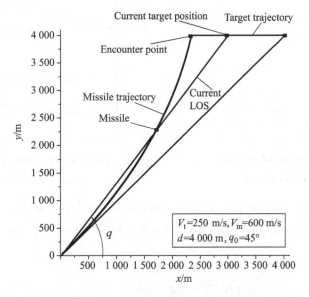

Fig. 7.3 – 8 Flight trajectories of the target and the missile

$q(t), \dot{q}(t)$, and $\ddot{q}(t)$ for this example are respectively shown in Fig. 7.3 – 9, Fig. 7.3 – 10, and Fig. 7.3 – 11. It can be seen from the $\dot{q}(t)$ plot that the LOS angular velocity is increasing when the missile is coming toward the guidance station, and its value reaches the maximum at the end of the missile-target engagement.

Fig. 7.3 – 12 displays the corresponding lateral Coriolis acceleration, translational acceleration, and total lateral acceleration as the missile follows the LOS. It indicates that the required lateral acceleration is primarily to accommodate the Coriolis acceleration associated with the LOS angular velocity $\dot{q}$.

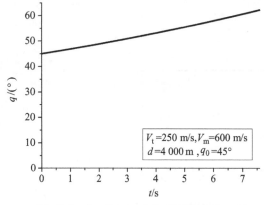

Fig. 7.3 – 9　Curve of the LOS angle q

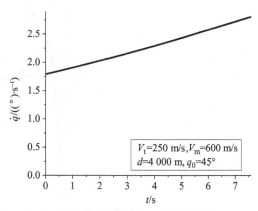

Fig. 7.3 – 10　Curve of the LOS angular velocity $\dot{q}$

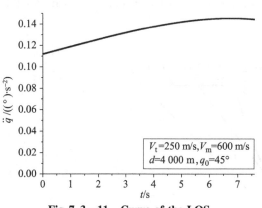

Fig. 7.3 – 11　Curve of the LOS angular acceleration $\ddot{q}$

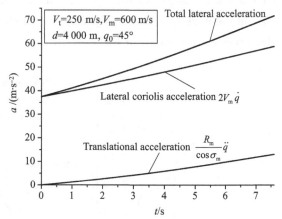

Fig. 7.3 – 12　Acceleration curve and acceleration components

Fig. 7.3 – 13 shows the variation of the missile trajectory-beam angle σ_m during this guidance process. It can be observed that this angle reaches its maximum at the end of missile-target engagement.

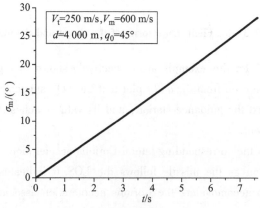

Fig. 7.3 – 13　Variation of the missile trajectory-beam angle σ_m

In the following, the LOS motion $q(t)$ corresponding to the above scenario is used as the command to the designed missile guidance loop (Fig. 7.3-14). Fig. 7.3-15 shows the definition of the position error h and the lateral acceleration a. Here, a positive h represents the missile lagging behind the LOS, and the positive a corresponds to a positive guidance error h. Fig. 7.3-16 shows the curve of the guidance process $h(t)$, and Fig. 7.3-17 shows the normal acceleration command $a_c(t)$ and the autopilot acceleration output $a(t)$.

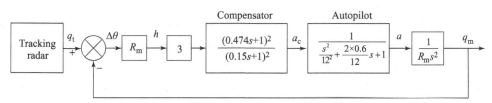

Fig. 7.3-14 Block diagram of the LOS guidance system

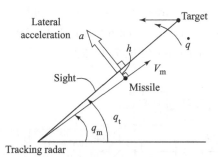

Fig. 7.3-15 Definition of relative variables of the LOS guidance

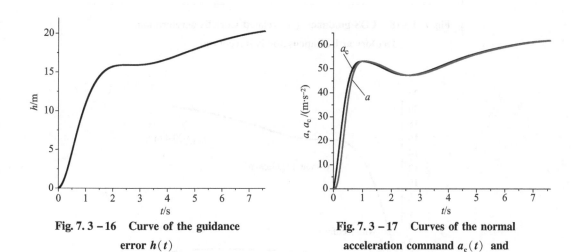

Fig. 7.3-16 Curve of the guidance error $h(t)$

Fig. 7.3-17 Curves of the normal acceleration command $a_c(t)$ and the autopilot acceleration output $a(t)$

It is seen from Fig. 7.3-17 that the required command $a_c(t)$ increases with the increase of the LOS angular velocity $\dot{q}$ during the guidance, and it reaches the maximum at the missile-target encounter with a value of 6.2 g. Due to the stability constraints of the type II system, the open loop gain value of the LOS guidance loop is low (in this case $K = 3$). For this reason, the required guidance acceleration command can only be generated by a large guidance error h. In this case, the miss distance at the encounter can reach 20.4 m.

Since the tracking radar can provide the LOS angular velocity $\dot{q}$ output when tracking the target, the LOS guidance system can adopt a feedforward strategy to improve its guidance accuracy. That is, the tracking radar can use its measured LOS angular velocity $\dot{q}$ and the missile velocity V_m to generate the Coriolis acceleration feedforward command and superimpose this feedforward command on the error command (Fig. 7.3 – 18). In Fig. 7.3 – 18, the time-varying block in the feedforward channel is used to smooth the initial sudden change in the feedforward signal. In this way, since most of the required acceleration commands for the missile are provided by the feedforward signal, it greatly reduces the contribution of the tracking error Δq in the command a_c. That is to say, the accuracy of guidance can be greatly improved. Fig. 7.3 – 19 shows the guidance error curves before and after the adoption of feedforward control, while Fig. 7.3 – 20 presents the corresponding acceleration command curve. These figures demonstrate that the terminal miss distance is significantly reduced from 20.4 m to 0.68 m with feedforward control (Fig. 7.3 – 19). However, the requirement for terminal acceleration remains unchanged (Fig. 7.3 – 20).

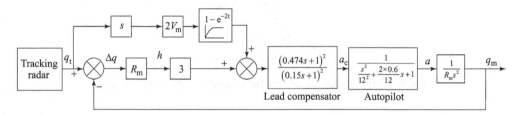

Fig. 7.3 – 18 LOS guidance ($\dot{q}$ related Coriolis acceleration feedforward compensator is introduced)

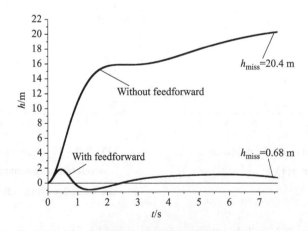

Fig. 7.3 – 19 Position error curves before and after the introduction of feedforward

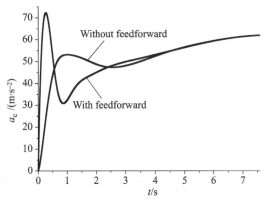

Fig. 7.3 – 20 Curves of the acceleration command before and after the introduction of feedforward

§7.4 Lead Angle Method

In LOS guidance, the missile trajectory becomes increasingly curved over time, with the lateral acceleration peaking towards the end of the missile-target engagement. Allowing the missile to follow a straighter path ahead of the LOS can reduce the required terminal acceleration. Since the guidance error is proportional to the acceleration command, reducing the terminal acceleration will simultaneously decrease the miss distance. This approach means that the missile will follow a virtual LOS which is ahead of the actual LOS, rather than the real LOS. With this approach, the missile will no longer fly along the actual LOS, but tracks a virtual LOS ahead of the real one. Let us denote the virtual LOS as q^* (Fig. 7.4 – 1),

$$q^* = q + \Delta q. \tag{7.4-1}$$

The expression of the lead angle Δq is taken as

$$\Delta q = c(R_t - R_m) = c \cdot \Delta R. \tag{7.4-2}$$

Clearly, $R_t - R_m = \Delta R = 0$ when the missile encounters the target. Therefore, the virtual LOS will coincide with the actual LOS automatically at the end of the missile-target engagement ($\Delta q = 0$). That is, this approach also ensures that the missile hits the target at the end of the engagement.

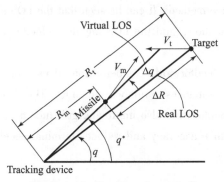

Fig. 7.4 – 1 Lead angle method

In addition, since the missile's required acceleration is mainly related to the LOS angular velocity, a constraint $\dot{q}^* = 0$ as $\Delta R = 0$ can be taken to derive the coefficient expression. In this way, the terminal acceleration required will certainly be reduced.

Since
$$\dot{q}^* = \dot{q} + \dot{c} \cdot \Delta R + c \cdot \Delta \dot{R}, \tag{7.4-3}$$
take $\dot{q}^* = 0$ as $\Delta R = 0$, then
$$\dot{q}^* = \dot{q} + c \cdot \Delta \dot{R} = 0. \tag{7.4-4}$$

As c is taken as $c = -\dfrac{\dot{q}}{\Delta \dot{R}} = \dfrac{\dot{q}}{|\Delta \dot{R}|}$, this will make $\dot{q}^* = 0$ when $\Delta R = 0$. The final lead angle expression will be

$$\Delta q = \frac{\dot{q}}{|\Delta \dot{R}|} \Delta R. \tag{7.4-5}$$

Since both $\dot{q}$ and ΔR can be obtained from the ground tracking device, this scheme is not very difficult to realize. However, due to the constraint of the limited beam angle width of the tracking device, a half-lead angle approach is more widely used. Specifically

$$q^* = q + \frac{1}{2}\left(\frac{\dot{q}}{|\Delta \dot{R}|}\right)\Delta R. \tag{7.4-6}$$

Differentiating the above expression with respect to t with $\Delta R = 0$ gives the following relation:

$$\dot{q}^* = \frac{1}{2}\dot{q}. \tag{7.4-7}$$

In other words, the half-lead angle approach can reduce the required acceleration command by approximately half when engaging the target.

Fig. 7.4-2 shows the variation of q and q^* with the change of ΔR in the three-point method, the half-lead angle method, and the lead angle method for the previous example. It can be seen that when $\Delta R = 0$, that is when the missile encounters the target, the value of q is the same for the three methods, and all the missiles will hit the target. However, the lead angle in the half-lead angle method is small, which reduces the risk of the missile deviating from the beam. Fig. 7.4-3 shows the variation of $\dot{q}$ and $\dot{q}^*$ in three methods. It can be seen that the LOS angular velocity $\dot{q}^*$ is zero when $\Delta R = 0$ using the lead angle method, and the LOS angular velocity of the half-lead angle method is half that of the three-point method.

As mentioned earlier, feedforward compensation does not alter the required terminal acceleration but significantly enhances guidance accuracy. The lead angle method can reduce terminal acceleration requirements while also improving guidance accuracy. Combining both methods simultaneously improves guidance accuracy and reduces terminal acceleration.

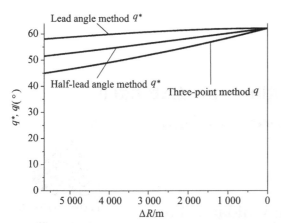

Fig. 7.4−2 Variation of q and q^* in the three-point method, the half-lead angle method, and the lead angle method

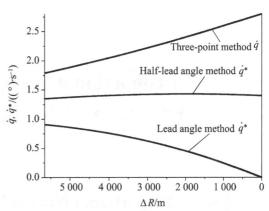

Fig. 7.4−3 Variation of $\dot{q}$ and $\dot{q}^*$ in the three-point method, the half-lead angle method, and the lead angle method

8

Proportional Navigation and Extended Proportional Navigation Guidance Laws

§ 8.1 Proportional Navigation Guidance Law

PROPORTIONAL NAVIGATION GUIDANCE

8.1.1 Proportional Navigation Guidance Law (PN Guidance Law)

Historical maritime experience shows that if a ship's navigator observes another ship and notices that the LOS remains stationary, a collision is inevitable if both ships continue on their current paths, as illustrated in Fig. 8.1 – 1. To avoid such collisions, the navigator must alter the course to ensure the observation line rotates, thereby preventing the two ships from meeting.

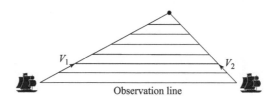

Fig. 8.1 – 1 Sketch of a sea accident

Missile guidance is essentially the inverse of the navigational problem described above. The guidance law designer must ensure that the missile-target line (or LOS) remains stationary in inertial space while the missile is in flight. By doing so, the missile is guaranteed to strike the target at the end of its trajectory. To implement this approach, people skillfully designed the seeker, which can measure the angular velocity $\dot{q}$ of the missile-target LOS in the inertial space. When the measured $\dot{q}$ is not zero, the missile changes its flight path direction through the PN guidance law $\dot{\theta} = N\dot{q}$ to control $\dot{q}$ towards zero so as to hit the target. This is the basic idea used in proportional navigation guidance (PNG).

Fig. 8.1 – 2 shows the schematic of the target and missile trajectories as the PN guidance law is implemented, in which the proportional guidance constant is taken as $N = 4$ and the ratio of the missile velocity and the target velocity is $2 (V_m/V_t = 2)$.

Years later, researchers discovered that proportional navigation guidance law is, in fact, an optimal guidance law, a fact that can be rigorously proven. The mathematical model for this optimal guidance problem is presented below.

8 Proportional Navigation and Extended Proportional Navigation Guidance Laws

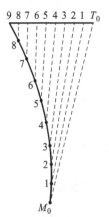

Fig. 8.1 – 2 Proportional guidance target and missile trajectories ($V_m/V_t = 2$, proportional guidance constant $N = 4$)

Suppose that the missile and the target have already been in the nominal engagement trajectories (Fig. 8.1 – 3), and V_m and V_t are the missile and the target velocities. At the time t moment, the missile will be at point A, while the target will be at point B. The origin of the coordinate system is placed at point B. The direction x is in the nominal direction of the missile-target LOS. Since the xBy coordinate system is moving in the inertial space with a constant velocity, it can be considered as an inertial system, where Newton's Law applies.

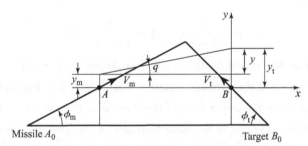

Fig. 8.1 – 3 Relative relationship between the missile and the target

Suppose that a disturbance deviation occurs in the missile and target flight in relation to the nominal trajectory coordinate system xBy, and the disturbances are y_t and y_m, in which y_t is the position disturbance of the target perpendicular to the LOS and y_m is the position disturbance of the missile perpendicular to the LOS. Therefore, the relative position error of the two is $y = y_t - y_m$, the relative velocity error is $V = \dot{y}$, and the relative acceleration error $\ddot{y}$ is $a_t - a_m$.

It should be noted that y and V are respectively the relative position and velocity of the missile and target, while a_t, a_m are the absolute accelerations of the target and the missile perpendicular to the LOS with respect to inertial space, and only their difference is the relative acceleration. Therefore, the dynamic model of this problem can be established as

$$\begin{cases} \dot{y} = V, \\ \dot{V} = a_t - a_m. \end{cases} \quad (8.1-1)$$

In studying the basic model, a simple situation like the following is often assumed: the target does not maneuver (i.e., $a_t = 0$), and the required missile maneuver acceleration a_m in the inertial space is taken as the guidance command a_c ($a_m = a_c$). Then the following basic dynamic model can be obtained

$$\begin{cases} \dot{y} = V, \\ \dot{V} = -a_c. \end{cases} \tag{8.1-2}$$

Set the objective function for deriving the optimal guidance law as $J = \left[S \dfrac{y^2(T)}{2} + \dfrac{1}{2} \int_0^T a_c^2(t) \, dt \right]$. Then, the optimization problem will be given as

$$\min J = \min \left[S \dfrac{y^2(T)}{2} + \dfrac{1}{2} \int_0^T a_c^2(t) \, dt \right], \tag{8.1-3}$$

where T is the engagement time, and $\phi(T) = S \dfrac{y^2(T)}{2}$ is the penalty function of the miss distance at the moment of T. When the value of the coefficient S is set as ∞, the miss distance will be forced to become zero at the end of the engagement. The integral term of the objective function represents the integrated square of the missile acceleration over the missile flight time interval of T, which means that the design tries to hit the target with minimum control costs.

After the above optimization problem is solved, the optimal control solution will give the following guidance law

$$a_c = \dfrac{3}{(T-t)^2} y(t) + \dfrac{3}{(T-t)} V(t). \tag{8.1-4}$$

Please refer to the end of the section for the detailed mathematical derivation process.

It is important to note that the dynamic equation discussed is a LTI model. It is known from the basic control theory that when the upper limit of the integral of the objective function is taken as ∞, the optimal control solution obtained is the state feedback of the normal optimal control. It is a LTI control law. However, when the integral time T of the objective function is finite, such an optimal control solution will become a linear time-varying control law because the guidance law is related to time $(T - t)$.

The above-mentioned closed-loop guidance law is expressed in terms of the state variables $y(t)$ and $V(t)$ requires the missile to provide the relative position and velocity of the current missile and target perpendicular to the LOS. The guidance law coefficient is related to time. Such a guidance law is not suitable for engineering applications. However, it has been found that when the seeker output $\dot{q}$ is used for guidance feedback, the guidance law can be greatly simplified.

In the case of small disturbances, the LOS angle of the missile to the target can be expressed as $q = \dfrac{y(t)}{V_r \cdot (T - t)}$, in which V_r is the relative velocity of the missile and target along the LOS, that is $V_r = V_m \cos\phi_m + V_t \cos\phi_t$ (Fig. 8.1-3).

Taking the derivative of q with respect to time. The following expression can be written as

8 Proportional Navigation and Extended Proportional Navigation Guidance Laws

$$\dot{q} = \frac{y(t)}{V_r \cdot (T-t)^2} + \frac{V(t)}{V_r \cdot (T-t)}. \tag{8.1-5}$$

Another form of the proportional navigation law can be acquired by substituting the above expression of $\dot{q}$ into the optimal guidance law, that is

$$a_c = 3V_r\dot{q}. \tag{8.1-6}$$

So, the general form of proportional navigation law that is more commonly used in engineering implementation is

$$a_c = NV_r\dot{q}. \tag{8.1-7}$$

Here. N is known as the proportional navigation constant.

The benefits of converting the guidance law feedback variable from the missile-target relative position y and the relative velocity V to the seeker output $t_{go} = T - t$ are:

(a) The number of the feedback variables is changed from y and V, which are hard to measure, to measurable seeker output $t_{go} = T - t$;

(b) The feedback guidance law is changed from a time-varying guidance law to a time-invariant guidance law, that is to say, to implement the PN guidance law, the missile does not need to know $t_{go} = (T - t)$. This significantly simplifies the engineering application of the PN guidance law.

Since the above acceleration command a_c of the PN guidance law refers to the acceleration command perpendicular to the LOS, when there is an angle of ϕ_m between the missile velocity direction and the LOS (Fig. 8.1-4), the acceleration autopilot command should be modified as

$$a_c = \frac{NV_c}{\cos\phi_m} \cdot \dot{q}. \tag{8.1-8}$$

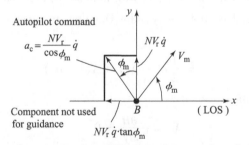

Fig. 8.1-4 Acceleration command modification when a seeker gimbal angle ϕ_m exists

When the guidance law is implemented, the angle ϕ_m can be approximated by the seeker gimbal angle.

The following section introduces the proof of the proportional navigation law with the help of optimal control theory.

Assuming small disturbances, the simplified model of a missile targeting a constant-velocity flight target can be represented as shown in Fig. 8.1-5. The corresponding state equation is

$$\begin{cases} \dot{y} = V, \\ \dot{V} = -a_c. \end{cases} \tag{8.1-9}$$

In Equation (8.1-9), y is the relative position of the missile-target perpendicular to the LOS, V is the relative velocity of the missile-target perpendicular to the LOS, and a_c is the absolute acceleration command of the missile-target perpendicular to the LOS. It is given that $a_t = 0, a_m = a_c$.

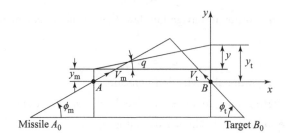

Fig. 8.1-5 Simplified model of the missile attacking a target

Taking the objective function for deriving the optimal guidance law as

$$J = S\frac{y(T)^2}{2} + \frac{1}{2}\int_0^T a_c^2(t)\,dt, \tag{8.1-10}$$

where T is the engagement time and $\phi(T) = S\dfrac{y^2(T)}{2}$ is the penalty function of the miss distance at the time moment T. The miss distance will be zero when $S \to \infty$. The integral term of the objective function involves minimizing the integral of the missile acceleration squared over the engagement time interval. requires the minimization of the integral of the missile acceleration square over the engagement time interval. Obviously, the purpose of this optimal model is to seek an optimal control law $a_c(t)$ so that the missile can hit the target with minimal control cost.

According to the optimal control theory, the Hamiltonian function of this problem should be given first

$$H = \frac{1}{2}a_c^2 + \lambda_y V + \lambda_V(-a_c), \tag{8.1-11}$$

followed by its adjoint equations

$$\begin{aligned}\dot{\lambda}_y &= -\frac{\partial H}{\partial y} = 0, \\ \dot{\lambda}_V &= -\frac{\partial H}{\partial V} = -\lambda_y.\end{aligned} \tag{8.1-12}$$

The optimal control a_c can be given by the stationarity of this problem

$$\frac{\partial H}{\partial a_c} = a_c - \lambda_V = 0, \tag{8.1-13}$$

$$a_c = \lambda_V. \tag{8.1-14}$$

It is known that the boundary value of the state equation is given as the initial value $y(0)$, $V(0)$ of the state. The boundary value of the adjoint equation is given by the final values $\lambda_y(T)$ and $\lambda_V(T)$ of λ_y and λ_V, and it can be derived from the penalty function $\phi(T)$.

8 Proportional Navigation and Extended Proportional Navigation Guidance Laws

$$\lambda_y(T) = \frac{\partial \phi(T)}{\partial y(T)} = S \cdot y(T),$$
$$\lambda_V(T) = \frac{\partial \phi(T)}{\partial V(T)} \equiv 0. \qquad (8.1-15)$$

The differential equations of the boundary value problem can be acquired as follows by substituting the optimal control expression $a_c = \lambda_V$ into the state and adjoint equations.

$$\begin{cases} \dot{y} = V, \\ \dot{V} = -\lambda_V, \\ \dot{\lambda}_y = 0, \\ \dot{\lambda}_V = -\lambda_y. \end{cases} \qquad (8.1-16)$$

The boundary conditions for solving this problem are:
the state initial values,

$$y(0), V(0);$$

and the adjoint final values,

$$\lambda_y(T) = S \cdot y(T),$$
$$\lambda_V(T) = 0. \qquad (8.1-17)$$

Generally, there are no analytical solutions for dual boundary value differential equations, but there are analytical solutions expressed with $y(0)$, $V(0)$, $\lambda_y(T)$ and t for this problem, and the open loop optimal control law for this problem can be obtained by substituting the expression of λ_V into a_c:

$$a_c(t) = (T-t)\frac{S}{1+\frac{S}{3}T^3}[y(0) + T \cdot V(0)]. \qquad (8.1-18)$$

Suppose $S \to \infty$, and the open loop guidance law with a zero miss distance is

$$a_c(t) = \frac{3(T-t)}{T^3}[y(0) + T \cdot V(0)]. \qquad (8.1-19)$$

Consider the current time t as the initial time, then the engagement time T becomes $(T-t)$, and the initial state value will be $y(t)$ and $V(t)$. The time variable after the initial time t can be expressed with τ (Fig. 8.1-6).

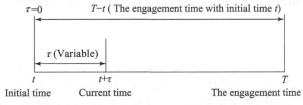

Fig. 8.1-6 Time variable transformation diagram

Then, the open loop guidance law with the initial time (the time t) and the time variable τ is

$$a_c(t+\tau) = \frac{3(T-t-\tau)}{(T-t)^3}[y(t) + (T-t) \cdot V(t)]. \qquad (8.1-20)$$

Let $\tau = 0$ for each moment t, the desired proportional navigation law is obtained as follows

$$a_c(t) = \frac{3}{(T-t)^2} y(t) + \frac{3}{(T-t)} V(t), \qquad (8.1-21)$$

where the missile-target LOS angle q is small, that is

$$q \approx \tan q = \frac{y(t)}{(T-t) \cdot V_r}. \qquad (8.1-22)$$

Take q by the derivative of time,

$$\dot{q} = \frac{y(t)}{(T-t)^2 \cdot V_r} + \frac{V(t)}{(T-t) \cdot V_r}. \qquad (8.1-23)$$

The PN guidance law with feedback $\dot{q}$ can be obtained as follows

$$a_c(t) = 3 V_r \cdot \dot{q}. \qquad (8.1-24)$$

8.1.2 Analysis of PN Guidance Law with No Guidance System Lag

1) Effect of the initial heading error on PNG

The optimal guidance law derived from the above model can be used to hit the target with minimal control cost under any initial state $(y(0), V(0))$ disturbance.

When there is an initial missile heading error ε, the relative velocity $V(0)$ disturbance (Fig. 8.1-7) that is perpendicular to the LOS will be

$$V(0) = V_m \varepsilon \cos\phi_m. \qquad (8.1-25)$$

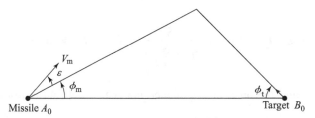

Fig. 8.1-7 Proportional guidance model under the $V(0)$ disturbance

Under the $V(0)$ disturbance, the block diagram of the PNG closed-loop control is shown in Fig. 8.1-8. Here again, it is emphasized that y_m and y_t represent the absolute displacement of the missile and the target perpendicular to the LOS, y is the relative displacement between the missile and the target in the same direction, and V_m is the flight velocity of the missile.

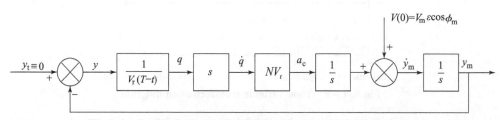

Fig. 8.1-8 Block diagram of PNG under the $V(0)$ disturbance

Fig. 8.1-8 can be simplified to an equivalent block diagram Fig. 8.1-9.

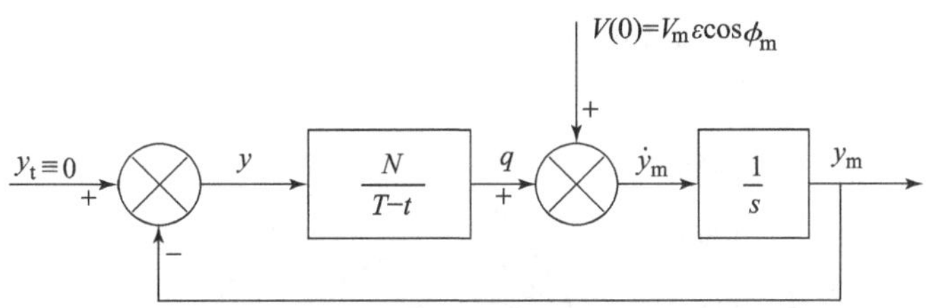

8 Proportional Navigation and Extended Proportional Navigation Guidance Laws

Fig. 8.1-9 Simplified block diagram of PNG under the $V(0)$ disturbance

The transfer function from $V_m \varepsilon \cos\phi_m$ to y_m is

$$\frac{y_m}{V_m \varepsilon \cos\phi_m} = \frac{\frac{1}{s}}{1 + \left(\frac{1}{s}\right)\left(\frac{N}{T-t}\right)} = \frac{1}{s + \frac{N}{T-t}}, \quad (8.1-26)$$

or it can be more precisely expressed as a linear time-varying differential equation

$$\dot{y}_m + \frac{N}{T-t} y_m = V_m \varepsilon \cos\phi_m. \quad (8.1-27)$$

The solution of the equation is

$$y_m = V_m \varepsilon \cos\phi_m \frac{T\left(1 - \frac{t}{T}\right)}{(N-1)} \left[1 - \left(1 - \frac{t}{T}\right)^{N-1}\right], \quad (8.1-28)$$

and the non-dimensional missile displacement will be

$$\frac{y_m}{V_m \varepsilon T \cos\phi_m} = \frac{\left(1 - \frac{t}{T}\right)}{(N-1)} \left[1 - \left(1 - \frac{t}{T}\right)^{N-1}\right]. \quad (8.1-29)$$

Obviously, for a simplified lag-free PNG model, there is $y = 0$ when $t = T$, i.e., the missile can certainly hit the target.

The dimensionless model of $\dot{y}_m$ is as follows by taking the derivative of y_m with respect to time t

$$\frac{\dot{y}_m}{V_m \varepsilon \cos\phi_m} = -\frac{1}{(N-1)} \left[1 - N\left(1 - \frac{t}{T}\right)^{N-1}\right]. \quad (8.1-30)$$

Then, taking the above equation by the derivative of time t, the following expression can be written

$$-\frac{\ddot{y}_m T}{V_m \varepsilon \cos\phi_m} = N\left(1 - \frac{t}{T}\right)^{N-2}. \quad (8.1-31)$$

Since $\ddot{y}_m = a_c \cos\phi_m$, the missile acceleration can be non-dimensionalized as

$$-\frac{a_c T}{V_m \varepsilon} = N\left(1 - \frac{t}{T}\right)^{N-2}. \quad (8.1-32)$$

Obviously, for the given N and t/T, there are $y \propto V_m \varepsilon T$, $\dot{y}_m \propto V_m \varepsilon$ and $a_c \propto \frac{V_m \varepsilon}{T}$.

Fig. 8.1-10, Fig. 8.1-11, and Fig. 8.1-12 show the non-dimensional relative position, velocity, and absolute missile acceleration time curves for PNG under the $V(0)$ disturbance with different proportional navigation constants N.

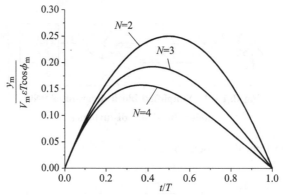

Fig. 8.1-10　Non-dimensional relative position time curves of the PNG system when $N = 2, 3, 4$

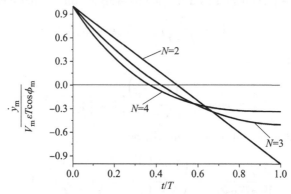

Fig. 8.1-11　Non-dimensional velocity time curves of the PNG system when $N = 2, 3, 4$

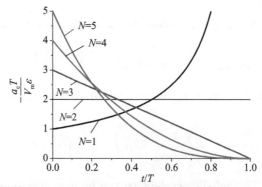

Fig. 8.1-12　Non-dimensional absolute missile acceleration time curves of the PNG system when $N = 1, 2, 3, 4, 5$

It is indicated from the missile acceleration curve in Fig. 8.1-12 that the required missile acceleration diverges and cannot be used when $N < 2$; it is a constant when $N = 2$; the missile acceleration curve is a straight line when $N = 3$. As $N > 3$, the initial required missile acceleration will increase with an increase of N, but the required missile acceleration will decrease later. To prevent the missile acceleration from reaching saturation during the terminal phase of guidance, the required acceleration in this phase should be minimized as much as possible. In addition, it is known from the following analysis that when the target is maneuvering, the proportional navigation constant N is required to be not less than 3, so in practical PNG implementation, N is often taken as 4. Therefore, in real applications, even if the variation of N could reach $\pm 25\%$, the N value used could still be in the range of 3 to 5, which will not greatly affect the guidance performance.

From the displacement curve of Fig. 8.1 – 10, the trajectory is a circular arc when $N = 2$. Since the required acceleration is large at the early stage of guidance with the increase of N and then it decreases afterward, the trajectory turns to be closer to a straight line before the interception of the target.

2) Effect of constant target maneuver on PNG performance

The performance of the proportional guidance control, when the target has a constant maneuver acceleration a_t perpendicular to the LOS, will be discussed in the following. Since the above-studied model of the proportional navigation law is supposed to be with no target maneuver, it is optimal only when the disturbance is the system's initial state $y(0)$ or $V(0)$. It should be noted that proportional navigation guidance is not considered an optimal guidance law for attacking maneuvering targets. However, because it is a closed-loop guidance law, it can still be used against maneuvering targets. Its effectiveness for such applications should be evaluated through rigorous theoretical analysis.

When $y(0)$ and $V(0)$ are zero and the target has an absolute maneuver acceleration a_t perpendicular to the LOS, the block diagram of the PNG control is shown in Fig. 8.1 – 13.

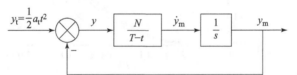

Fig. 8.1 – 13 Block diagram of PNG under the disturbance of constant target maneuver $y_t = \dfrac{1}{2} a_t t^2$

Its equivalent control block diagram is illustrated in Fig. 8.1 – 14.

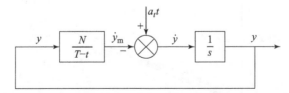

Fig. 8.1 – 14 Simplified block diagram of PNG under the disturbance of the constant target maneuver $y_t = \dfrac{1}{2} a_t t^2$

Then, the corresponding linear time-varying differential equation will be

$$\dot{y} + \left(\frac{N}{T-t}\right)y = a_t t. \tag{8.1-33}$$

The following is expressed as

$$y_m = \frac{1}{2}a_t t^2 - y. \tag{8.1-34}$$

By solving the above differentiation equation and substituting y with the expression of y_m, the following expression can be written as

$$\frac{y_m}{a_t T^2} = \frac{1}{2}\left(\frac{t}{T}\right)^2 - \frac{\left(1-\frac{t}{T}\right)}{(N-1)(N-2)}\left[(N-1)\frac{t}{T} - 1 + \left(1-\frac{t}{T}\right)^{N-1}\right]. \tag{8.1-35}$$

It is clear that as $t = T$, $y_m = \frac{1}{2}a_t T^2$ and $y = 0$.

The solution of its non-dimensional acceleration command will be

$$\frac{a_c}{a_t} = \left(\frac{N}{N-2}\right)\left[1 - \left(1-\frac{t}{T}\right)^{N-2}\right]. \tag{8.1-36}$$

Fig. 8.1-15 and Fig. 8.1-16 are the non-dimensional trajectory and acceleration curves when the missile engages a maneuvering target with constant acceleration a_t.

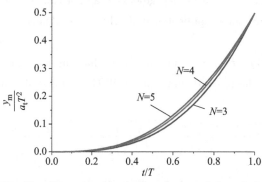

Fig. 8.1-15 Non-dimensional trajectory curves of the missile targeting a constantly maneuvering target when $N = 3, 4, 5$

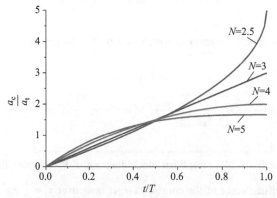

Fig. 8.1-16 Non-dimensional acceleration curves of the missile targeting a constantly maneuvering target when $N = 2.5, 3, 4, 5$

As shown in Fig. 8.1 – 16, it is known from the non-dimensional acceleration curve of the missile when attacking a constantly maneuvering target that the required acceleration will diverge when $N < 3$. Therefore, when attacking a maneuvering target, the proportional navigation constant N should not be less than 3. The missile's initial acceleration will increase with the increase of m^2 when $N > 3$, but the terminal acceleration will decrease, e.g., $\dfrac{a_c}{a_t} = 3$ at the end of guidance when $N = 3$; $\dfrac{a_c}{a_t} = 2$ at the end of guidance when $N = 4$.

To ensure that the missile has sufficient acceleration to intercept a maneuvering target, a US missile designer recommended that the missile's available acceleration be designed to be five times the target's maneuvering acceleration, that is $a_c = 5a_t$. While Russian references suggested it to be $a_c = 3a_t + 10g$.

It is known from the equation $y_m = \dfrac{1}{2}aT^2$ when $t = T$. In other words, although proportional navigation guidance is not an optimal guidance law for attacking maneuvering targets, it can still be effective if the missile has sufficient acceleration capability. As long as the missile's acceleration is adequate, it can successfully intercept a maneuvering target.

8.1.3 PNG Characteristics with the Missile Guidance Dynamics Included

In the previous section, it was assumed that the entire signal flow from the measurement of the LOS angular velocity $\dot{q}$ to the generation of the missile acceleration a_m is instantaneous without any dynamics involved. However, in real engineering implementation, the guidance loop includes various hardware elements (e.g., seeker, guidance filter, and autopilot), each of which has its own dynamics. To simplify the system analysis, it is often supposed that the seeker and guidance filter both have a first-order dynamics, the autopilot has a second-order dynamics and the four first-order dynamic components all have the same time constant $T_g/4$ (Fig. 8.1 – 17). Therefore, the whole guidance dynamics will have the form of $\left(\dfrac{T_g}{4}s + 1\right)^{-4}$.

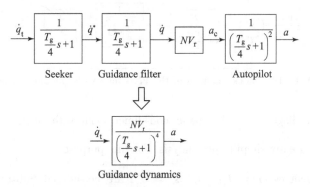

Fig. 8.1 – 17 **Simplified model of the guidance dynamics**

It can be seen from the Bode diagram in Fig. 8.1 – 18 (here taking $T_g = 0.5$ s) of the guidance dynamics transfer function $\dfrac{1}{\left(\dfrac{T_g}{4}s+1\right)^4} = \dfrac{1}{\left(\dfrac{T_g^4}{256}s^4 + \dfrac{T_g^3}{16}s^3 + \dfrac{3T_g^2}{8}s^2 + T_g s + 1\right)}$ that its low-frequency dynamics characteristics are quite similar to a first-order system $\dfrac{1}{T_g s + 1}$, in which T_g is often known as the guidance time constant. However, it should be noted that the phase shift and gain roll-off rate at higher frequencies are significantly different from those of a lower-order system. Their effects on guidance loop stability and guidance miss distance must be carefully considered.

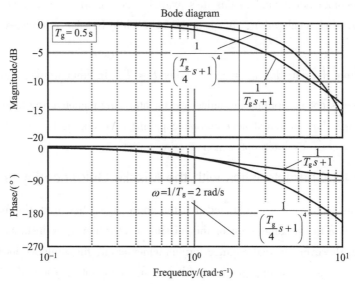

Fig. 8.1 – 18 Bode diagram of the guidance dynamics transfer function $\left(\dfrac{T_g}{4}s+1\right)^{-4}$ and first-order system $(T_g s + 1)^{-1}$

The block diagram of the PNG loop, considering guidance dynamics, is shown in Fig. 8.1 – 19.

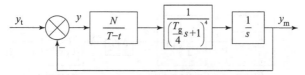

Fig. 8.1 – 19 Block diagram of the PNG loop with guidance dynamics

The above block diagram can be made non-dimensional with respect to the guidance time constant T_g. This will greatly simplify its analysis. For this purpose, $\bar{t} = \dfrac{t}{T_g}$ and the corresponding non-dimensional frequencies $\bar{\omega} = T_g \omega$ and $\bar{s} = T_g s$ are considered. Substituting the above non-dimensional time domain and frequency domain expressions into Fig. 8.1 – 19, the PNG loop can be simplified as Fig. 8.1 −20 and Fig. 8.1 −21, in which $\bar{t}_{go} = \dfrac{T-t}{T_g} = \dfrac{t_{go}}{T_g}$.

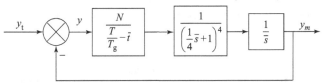

Fig. 8.1 −20 Simplified PNG loop with guidance dynamics (I)

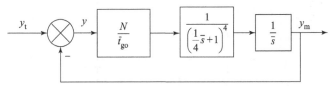

Fig. 8.1 −21 Simplified PNG loop with guidance dynamics (II)

It can be seen from Fig. 8.1 − 21 that with the missile approach to the target, the $\bar{t}_{go}$ will decrease, and the guidance loop gain $K = \dfrac{N}{\bar{t}_{go}}$ could finally go to ∞.

Fig. 8.1 − 22 shows the changes of the guidance loop gain $K = \dfrac{N}{\bar{t}_{go}}$ with the decrease of $\bar{t}_{go}$ when $N = 3, 4$, and 5.

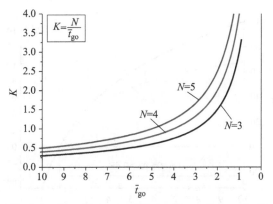

Fig. 8.1 − 22 Changes of the guidance loop gain K with the decrease of $\bar{t}_{go}$

Take $N = 4$ and draw the open-loop Bode diagram (Fig. 8.1 − 23) of PNG with different $\bar{t}_{go}$. It can be seen that the gain margin and phase margin of PNG decrease when the missile approaches the target ($\bar{t}_{go}$ decreases). Fig. 8.1 − 24 shows the variation of the gain margin and phase margin with the decrease of $\bar{t}_{go}$. From Fig. 8.1 − 23 and Fig. 8.1 − 24, it can be seen that as $N = 4$ and $\bar{t}_{go} = \dfrac{t_{go}}{T_g} = 1.76$, the phase shift of the loop has reached − 180° and the actual control direction has been reversed in comparison with the required feedback control direction. This means that the guidance loop tends to be unstable after $\bar{t}_{go} < 1.76$.

It should be noted that since the proportional guidance is a linear time-varying system, the stability analysis conducted using a LTI model is not entirely accurate in theory. However, qualitatively, the conclusion that stability issues may arise in the guidance loop during the later stages of PNG remains valid.

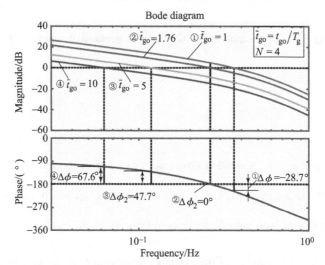

Fig. 8.1 – 23 Open-loop Bode diagram of PNG with different $\bar{t}_{go}$ parameters

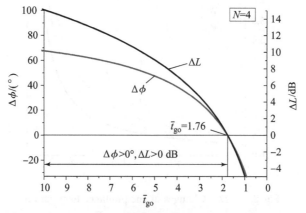

Fig. 8.1 – 24 Variation of the gain margin and phase margin with the decrease of $\bar{t}_{go}$

The effects of various disturbance sources on the performance of proportional navigation guidance will be examined in the following.

1) Analysis of the effect of the initial heading error ε on PNG

Fig. 8.1 – 25 shows the block diagram of the non-dimensional PNG control with guidance dynamics under the initial heading error disturbance ε.

The variation curve (Fig. 8.1 – 26) of the non-dimensional miss distance $\dfrac{y_{miss}}{V_m T_g \varepsilon \cos\phi_m}$ with the

8 Proportional Navigation and Extended Proportional Navigation Guidance Laws

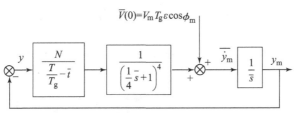

Fig. 8.1−25 Block diagram of the non-dimensional PNG control with guidance dynamics under the initial heading error disturbance ε

non-dimensional engagement time T/T_g under the initial heading error disturbance ε can be obtained by using the adjoint method (Section 8.1.4). Note that y_{miss} is defined as the guidance terminal miss distance.

As shown in Fig. 8.1−26, the presence of fourth-order guidance dynamics significantly affects the miss distance, with the duration of the engagement time having a considerable impact. To ensure the convergence of the miss distance, the engagement time should not be less than 10 times the guidance time constant T_g.

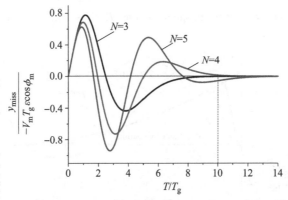

Fig. 8.1−26 Miss distance curves for the guidance system with guidance dynamics considered for different engagement time T/T_g and N

Next, the PNG acceleration under the initial heading error disturbance ε will be discussed. Fig. 8.1−27 shows the block diagram for this investigation. Here, the missile acceleration is taken as the output.

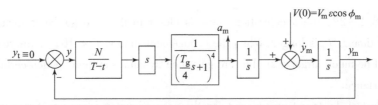

Fig. 8.1−27 Block diagram of the guidance system with fourth-order lag dynamics

When studying the missile acceleration, it is more reasonable to take the engagement time T to non-dimensionalize the loop. For this reason, take $\bar{\bar{t}} = \dfrac{t}{T}, \bar{\bar{\omega}} = T\omega$, and $\bar{\bar{s}} = Ts$. That is, $t = T\bar{\bar{t}}$ and $s = \dfrac{1}{T}\bar{\bar{s}}$. Therefore, the block diagram after non-dimensionalization is shown in Fig. 8.1 – 28.

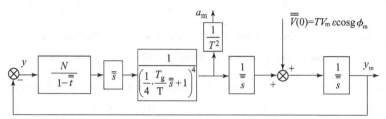

Fig. 8.1 – 28 Block diagram after non-dimensionalization

At this time, the expression of the non-dimensional acceleration is

$$\bar{\bar{a}}_m = \dfrac{T}{V_m \varepsilon \cos\phi_m} a_m. \qquad (8.1-37)$$

Fig. 8.1 – 29 shows the variation curves of the non-dimensional acceleration with the non-dimensional time t/T, when different engagement time T/T_g is taken as the parameter and $N = 4$.

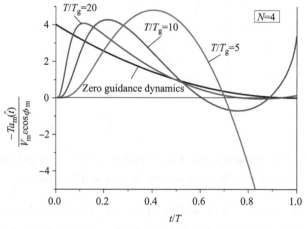

Fig. 8.1 – 29 Variation curves of the non-dimensional acceleration with the non-dimensional time for different engagement time T/T_g when $N = 4$.

From Fig. 8.1 – 29, the acceleration curve is closer to the zero guidance dynamics system as the engagement time T/T_g is longer. When the engagement time is $T < 10T_g$, the terminal guidance acceleration will be much larger than the zero guidance dynamics system, which certainly will lead to a larger miss distance.

2) The effect of the target maneuver a_t on the proportional guidance performance

The miss distance caused by the target maneuver a_t will be analyzed in the following sections. Still taking the guidance time constant T_g to make the non-dimensionalization of the time t,

$\bar{t} = \dfrac{t}{T_g}$ ($\bar{\omega} = T_g \omega, \bar{s} = T_g s$), and the resulting non-dimensional block diagram is shown in Fig. 8.1 – 30.

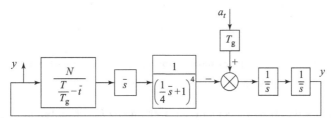

Fig. 8.1 – 30 Block diagram of the non-dimensional guidance dynamics system under the influence of the target maneuver

The non-dimensional miss distance is defined as

$$\bar{y}_{\text{miss}} = \dfrac{y_{\text{miss}}}{a_t T_g^2}. \qquad (8.1-38)$$

Fig. 8.1 – 31 shows the variation curves of the non-dimensional miss distance over the non-dimensional engagement time $\bar{T} = T/T_g$. It can be seen from the figure that when a maneuvering target is involved, the PNG constant N should not be lower than 3 to ensure the convergence of miss distance. At the same time, the missile's engagement time T should not be less than 10 times the missile's guidance time constant T_g.

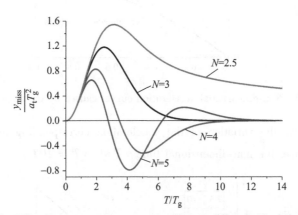

Fig. 8.1 – 31 Miss distance curves corresponding to different N with the guidance dynamics taken into account (under the influence of the target maneuver)

Similarly, in analyzing the missile acceleration, the missile's engagement time is used to non-dimensionalize t by defining T. Taking $\bar{\bar{t}} = \dfrac{t}{T}$, $\bar{\bar{\omega}} = T\omega$, and $\bar{\bar{s}} = Ts$, the guidance block diagram after non-dimensionalization is shown in Fig. 8.1 – 32.

From Fig. 8.1 – 32, the expression of the non-dimensional acceleration can be given as follows

$$\bar{\bar{a}}_m = \dfrac{a_m}{a_t}. \qquad (8.1-39)$$

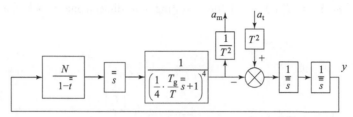

Fig. 8.1 – 32 Guidance block diagram after non-dimensionalization with respect to the engagement time T

Fig. 8.1 – 33 shows the missile acceleration over time t/T with different engagement times T/T_g when $N=4$, where one can observe that the missile acceleration gradually deviates from the ideal PN guidance law when the engagement time T is less than 10 times the guidance time constant T_g, which also leads to a larger miss distance. Therefore, when attacking a maneuvering target in the presence of guidance dynamics, the engagement time T should not be less than 10 times the guidance time constant T_g to obtain an acceptably small miss distance.

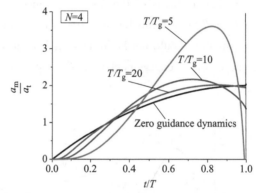

Fig. 8.1 – 33 Non-dimensional acceleration curves caused by the target maneuver

Fig. 8.1 – 34 shows the variation of the acceleration corresponding to different proportional navigation constant N over the non-dimensional time t/T when $T = 10T_g$.

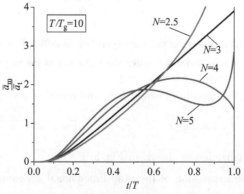

Fig. 8.1 – 34 Variation of the acceleration corresponding to different proportional navigation constant N under the influence of the target maneuver

8 Proportional Navigation and Extended Proportional Navigation Guidance Laws

3) Analysis of the influence of the radar seeker receiver thermal noise

Since the thermal noise of the seeker receiver has a wide bandwidth, it can be regarded as a white noise input in comparison with the guidance loop bandwidth. Suppose that the power spectrum density of the white noise signal u_{RN} is $S(\omega) = \phi_{RN}$ rad^2/Hz $= \phi_{RN}$ rad$^2 \cdot$ s when the missile-target distance is R_0. Since the echo signal from the target increases as the missile-target distance decreases, the seeker is equipped with an automatic gain adjustment function. This feature ensures that a constant signal level is maintained for the subsequent circuitry as the missile approaches the target and the echo signal intensifies. For this, the automatic gain adjustment coefficient of the seeker is R^2/R_0^2, in which R is the current missile-target distance. Therefore, the actual effective power spectrum of the thermal noise u_{RN} is $S(\omega,R) = \dfrac{R^2}{R_0^2}\phi_{RN} = \left(\dfrac{(T-t)V_r}{R_0}\right)^2 \phi_{RN}$ (rad$^2 \cdot$ s). Accordingly, the guidance loop diagram under the influence of thermal noise is shown in Fig. 8.1 – 35.

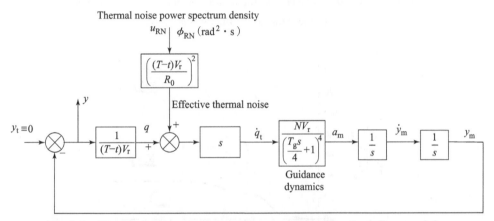

Fig. 8.1 – 35 Block diagram of the guidance system under the influence of the thermal noise (including the fourth-order guidance dynamics)

Take the above block diagram to make the non-dimensionalization with respect to the guidance time constant T_g. The result is shown in Fig. 8.1 – 36, in which $\bar{u}_{RN}$ is the non-dimensional thermal noise signal, and the non-dimensional power spectrum of its corresponding disturbance white noise power spectrum is $\bar{\phi}_{RN}$ $\left(\bar{\phi}_{RN} = \dfrac{\phi_{RN}}{T_g}\right)$. Its unit is rad^2.

Next, the adjoint function method can be employed to solve this problem. For a detailed solution procedure, please refer to Section 8.1.4 of this book. The expression of the non-dimensional miss distance standard deviation $\bar{\sigma}_{miss(RN)}$ caused by the thermal noise is as follows

$$\bar{\sigma}_{miss(RN)} = \dfrac{\sigma_{miss(RN)} R_0^2}{V_r^3 T_g^3 \sqrt{\bar{\phi}_{RN}}}, \qquad (8.1-40)$$

where $\sigma_{miss(RN)}$ is the miss distance standard deviation caused by the thermal noise, and its unit is in meters. The non-dimensional miss distance of the system under the influence of the thermal noise is shown in Fig. 8.1 – 37.

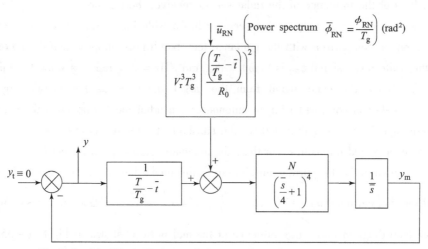

Fig. 8.1-36 Block diagram of the non-dimensional guidance system under the influence of the thermal noise

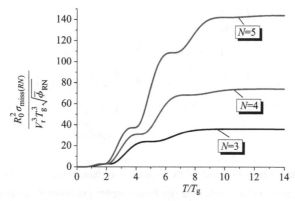

Fig. 8.1-37 Non-dimensional miss distance of the system under the influence of the thermal noise

It can be seen that the miss distance caused by the thermal noise does not converge to "0" no matter how long the engagement time T is. However, in the case where the usual engagement time $T/T_g > 10$ is adopted, the miss distance caused by the thermal noise is basically stable and flat. However, when the proportional guidance coefficient N increases, the bandwidth of the guidance loop will also increase, which will certainly increase the miss distance response to the high-frequency thermal noise input.

4) The influence of target glint

As mentioned in Chapter 4, the target glint noise is a low-frequency disturbance noise. Generally, it can be simulated with a white noise with a power spectrum density ϕ_{GL} ($m^2 s$) passing a first-order low pass filter with the time constant T_{gl}. It should be noted that the output colored noise signal represents the target position glint, which can be converted into the target angle glint and incorporated into the guidance system block diagram, as shown in Fig. 8.1-38.

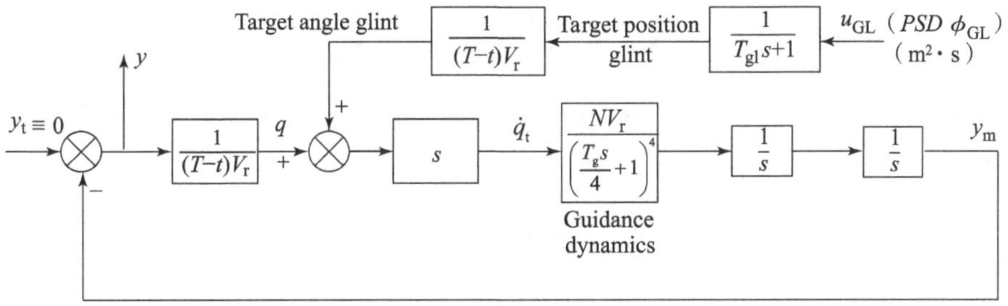

8 Proportional Navigation and Extended Proportional Navigation Guidance Laws

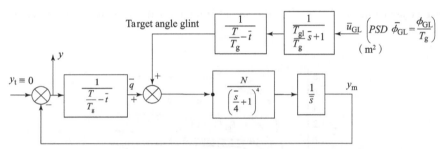

Fig. 8.1-38 Block diagram of the guidance system under the influence of the target glint

Taking non-dimensionalization with respect to the guidance time constant T_g, the system can be simplified as Fig. 8.1-39, in which $\bar{u}_{GL}$ is the non-dimensional white noise signal, and its corresponding non-dimensional power spectrum is $\bar{\phi}_{GL}$ $\left(\bar{\phi}_{GL} = \dfrac{\varphi_{GL}}{T_g}\right)$, with a unit m^2.

Fig. 8.1-39 Block diagram of the non-dimensional guidance system under the influence of the target glint

Using the adjoint method (see Section 8.1.4 for details), the expression for the non-dimensional miss distance standard deviation caused by the target glint can be obtained as

$$\bar{\sigma}_{miss(GL)} = \dfrac{\sigma_{miss(GL)}}{\sqrt{\bar{\phi}_{GL}}}. \qquad (8.1-41)$$

According to the theory of random process, when the white noise with the power spectrum $\phi_{GL}(m^2 \cdot s)$ passes through a first-order low pass filter with a time constant T_{gl} (Fig. 8.1-40), the output colored noise variance will be

$$\sigma_{GL}^2 = \dfrac{\phi_{GL}}{2T_{gl}}(m^2). \qquad (8.1-42)$$

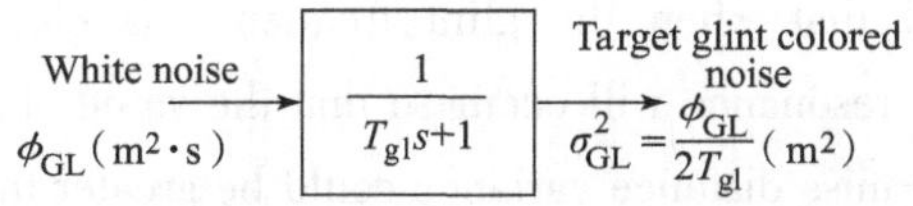

Fig. 8.1-40 Model of the colored noise

Therefore, the relationship between the power spectrum of the target glint power spectrum density and its standard deviation is

$$\sqrt{\overline{\phi}_{GL}} = \sqrt{\frac{\phi_{GL}}{T_g}} = \sqrt{\frac{2T_{gl}}{T_g}} \sigma_{GL}. \quad (8.1-43)$$

Taking the above equation into the expression $\overline{\sigma}_{miss(GL)}$ obtained via the adjoint method, the following expression can be written as

$$\overline{\sigma}_{miss(GL)} = \frac{\sigma_{miss(GL)}}{\sqrt{2\frac{T_{gl}}{T_g}}\sigma_{GL}}. \quad (8.1-44)$$

Thus, the ratio $\sigma_{miss(GL)}/\sigma_{GL}$ of the root-mean-squares (RMS) of the miss distance to the glint noise is

$$\frac{\sigma_{miss(GL)}}{\sigma_{GL}} = \sqrt{2\frac{T_{gl}}{T_g}}\,\overline{\sigma}_{miss(GL)}. \quad (8.1-45)$$

Fig. 8.1 – 41 shows the variation of the ratio $\sigma_{miss(GL)}/\sigma_{GL}$ of the miss distance to the glint noise over the non-dimensional engagement time. It can be seen from the figure that in the case of the common engagement time $T/T_g > 10$, the target glint miss distance has converged to a constant error.

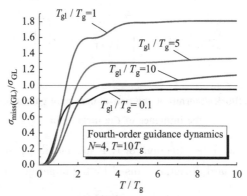

Fig. 8.1 – 41 Variation of the ratio $\sigma_{miss(GL)}/\sigma_{GL}$ of the miss distance to the glint noise over the non-dimensional engagement time

Fig. 8.1 – 42 shows the effect of T_{gl}/T_g on $\sigma_{miss(GL)}/\sigma_{GL}$. From this figure, it can be known that when the glint frequency is lower in comparison with the guidance dynamics frequency (i.e., T_{gl}/T_g is high), the missile can completely track the glint, and thus $\sigma_{miss(GL)}/\sigma_{GL} \approx 1$. When the glint frequency is high, the guidance loop fails to respond to the target glint (T_{gl}/T_g is very small), and the value of $\sigma_{miss(GL)}/\sigma_{GL}$ can be much less than 1.

It also should be noted that when the glint frequency is close to the guidance dynamics frequency ($T_{gl}/T_g \approx 1$), a resonance will occur so that the value of $\sigma_{miss(GL)}/\sigma_{GL}$ could be greater than 1; this is to say that the miss distance variance could be greater than the target glint variance in certain cases.

8 Proportional Navigation and Extended Proportional Navigation Guidance Laws

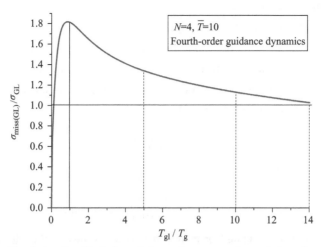

Fig. 8.1-42 Variation of the ratio $\sigma_{\text{miss(GL)}}/\sigma_{\text{GL}}$ of the miss distance with respect to T_{gl}/T_g when $\overline{T}=10$

8.1.4 Adjoint Method

The adjoint method is an effective tool for analyzing the linear time-varying system's final state $y(T)$ (here T is the final control time) under deterministic disturbance or random disturbance.

The linear time-varying system loops in which the adjoint method can be used should have the following characteristics: its time-varying blocks should only contain $(T-t)$ form elements, that is t_{go}, where T is the total control time.

The transformation of the above linear time-varying system loop to the adjoint system should follow the steps below.

(1) Replace the original time variable t with the new time variable $\tau = T - t$. So that all time-varying structures $(T-t)$ can be converted to τ, and the new system no longer contains T.

(2) Invert the signal flow of the original loop. Convert the branch nodes of the original system into summing junctions and the original summing junctions into branch nodes, as illustrated in Table 8.1-1.

Table 8.1-1 Interchange the branch nodes and summing junctions for the adjoint method

Items	Original systems	Adjoint systems
Summing junctions	$G_2(s)$, $G_1(s)$	$G_2(s)$, $G_1(s)$
Branch nodes	$G_2(s)$, $G_1(s)$	$G_2(s)$, $G_1(s)$

(3) Transform the original system.

(a) For deterministic step disturbance input.

Transform the original system output into the impulse input for the adjoint system, and convert the original step input into the impulse response output of the adjoint system. With deterministic step disturbance input, the linear time-varying system's final state $y(T)$ is the integration of the impulse response output of the adjoint system (Fig. 8.1 – 43).

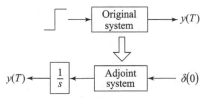

Fig. 8.1 – 43 Transformation of the original system to the adjoint system with a deterministic step disturbance input

(b) For white noise input with a power spectrum of ϕ.

It should be noted that the unit of power spectrum density ϕ for angle noise is rad^2/Hz, and the unit of power spectrum density ϕ for position noise is m^2/Hz.

Transform the output of the original system into the impulse input for the adjoint system. Treat the input of the original system as the impulse response of the adjoint system. To obtain the linear time-varying system final state variance $E[y^2(T)]$, the adjoint system impulse response should be squared, integrated, and multiplied by the power spectrum density ϕ (Fig. 8.1 – 44).

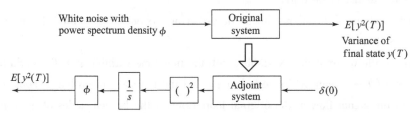

Fig. 8.1 – 44 Transformation of the original system to the adjoint system under a white noise input

Notice that the adjoint system time variable is taken as τ. When $\tau = T$, the adjoint system output $y(\tau)$ will be $y(T)$ or $E[y^2(T)]$. The benefit of the adjoint method is that the original system's final state $y(T)$ or its variance $E[y^2(T)]$ for different disturbances and different control time T can be obtained by only one simulation.

In the following, two examples are given to show the applicability of the adjoint method.

1) Example for deterministic disturbance

The original PNG loop under deterministic disturbance of target maneuver with acceleration $a_T(m/s^2)$ is given as Fig. 8.1 – 45, where a fourth-order guidance dynamic $\left(\dfrac{T_g}{4}s + 1\right)^{-4}$ is considered.

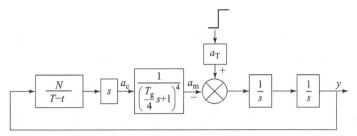

Fig. 8.1-45 Original block diagram with target maneuver a_T

The adjoint block diagram of the proportional guidance system under target maneuvering acceleration a_T disturbance is shown in Fig. 8.1-46.

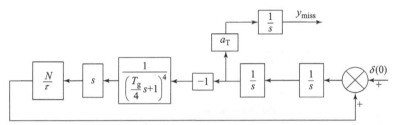

Fig. 8.1-46 Adjoint block diagram of the corresponding system

Suppose the target maneuvering acceleration a_T is $5g(50 \text{ m/s}^2)$ and the system guidance time constant T_g is 0.5 s. The miss distance y_{miss} obtained by the adjoint method is given in Fig. 8.1-47 for different proportional constant N and control time T.

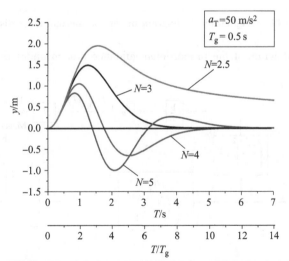

Fig. 8.1-47 Miss distance curves corresponding to different N for the proportional guidance system (under the target maneuver disturbance)

The adjoint method analysis result shows that to cope with target maneuver, the proportional navigation constant N should be greater than 3 and the non-dimensional guidance time T/T_g more than 10.

2) Example of random disturbance

Due to the seeker's automatic gain control, the thermal noise generated effective angular variation will be proportional to the square of the missile-target distance R^2, and its effective angular disturbance power spectrum density will be $\dfrac{R^2}{R_0^2}\phi_{RN} = \left(\dfrac{(T-t)V_r}{R_0}\right)^2 \phi_{RN}$ (rad^2/Hz), where ϕ_{RN} is the thermal noise power spectrum density at a distance R_0, and V_r is the relative missile-target velocity.

Fig. 8.1-48 shows the original PNG loop under random disturbance of seeker receiver thermal noise, where a fourth-order guidance dynamic $\left(\dfrac{T_g}{4}s+1\right)^{-4}$ is considered.

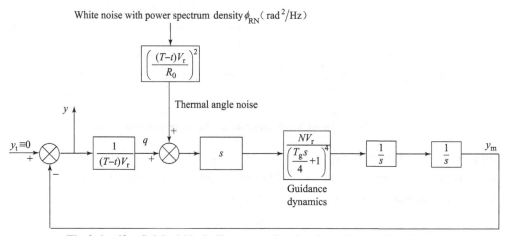

Fig. 8.1-48 Original block diagram under the thermal noise disturbance

The adjoint block diagram of the proportional guidance system under thermal noise disturbance is shown in Fig. 8.1-49.

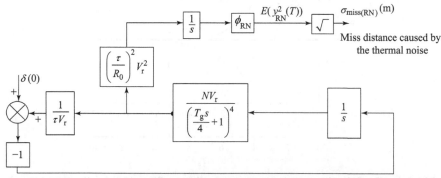

Fig. 8.1-49 Adjoint block diagram of the corresponding system under the thermal noise disturbance

Suppose the initial power spectrum density of the thermal noise signal at R_0 is R_0 is $\phi_{RN} = 1.20 \times 10^{-9}\,rad^2/Hz$, the relative velocity of the missile and target is $V_r = 800$ m/s and the initial missile-target distance is $R_0 = 10$ km. The miss distance $\sigma_{miss(RN)}$ generated by the adjoint method is given in Fig. 8.1 – 50 for different proportional constant N and control time T.

The adjoint method analysis result shows that the thermal noise introduced by the miss distance is small and its value approaches constant for normal guidance time $T > 10T_g$.

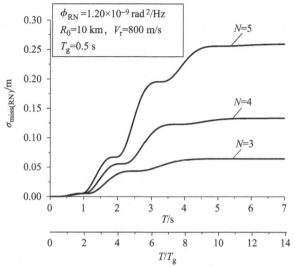

Fig. 8.1 – 50 Miss distance curves corresponding to different N for the proportional guidance system (under the thermal noise disturbance)

§ 8.2 Extended Proportional Navigation (Optimal Proportional Navigation, OPN) Guidance Laws

EXTENDED GUIDANCE LAW
WITH IMPACT ANGLE CONSTRAINT

8.2.1 Optimal Proportional Navigation Guidance Law (OPN1): Consideration of Missile Guidance Dynamics

It is important to note that the PN guidance law (described in Section 8.1) was derived under the assumption of zero guidance dynamics. However, because it is a feedback guidance law with strong robustness, it remains applicable even when guidance dynamics are not zero, although it is no longer optimal in such cases. The application restriction normally is $T/T_g > 10$. This limitation will restrict the missile's minimum operational range. When the actual guidance dynamics, including seeker, guidance filter, and autopilot dynamics, are known, a modified proportional navigation guidance law can be derived. This revised guidance law optimizes the guidance process while significantly reducing the missile's minimum operational range.

Suppose the higher-order guidance dynamics can be simplified to a first-order model, i.e.,

$$\frac{a_m(s)}{a_c(s)} = \frac{1}{T_g s + 1}. \tag{8.2-1}$$

The extended state equations with this simplified guidance dynamics will be

$$\begin{bmatrix} \dot{y} \\ \dot{V} \\ \dot{a}_m \end{bmatrix} = \begin{bmatrix} 0 & 1 & 0 \\ 0 & 0 & -1 \\ 0 & 0 & -1/T_g \end{bmatrix} \begin{bmatrix} y \\ V \\ a_m \end{bmatrix} + \begin{bmatrix} 0 \\ 0 \\ 1/T_g \end{bmatrix} a_c. \qquad (8.2-2)$$

The three state variables of this state equation are the relative distance y between the missile and the target perpendicular to the LOS, the relative velocity V, and the absolute acceleration a_m of the missile in inertial space. The control variable is the missile's autopilot command a_c. Take the objective function of this optimization problem as $J = S\dfrac{y(T)^2}{2} + \dfrac{1}{2}\int_0^T a_c^2(t)\,dt$, and take $S \to \infty$.

By solving the above optimal control problem, the optimal OPN1 guidance law is given as

$$a_c = N'\left[\frac{1}{t_{go}^2}y(t) + \frac{1}{t_{go}}V(t) + \frac{1}{t_{go}^2}(1 - e^{-\bar{t}_{go}} - \bar{t}_{go})a_m(t)\right]. \qquad (8.2-3)$$

In Equation $(8.2-3)$, $t_{go} = T - t$ is the remaining engagement time, $\bar{t}_{go} = \dfrac{T-t}{T_g} = \dfrac{t_{go}}{T_g}$ is the non-dimensional remaining engagement time, N' is the effective navigation ratio, and its expression is

$$N' = \bar{t}_{go}^2(e^{-\bar{t}_{go}} + \bar{t}_{go} - 1)\left(-\frac{1}{2}e^{-2\bar{t}_{go}} - 2\bar{t}_{go}e^{-\bar{t}_{go}} + \frac{1}{3}\bar{t}_{go}^3 - \bar{t}_{go}^2 + \bar{t}_{go} + \frac{1}{2}\right)^{-1}. \qquad (8.2-4)$$

It is known that under the condition of small disturbance, the LOS angle can be approximated as $q = \dfrac{y(t)}{V_r \cdot (T-t)} = \dfrac{y(t)}{V_r \cdot t_{go}}$. Taking the derivative of q with respect to time t will give

$$\dot{q} = \frac{y(t)}{V_r \cdot t_{go}^2} + \frac{V(t)}{V_r \cdot t_{go}}. \qquad (8.2-5)$$

A more practical proportional navigation law for engineering applications can be obtained as follows by substituting the expression of $\dot{q}$ into the above guidance law expression

$$a_c = N'V_r\dot{q} + C_1 a_m, \qquad (8.2-6)$$

where

$$N' = \bar{t}_{go}^2(e^{-\bar{t}_{go}} + \bar{t}_{go} - 1)\left(-\frac{1}{2}e^{-2\bar{t}_{go}} - 2\bar{t}_{go}e^{-\bar{t}_{go}} + \frac{1}{3}\bar{t}_{go}^3 - \bar{t}_{go}^2 + \bar{t}_{go} + \frac{1}{2}\right)^{-1}$$

and

$$C_1 = N'\frac{1}{t_{go}^2}(1 - e^{-\bar{t}_{go}} - \bar{t}_{go}).$$

The differences between the OPN1 guidance law and the standard PN guidance law are:
(1) The OPN1 guidance law has one more feedback term (missile acceleration a_m);
(2) The feedback coefficients of the state $\dot{q}$ and state a_m have become time varying, which are related to $\bar{t}_{go} = \dfrac{t_{go}}{T_g} = \dfrac{T-t}{T_g}$.

When there is a initial heading error disturbance ε, the block diagram of the OPN1 law is shown in Fig. 8.2-1.

According to the proposition of this problem, when the guidance dynamics is a first-order model, this extended proportional navigation guidance law will hit the target with minimum control cost and no limit on t_{go} as long as the required acceleration condition is not saturated. Next, the effect of initial heading error ε on this guidance law will be studied.

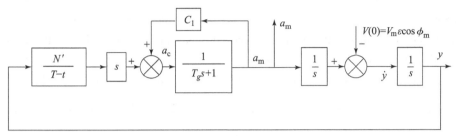

Fig. 8.2 –1 Block diagram of the OPN1 guidance law under a initial heading error disturbance ε

Make the loop non-dimensionalized with respect to the engagement time T. For this purpose, take $\bar{t} = \dfrac{t}{T}$, $\bar{s} = Ts$, and the non-dimensional time to go will be $\bar{t}_{go} = \dfrac{T-t}{T_g} = \dfrac{T}{T_g}(1 - \bar{t})$. The block diagram after non-dimensionalization is shown in Fig. 8.2 –2.

Fig. 8.2 – 3 compares the OPN1 guidance law with the standard PN guidance law. It shows that, during the initial phase of guidance, OPN1 requires more acceleration than PN. This increased acceleration command in the early guidance stage helps compensate for the lag in guidance dynamics, ensuring that the miss distance remains zero even in the presence of first-order guidance dynamics. Similar to the proportional guidance, OPN1 also has the feature that the smaller the engagement time T/T_g, the greater the required acceleration.

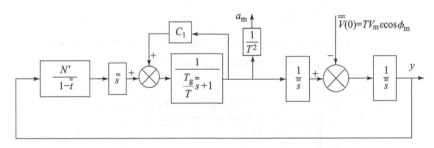

Fig. 8.2 –2 Block diagram of the non-dimensional OPN1 guidance law under a initial heading error disturbance ε

Fig. 8.2 –4 shows the acceleration command a_c and the acceleration output a_m for the PN and OPN1 guidance laws. It can be seen from the figure that the acceleration command a_c of the OPN1 guidance law at the end of the guidance is the same as the normal proportional navigation guidance. Both are zero, but the acceleration output a_m is not zero for OPN1 at the end of the guidance. This is because a_m is the output of a_c through first-order dynamics lag.

Next, the implementation approach of OPN1 in real engineering applications is studied. Suppose that the guidance dynamics model includes the dynamics of the seeker and the guidance filter $\left(\frac{T_g}{4}s + 1\right)^{-2}$, as well as the autopilot second-order dynamics $\left(\frac{T_g}{4}s + 1\right)^{-2}$. In actual engineering applications, the feedback signal $\dot{q}$ can only be taken as the output of the guidance filter, and the feedback signal a_m can be taken as the output of the autopilot accelerometer. When the feedback signals are taken accordingly, the guidance time constants in the calculation of the guidance law coefficient N' and C_1 can be taken approximately as T_g (Model I) or $T_g/2$ (Model II); their corresponding block diagrams are shown in Fig. 8.2-5 and Fig. 8.2-6. In Fig. 8.2-5, $T_g = 4 \times \left(\frac{1}{4}T_g\right)$ is taken in the calculation of N' and C_1. In Fig. 8.2-6, the second-order dynamics lag of the autopilot alone is taken in the calculation of the guidance law coefficient N'^*, C_1^*, that is to say, that here $T_g^* = 2 \times \left(\frac{1}{4}T_g\right) = \frac{1}{2}T_g$ is taken in the calculation of the time-varying feedback coefficient N'^* and C_1^*.

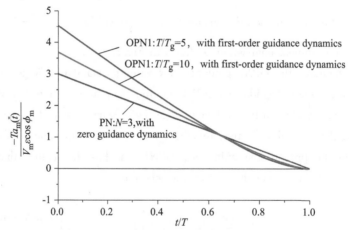

Fig. 8.2-3 Non-dimensional acceleration of standard PN guidance law and the OPN1 guidance law

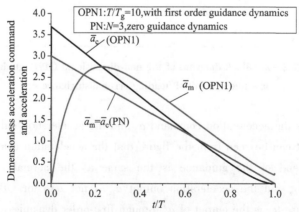

Fig. 8.2-4 Comparison between the non-dimensional acceleration command a_c and the acceleration output a_m for the OPN1 guidance law and the standard PN guidance law

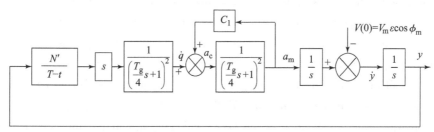

Fig. 8.2-5 Fourth-order dynamics OPN1 guidance law model Ⅰ in engineering applications (case ③)
(The guidance time constant is taken as T_g in the calculation of N' and C_1)

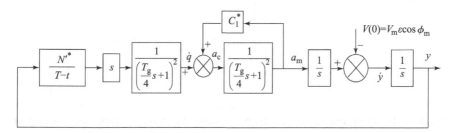

Fig. 8.2-6 Fourth-order dynamics OPN1 guidance law model Ⅱ in engineering applications (case ④)
(The guidance time constant is taken as $T_g^* = \frac{1}{2}T_g$ in the calculation of N'^* and C_1^*; that is, the seeker dynamics is ignored in the guidance law, and only the pilot dynamics is taken into account)

Fig. 8.2-7 displays the non-dimensional miss distance curves for different guidance laws under the initial heading error disturbance. Case ① shows the results for the standard PN guidance law applied to a fourth-order dynamics system. Case ② illustrates the performance of the OPN1 guidance law with a first-order dynamics system. Cases ③ and ④ depict the outcomes of engineering implementation approaches ε.

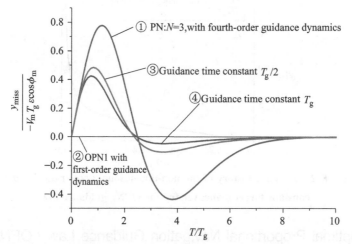

Fig. 8.2-7 Non-dimensional miss distance curves for different guidance laws under the initial heading error disturbance ε

Obviously, under the action of the initial heading error disturbance ε, the OPN1 guidance law is optimal for first-order guidance dynamics, and its miss distance is zero for all engagements T/T_g. When this guidance law is applied with engineering implementation approaches case ③ and ④ with two engineering practical "second order seeker + second order pilot" dynamics models, the convergence time of the miss distance converges for $T/T_g > 6$ instead of $T/T_g > 10$ for normal PN, and its miss distance is smaller with the same T/T_g.

In the following, the acceleration of the OPN1 guidance law under the target maneuver a_t is analyzed, and its block diagram is shown in Fig. 8.2 – 8.

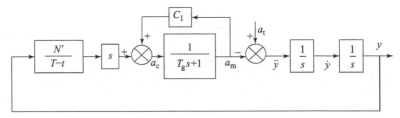

Fig. 8.2 – 8 Block diagram of the OPN1 guidance law with constant target maneuver

Fig. 8.2 – 9 shows the non-dimensional acceleration time history for the OPN1 guidance law under constant target maneuver. The terminal acceleration diverges due to the target maneuver, indicating that the guidance law becomes ineffective when dealing with a constant target maneuver. The reason is that this guidance law a_m feedback is used to compensate for the influence of the time constant T_g of the guidance dynamics, but not the target maneuver. In other words, this guidance law is not suitable for scenarios involving target maneuvers. To address this issue, it is necessary to incorporate target maneuver compensation terms into the guidance law design, as will be discussed in the subsequent OPN2 and OPN3 approaches.

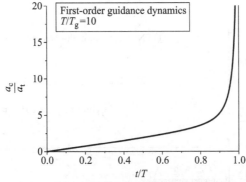

Fig. 8.2 – 9 Time history of the non-dimensional acceleration under constant target maneuver for the OPN1 guidance law

8.2.2 Optimal Proportional Navigation Guidance Law (OPN2): Consideration of the Constant Target Maneuver

From the analysis in Section 8.1, it is understood that when using proportional navigation for targeting maneuvering objects, the missile's required acceleration should be 3 to 5 times greater than the target's maneuver acceleration. If the missile's available acceleration is insufficient to meet this requirement, acceleration saturation will result in an excessive miss distance. To address this issue, if the missile is capable of estimating the target's maneuver acceleration, this estimated value can be incorporated into the guidance law as compensation. This adjustment will reduce the required missile acceleration and enhance the accuracy of guidance.

When the input to the guidance system is constant target maneuver acceleration a_t alone, the dynamics model will be

$$\begin{cases} \dot{y} = V, \\ \dot{V} = a_t - a_c. \end{cases} \quad (8.2-7)$$

Take the objective function for deriving the optimal guidance law as

$$\min J = \min \left[S \frac{y(T)^2}{2} + \frac{1}{2} \int_0^T a_c^2(t) \, dt \right]. \quad (8.2-8)$$

In Equation (8.2-8), T is the engagement time and $\phi(T) = S \frac{y(T)^2}{2}$ is the penalty function of the miss distance at the time moment T. When the value of S approaches ∞, the miss distance is zero. The integration term of the objective function represents the minimum integral of the squared missile acceleration over the time interval T, which means that it tries to hit the target with minimum control cost.

By solving the above optimum control problem, the optimum control solution will be

$$a_c = N \left(\frac{y}{t_{go}^2} + \frac{\dot{y}}{t_{go}} + \frac{1}{2} a_t \right), \quad (8.2-9)$$

where t_{go} is the remaining flight time and $t_{go} = T - t$.

Based on the previous discussion, the guidance law can be revised to a more practical form:

$$a_c = N \left(V_r \dot{q} + \frac{1}{2} a_t \right). \quad (8.2-10)$$

When there are no guidance dynamics, this revised guidance law ensures that the miss distance when targeting a constantly maneuvering object is zero. The corresponding guidance block diagram is illustrated in Fig. 8.2-10.

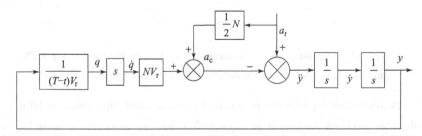

Fig. 8.2-10 Guidance block diagram of the enhanced proportional navigation law OPN2 with the effect of constant target maneuver

Make the loop non-dimensionalized with respect to the engagement time T. For this purpose, take $\bar{\bar{t}} = \dfrac{t}{T}$, $\bar{\bar{s}} = Ts$, so the non-dimensional remaining engagement time becomes $\bar{t}_{go} = \dfrac{T-t}{T_g} = \dfrac{T}{T_g}(1 - \bar{\bar{t}})$. The block diagram after the non-dimensionalization is shown in Fig. 8.2-11.

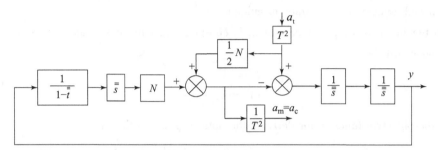

Fig. 8.2-11 Non-dimensional block diagram of the OPN2 guidance law

From the analysis of the non-dimensional acceleration time history in Fig. 8.2-12, it is known that using this guidance law to compensate a_t can ensure that the required acceleration of the missile at the end of engagement is zero and the maximum required acceleration in the initial stage of the guidance be only 1.5 times that of the target maneuver acceleration.

The guidance accuracy of this guidance law for a system with fourth-order guidance dynamics will be analyzed in the following in this case. The guidance block diagram is shown in Fig. 8.2-13.

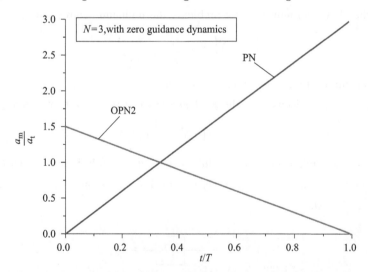

Fig. 8.2-12 Non-dimensional acceleration time history for the PN and OPN2 guidance laws, with target maneuver and guidance dynamics ignored

Fig. 8.2-14 illustrates the miss distance over the engagement time when applying this guidance law to a fourth-order guidance system. It is evident that this law does not account for the effects of guidance dynamics. Therefore, although the required minimum engagement time T remains 10 times that of the guidance dynamics time, the required terminal acceleration is reduced, and the miss distance improves. To further decrease both the minimum engagement time and the required terminal acceleration, compensation for both target maneuver and guidance dynamics is necessary (see the analysis of guidance law OPN3 in the following section).

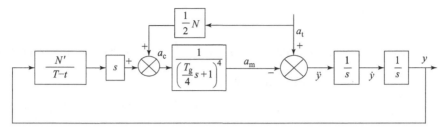

Fig. 8.2 – 13 Block diagram of the OPN2 guidance law with the fourth-order guidance dynamics

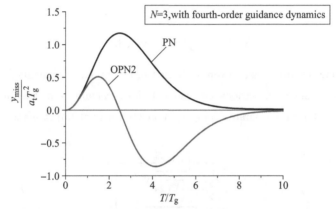

Fig. 8.2 – 14 Non-dimensional miss distance for the PN and OPN2 guidance laws with constant target maneuver a_t

8.2.3 Optimal Proportional Navigation Guidance Law (OPN3): Consideration of Both Constant Target Maneuvers and Missile Guidance Dynamics

According to the analysis in the previous section, to reduce the required terminal acceleration and further shorten the minimum allowable engagement time, compensations for both the target maneuver a_t and the guidance dynamics T_g in the guidance law are necessary.

To accomplish this task, the required extended state equation is

$$\begin{bmatrix} \dot{y} \\ \dot{V} \\ \dot{a}_m \end{bmatrix} = \begin{bmatrix} 0 & 1 & 0 \\ 0 & 0 & -1 \\ 0 & 0 & -1/T_g \end{bmatrix} \begin{bmatrix} y \\ V \\ a_m \end{bmatrix} + \begin{bmatrix} 0 \\ 1 \\ 0 \end{bmatrix} a_t + \begin{bmatrix} 0 \\ 0 \\ 1/T_g \end{bmatrix} a_c. \qquad (8.2-11)$$

The three states are the relative distance y between the missile and the target perpendicular to the LOS, the relative velocity V, and the absolute acceleration a_m of the missile in inertial space. The system disturbance input is the estimated target acceleration a_t, and its control variable is the acceleration autopilot command a_c.

The objective function for deriving the optimal guidance law is given in the same form as before

$$\min J = \min\left[S\frac{y(T)^2}{2} + \frac{1}{2}\int_0^T a_c^2(t)\,dt \right]. \tag{8.2-12}$$

By solving the optimal control problem, the optimal guidance law will be

$$a_c = N'\left[\frac{y}{t_{go}^2} + \frac{1}{t_{go}}V - \frac{1}{t_{go}^2}(e^{-\bar{t}_{go}} - 1 + \bar{t}_{go})a_m + 0.5a_t \right], \tag{8.2-13}$$

where $t_{go} = T - t$ is the remaining flight time, $\bar{t}_{go} = \dfrac{T-t}{T_g} = \dfrac{t_{go}}{T_g}$ is the non-dimensional remaining flight time, and N' is the effective navigation ratio.

$$N' = \bar{t}_{go}^2(e^{-\bar{t}_{go}} - 1 + \bar{t}_{go})\left(-\frac{1}{2}e^{-2\bar{t}_{go}} - 2Te^{-\bar{t}_{go}} + \frac{1}{3}\bar{t}_{go}^3 - \bar{t}_{go}^2 + \bar{t}_{go} + \frac{1}{2} \right)^{-1}. \tag{8.2-14}$$

Under the condition of small disturbance, the proportional navigation law in another form can be given by substituting the expression of $\dot{q}$ into the OPN3 guidance law, that is:

$$a_c = N'V_r\dot{q} + C_1 a_m + 0.5a_t, \tag{8.2-15}$$

The definition of $C_1 = N'\dfrac{1}{\bar{t}_{go}^2}(1 - e^{-\bar{t}_{go}} - \bar{t}_{go})$ is provided.

This guidance law ensures that, even with first-order guidance dynamics and constant target maneuver, the miss distances at various engagement times remain zero. The block diagram of the OPN3 guidance law, which incorporates compensation for constant target maneuver a_t, is shown in Fig. 8.2-15.

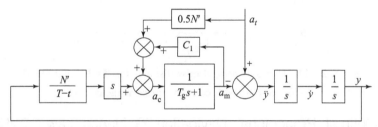

Fig. 8.2-15 Block diagram of the OPN3 guidance law with constant target maneuver a_t

Analysis of the non-dimensional acceleration time history in Fig. 8.2-16 shows that the OPN3 guidance law ensures that the missile's required acceleration at the end of the engagement is zero, even with a constant target maneuver. While the maximum required acceleration is higher during the early stages of guidance, this approach effectively addresses the acceleration divergence issue present in the OPN1 guidance law. In addition, the required acceleration under the OPN3 guidance law decreases with the increase in the engagement time T/T_g. As shown in Fig. 8.2-16, the maximum required accelerations are respectively 2.25 and 1.8 times the target maneuver acceleration when $T/T_g = 5$ and 10.

Fig. 8.2-17 shows the acceleration command a_c and missile acceleration a_m curves under the action of the PN and OPN3 guidance laws. It can be known from the figure that the acceleration command a_c under the OPN3 guidance law converges to zero at the end of the engagement, but its acceleration a_m is not zero. This is because a_m is the output of a_c when passing the first-order guidance lag dynamics.

8 Proportional Navigation and Extended Proportional Navigation Guidance Laws

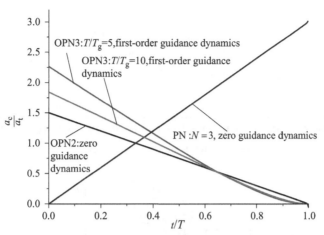

Fig. 8.2 – 16 Non-dimensional acceleration time history for the PN and OPN3 guidance laws under constant target maneuver a_t

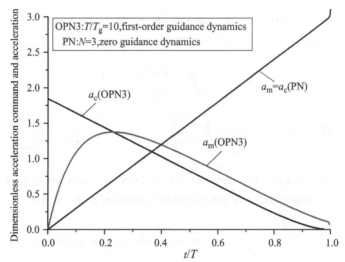

Fig. 8.2 – 17 Comparison between the non-dimensional acceleration command a_c and guidance dynamics output a_m for the OPN3 and PN guidance laws under constant target maneuver a_t

Fig. 8.2 – 18 shows the fourth-order dynamics guidance system block diagram of the PN guidance law in the presence of constant target maneuver a_t (case ①).

For practical applications, the feedback variables are limited to the guidance filter output $\dot{q}$ and the autopilot accelerometer output, as discussed in Section 8.2.1. The time T_g^* is then considered as $4 \times \left(\frac{1}{4}T_g\right) = T_g$ (case ③) or $2 \times \left(\frac{1}{4}T_g\right) = \frac{T_g}{2}$ (case ④), as illustrated in Fig. 8.2 – 19 and Fig. 8.2 – 20.

Fig. 8.2 – 21 shows the non-dimensional miss distance of the PN guidance law under the fourth-order dynamics guidance system (see Fig. 8.2 – 18 and case ①), the OPN3 guidance law under the first order dynamics guidance system (case ②), and the OPN3 guidance law adopting two engineering application schemes (cases ③ and ④) with constant target maneuver a_t.

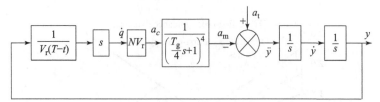

Fig. 8.2 – 18 Fourth-order dynamics for the PN guidance law with constant target maneuver a_t (case ①)

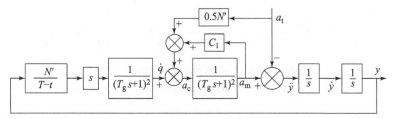

Fig. 8.2 – 19 Fourth-order dynamics for the OPN3 guidance law in engineering applications with constant target maneuver a_t (case ③)

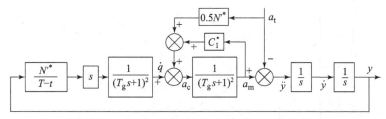

Fig. 8.2 – 20 Fourth-order dynamics for the OPN3 guidance law in engineering applications with constant target maneuver a_t (case ④)

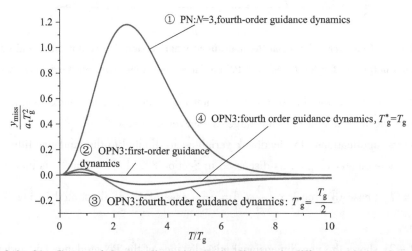

Fig. 8.2 – 21 Curves of the non-dimensional miss distance in different guidance law cases

Obviously, under constant target maneuver a_t, the OPN3 guidance law is optimal when the first-order guidance dynamics are taken into account, and the miss distance is zero. When the guidance law is applied to the dynamics model with two types of engineering implementation approaches, the miss distance performance can be enhanced from the required $T/T_g > 10$ in the normal proportional guidance to $T/T_g > 7$. Moreover, since the target maneuver acceleration is compensated, this guarantees that the required acceleration at the end of the engagement is smaller and the miss distance is reduced with the same T/T_g.

8.2.4 Estimation of Target Maneuver Acceleration

An active radar seeker measures the relative missile-target distance ΔR when working in medium to low pulse repetition modes. The direction angle q of the target in inertial space can be acquired by adding together the missile attitude angle ϑ, the seeker gimbal angle ϕ, and the beam angle measurement error ε (Fig. 8.2 – 22).

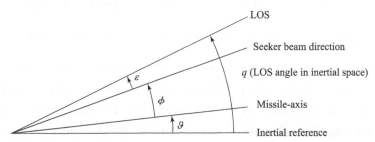

Fig. 8.2 – 22 Definition of related angles for LOS angle calculation

Given the LOS direction and the missile-target relative distance ΔR, which represents the missile-target relative distance vector ΔR in the three-dimensional (3D) space, the position vector $R_t = R_m + \Delta R$ of the target in the inertial space can be obtained from the position vector R_m of the missile in inertial space which are provided by the missile's navigation system and ΔR measured by the seeker (Fig. 8.2 – 23).

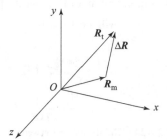

Fig. 8.2 – 23 Target position vector R_t in the inertial space

Obviously, the target maneuver acceleration $a_t(t)$ can be estimated by using the known target position vector $R_t(t)$ in inertial space with the help of the Kalman filtering technique. (Refer to Section 8.4 for the Kalman filter theory on acceleration estimation.)

When the seeker operates in medium pulse repetition frequency (PRF) mode, the missile can measure both the relative distance and radial velocity between the missile and the target. Utilizing this additional information can enhance the accuracy of target acceleration estimation.

With advancements in radar seeker technology allowing for more precise velocity and distance measurements, real-time estimation of target maneuver acceleration becomes feasible. Consequently, the proportional navigation guidance law with target maneuver compensation discussed earlier could be effectively implemented in future engineering applications.

8.2.5 Estimation of t_{go}

(1) In practice, the required accuracy for t_{go} in the extended proportional navigation guidance law is not high. It can be estimated by dividing the missile-target relative distance measured by the seeker by the current missile velocity.

(2) When t_{go} is so small that the calculated acceleration command is close to the missile's acceleration capacity limit, it can be taken as a constant and does not have to be further decreased.

8.2.6 Proportional Navigation Guidance Law with Impact Angle Constraint

In general, air-to-ground missiles often require a large impact angle when targeting stationary objects. Therefore, it is necessary to employ a proportional navigation guidance law that incorporates constraints to achieve the desired impact angle. Assuming that the flight velocity of the missile $V_{missile}$ is constant and the target has no maneuver, a_c is the acceleration autopilot command. Since the guidance dynamics are ignored, there is a missile lateral acceleration $a_m = a_c$. The state equation used to solve this problem remains

$$\begin{cases} \dot{y} = V, \\ \dot{V} = -a_c, \end{cases} \quad (8.2-16)$$

where the state variable y is the relative distance of the missile and the target perpendicular to the LOS, V is the relative velocity ($V = V_t - V_m$) in the direction perpendicular to the LOS, and V_m is the missile velocity in the direction perpendicular to the missile-target line (Fig. 8.2 – 24). Given that $V = V_t - V_m$, it follows that $V_m = -V$ will be $V_t = 0$.

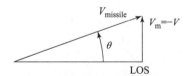

Fig. 8.2 – 24 Relationship between V and θ

Under the assumption of small disturbances and $V_t = 0$, it is known from the geometric relationship in Fig. 8.2 – 24 that

$$\theta = \frac{V_m}{V_{missile}} = \frac{-V}{V_r}. \quad (8.2-17)$$

Therefore, the expected impact angle θ_F can be given as

$$\theta_F = \frac{-V^*(T)}{V_r}, \quad (8.2-18)$$

where θ_F is the expected impact angle and $V^*(T)$ is the expected terminal relative velocity.

The objective function for deriving the optimal guidance law is

$$\min J = \min\left[S_1 \frac{y(T)^2}{2} + S_2 \frac{(V(T) - V^*(T))^2}{2} + \frac{1}{2}\int_0^T a_c^2(t)\,dt\right], \quad (8.2-19)$$

where $V^*(T)$ is the expected value of $V(T)$; $\phi_1(T) = S_1 \frac{y(T)^2}{2}$ is the penalty function of the miss distance at the moment T; $\phi_2(T) = S_2 \frac{(V(T) - V^*(T))^2}{2}$ is the penalty function for the velocity perpendicular to the LOS at the moment T and S_1, S_2 are the related weighting coefficients. When S_1 approaches ∞, the miss distance is zero; when S_2 approaches ∞, the relative velocity $V(T)$ reaches the expected value $V^*(T)$, i. e., the terminal impact angle will be the expected impact angle $\theta_F\left(\theta_F = -\frac{V^*(T)}{V_r}\right)$.

By solving the above optimal problem, the optimal guidance law for the zero guidance dynamics system will be

$$a_c(t) = -\frac{1}{t_{go}^2}(6y + 4t_{go}V + 2t_{go}V(T)^*), \quad (8.2-20)$$

or

$$a_c(t) = 4V_r\left(\frac{y(t) + V(t)t_{go}}{V_r t_{go}^2}\right) + \frac{2V_r}{t_{go}}\left(\frac{y(t)}{V_r t_{go}} + \frac{V(T)^*}{V_r}\right). \quad (8.2-21)$$

In Equation (8.2-20) and Equation (8.2-21), t_{go} is the remaining flight time. With small disturbance assumption, it is known as

$$q = \frac{y(t)}{V_r \cdot (T-t)}, \quad (8.2-22)$$

and

$$\dot{q} = \frac{y(t)}{V_r \cdot (T-t)^2} + \frac{V(t)}{V_r \cdot (T-t)}. \quad (8.2-23)$$

Using Equation (8.2-22), Equation (8.2-23), and $V^*(T) = -V_r\theta_F$, a more practical form of the proportional navigation guidance law with the impact angle constraint θ_F can be obtained as

$$a(t) = 4V_r\dot{q}(t) + \frac{2V_r}{t_{go}}(q(t) - \theta_F). \quad (8.2-24)$$

From Equation (8.2-24), it is known that the proportional navigation guidance law with the impact angle constraint is composed of two terms: the proportional guidance $\dot{q}(t)$ feedback term ($4V_r\dot{q}(t)$) that ensures the hitting of the target and the feedback term ($2V_r(q(t) - \theta_F)/t_{go}$) of ($q(t) - \theta_F$) that satisfies the impact angle constraint.

It should be noted that although the above guidance law is derived under the assumption of small disturbances, the result is a guidance law with both the miss distance feedback (feedback $\dot{q}$) and the impact angle control feedback (feedback ($q(t) - \theta_F$)). Therefore, this guidance law can be applied to a nonlinear system that requires both target hitting and impact angle control. However, achieving the absolute optimum cannot be guaranteed at this stage.

To validate the applicability of this guidance law for nonlinear systems and high-impact angle scenarios, a guided bomb example is used. In this analysis, it is assumed that the missile's guidance dynamics are negligible. The system in the pitch plane is described as follows

$$\begin{cases} \dot{V} = (-X - mg\sin\theta)/m, \\ \dot{\theta} = (Y - mg\cos\theta)/(mV), \\ \dot{x} = V\cos\theta, \\ \dot{y} = V\sin\theta, \end{cases} \quad (8.2-25)$$

where the drag is $X = c_x qS$, $c_x = 0.5$; the normal force is $Y = ma_c$, a_c is the above proportional navigation guidance law command; the reference area of the missile is taken as $S = 0.1075 \text{ m}^2$, diameter is $D = 370$ mm and the mass is $m = 500$ kg.

Suppose that the guided bomb's initial releasing condition is altitude $H_0 = 1\,000$ m and velocity $V_0 = 250$ m/s. The bomb is released horizontally. The target position is: $X_t = 5\,000$ m, $Y_t = 0$ m and the expected impact angles θ_F are respectively $0°$, $-30°$, $-60°$, and $-90°$.

Fig. 8.2-25 shows the guided trajectory curves with different expected impact angles θ_F. It can be observed that the trajectory-shaping guidance law remains effective even under conditions of large impact angles and nonlinear dynamics.

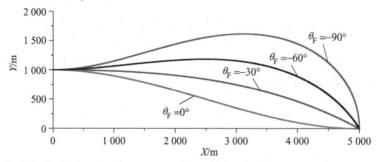

Fig. 8.2-25 Guided trajectory curves of the proportional navigation law with impact angle constraints under different expected impact angles

In engineering applications, a more general trajectory-shaping guidance law with a weighting coefficient N_y and a weighting coefficient N_q is often used.

$$a_c(t) = N_y V_r \dot{q}(t) + N_q V_r (q(t) - \theta_F)/t_{go}, \quad (8.2-26)$$

where the weighting coefficients N_y and N_q can be used to adjust the weight ratio of the impact position constraint and the impact angle constraint. Generally, the constraint on impact position should be given higher priority than the constraint on impact angle.

§8.3 Other Types of Proportional Navigation Laws

8.3.1 Gravity Over-Compensated Proportional Navigation Law

Normal guidance bombs and some tactical air-to-ground missiles flying at subsonic speed do not have high flight velocity in their terminal guidance phase, and their lateral acceleration capacity is limited (usually $(2-4)g$). When normal proportional navigation guidance law is adopted, an additional aerodynamic acceleration of about one g will have to be provided to counter balance the gravity. As this is done, the available missile lateral acceleration maneuvering upward will be $a_{\text{inertial available}} = a_{\text{aerodynamic available}} - g$, and the downward maneuvering capability will be $a_{\text{inertial available}} = a_{\text{aerodynamic available}} + g$. Due to the missile's lower upward maneuverability compared to its downward maneuverability—potentially differing by up to $2g$—the gravity over-compensated proportional navigation guidance law aims to address this imbalance. This law can reduce the lateral acceleration required at the end of the engagement to less than the standard value of $1g$, thereby equalizing the missile's upward and downward maneuverability. Additionally, it can enhance the impact angle to improve warhead effectiveness.

The general form of the gravity over-compensated proportional navigation guidance law is as follows

$$a_c = NV_r\dot{q} + (c-1)g. \qquad (8.3-1)$$

(normal gravity compensation: $c = 1$; over-gravity compensation: $c > 1$)

Fig. 8.3-1 shows its guidance block diagram, in which $a_{\text{m aerodynamic}}$ represents the missile's aerodynamic lateral acceleration, and $a_{\text{m inertial}}$ indicates the missile's maneuverability acceleration in inertial space (gravity force included). The relationship between the two is

$$a_{\text{m inertial}} = a_{\text{m aerodynamic}} - g \,(a \text{ is positive for upward acceleration}). \qquad (8.3-2)$$

Therefore, as shown in Fig. 8.3-2, the guidance law of the missile in the inertial space is

$$\begin{cases} a_{\text{m inertial}} = NV_r\dot{q} + (c-1)g, \\ a_{\text{m aerodynamic}} = NV_r\dot{q} + cg, \end{cases} \qquad (8.3-3)$$

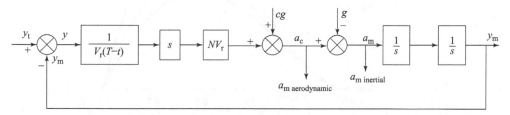

Fig. 8.3-1 Block diagram of the gravity over-compensated proportional navigation guidance law

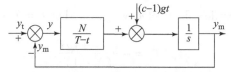

Fig. 8.3-2 Simplified block diagram of the gravity over-compensated proportional navigation guidance law

It is known from Fig. 8.3-2 that the linear time-varying differential equation of this guidance law is

$$\dot{y}_m + \frac{N}{T-t}y_m = (c-1)gt. \qquad (8.3-4)$$

By solving this differential equation, the non-dimensional position, velocity, and inertial acceleration under the guidance law can be determined

$$\frac{y_m}{(c-1)gT^2} = \frac{1-\bar{\bar{t}}}{(N-1)(N-2)}[(N-1)\bar{\bar{t}} - 1 + (\bar{\bar{t}}_{go})^{N-1}], \qquad (8.3-5)$$

$$\frac{V_m}{(c-1)gT} = \frac{N}{(N-1)(N-2)} - \frac{2}{(N-2)}\bar{\bar{t}} - \frac{N(\bar{\bar{t}}_{go})^{N-1}}{(N-1)(N-2)}, \qquad (8.3-6)$$

$$\frac{a_{m\ inertial}}{(c-1)g} = \frac{\dot{V}_m}{(c-1)g} = \frac{1}{(N-2)}[N(\bar{\bar{t}}_{go})^{N-2} - 2], \qquad (8.3-7)$$

where

$$\bar{\bar{t}} = \frac{t}{T}, \quad \bar{\bar{t}}_{go} = \frac{T-t}{T} = \frac{t_{go}}{T}.$$

Fig. 8.3 – 3, Fig. 8.3 – 4, and Fig. 8.3 – 5 respectively show the non-dimensional displacement, velocity, and lateral acceleration curves corresponding to the guidance law when the initial heading error is zero.

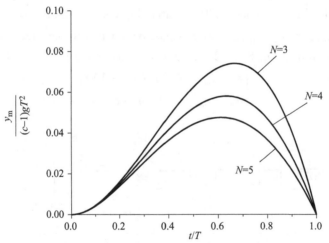

Fig. 8.3 – 3 Non-dimensional displacement curves

The expression of the missile's final velocity $V_m(T)$ when $t = T$ is

$$V_m(T) = -\frac{(c-1)gT}{(N-1)}. \qquad (8.3-8)$$

It is clear that when c is larger than 1, the missile's final velocity perpendicular to the ground is negative. That is, the gravity over-compensation can increase the impact angle of the missile.

The terminal acceleration values of the missile when $t = T$ are

$$a_{m\ inertial} = \frac{-2(c-1)g}{(N-2)}, \qquad (8.3-9)$$

$$a_{m\ aerodynamic} = \frac{-2(c-1)g}{(N-2)} + g. \qquad (8.3-10)$$

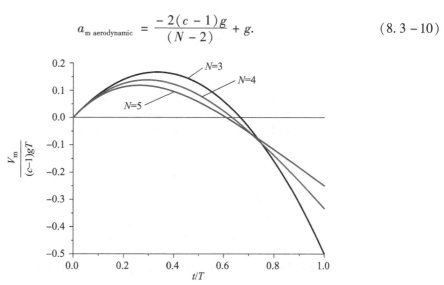

Fig. 8.3-4 Non-dimensional velocity curves

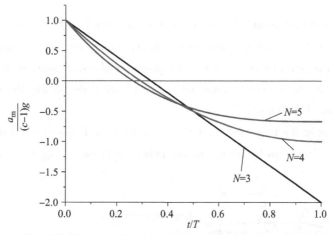

Fig. 8.3-5 Non-dimensional lateral acceleration curves

Fig. 8.3-6 shows the aerodynamic acceleration values $a_{m\ aerodynamic}$ at the end of the engagement for gravity normal-compensation ($c = 1$) and gravity over-compensation ($c = 2$) with different proportional navigation constants $-\frac{a_c T}{V_m \varepsilon} = N\left(1 - \frac{t}{T}\right)^{N-2}$. It can be seen from the figure that for normal proportional navigation ($c = 1$), the missile always requires one g aerodynamic acceleration to balance the gravity force. If $c = 2$ and the proportional guidance coefficient $N = 4$, the aerodynamic acceleration required at the end of engagement could be zero. This approach addresses the issue of asymmetry in the missile's upward and downward acceleration capabilities inherent in standard PNG. It effectively mitigates the problem of reduced aerodynamic performance during upward maneuvers.

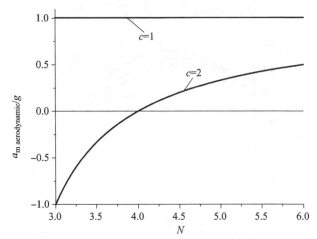

Fig. 8.3-6 Variation curve of $a_{m\ aerodynamic}$ over N ($c=1$, 2)

8.3.2 Lead Angle Proportional Navigation Guidance Law

When a passive infrared-guided missile targets an aircraft, the infrared seeker focuses on tracking the heat signature from the aircraft's engine, which emits the highest temperature. To maximize damage, it is advantageous to adapt the proportional navigation law to account for the size of the aircraft, enabling the seeker to track the engine's heat while guiding the missile to strike the more vulnerable areas of the target.

Based on the size of the target to be attacked, the forward displacement d of the attack point P relative to the engine flame can be set before launching the missile (Fig. 8.3-7).

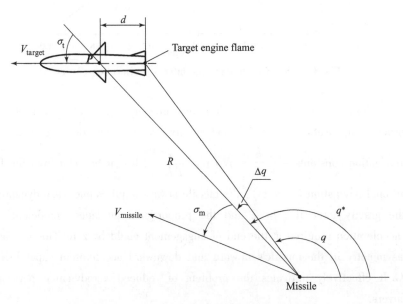

Fig. 8.3-7 Lead angle proportional navigation law

8 Proportional Navigation and Extended Proportional Navigation Guidance Laws

Suppose P represents the target point, and consider introducing an extended proportional navigation guidance law in the following form

$$a_c = NV_r \dot{q}^*, \qquad (8.3-11)$$

where q^* is the virtual missile-target LOS. Its definition is

$$q^* = q + \Delta q, \qquad (8.3-12)$$

$$\dot{q}^* = \dot{q} + \Delta \dot{q}. \qquad (8.3-13)$$

Since the detector tracks the heat source at the end of the engine, its output is naturally the seeker's output q. The question now becomes how to get the value of $\Delta \dot{q}$. It is known from Fig. 8.3-7

$$\Delta q = \frac{d \sin\sigma_t}{R} = \frac{d \sin\sigma_t}{V_r(T-t)} = \frac{d \sin\sigma_t}{V_r t_{go}}, \qquad (8.3-14)$$

$$\Delta \dot{q} = \frac{d \sin\sigma_t}{V_r}\left(-\frac{1}{t_{go}^2}\right)(-1) = \frac{d \sin\sigma_t}{V_r t_{go}^2}. \qquad (8.3-15)$$

During the process of approaching the target, there is

$$V_{\text{missile}} \sin\sigma_m = V_{\text{target}} \sin\sigma_t. \qquad (8.3-16)$$

Therefore

$$\sin\sigma_t = \frac{V_{\text{missile}}}{V_{\text{target}}} \sin\sigma_m, \qquad (8.3-17)$$

$$\Delta \dot{q} = \frac{V_{\text{missile}} d}{V_{\text{target}} V_r}\left(\frac{1}{t_{go}^2}\right) \sin\sigma_m. \qquad (8.3-18)$$

It is known that the relationship between the infrared heat source energy W measured by the seeker detector and the missile-target relative distance R are as follows

$$W = K\left(\frac{1}{R^\alpha}\right) \;(\alpha \text{ is about } 2), \qquad (8.3-19)$$

$$\ln W = \ln K - \alpha \ln R, \qquad (8.3-20)$$

$$\frac{d}{dt}(\ln W) = -\alpha \frac{\dot{R}}{R} = -\alpha \frac{V_r}{V_r t_{go}} = -\alpha\left(\frac{1}{t_{go}}\right), \qquad (8.3-21)$$

$$\frac{1}{t_{go}^2} = \frac{1}{\alpha^2}\left[\frac{d}{dt}(\ln W)\right]^2. \qquad (8.3-22)$$

Therefore, the expression of $\Delta \dot{q}$ is

$$\Delta \dot{q} = \left(\frac{V_{\text{missile}} d}{V_{\text{target}} V_r \alpha^2}\right)\left[\frac{d}{dt}(\ln W)\right]^2 \sin\sigma_m. \qquad (8.3-23)$$

Furthermore, the guidance equation of the extended proportional navigation law can be given as

$$a_c = NV_r\left\{\dot{q} + \frac{V_{\text{missile}} d}{V_{\text{target}} V_r \alpha^2}\left[\frac{d}{dt}(\ln W)\right]^2 \sin\sigma_m\right\}. \qquad (8.3-24)$$

In engineering applications, V_{missile}, V_{target}, and V_r can be approximately known. d and α are known, σ_m can be acquired from the output of the seeker gimbal angle, and $[d(\ln W)/dt]^2$ can be obtained from the change rule of the infrared heat source energy W measured by the seeker detector. This approach provides all the necessary information to implement the lead angle proportional navigation law, thereby ensuring the maximum damage to the target by utilizing this extended proportional guidance law.

§ 8.4 Target Maneuver Acceleration Estimation

As discussed in Section 8.2 regarding the extended proportional navigation guidance law, accurately estimating both the magnitude and direction of the target's maneuver acceleration $y_t = \frac{1}{2}at^2$ in inertial space allows for the use of a proportional navigation guidance law with target acceleration compensation. This approach significantly reduces the required missile acceleration at impact, thereby enhancing guidance accuracy.

Currently, the inertial navigation system of air-to-air missiles can provide the missile's coordinates and velocity at any given moment in inertial space. However, target information is typically obtained through the seeker system. Active radar seekers operate in various modes. Medium Distance and Medium Pulse Repetition Frequency Mode: This mode allows the seeker to provide data on missile-target distance, relative radial velocity, and LOS direction. With this data, the 3D relative velocity and position of the missile and target can be calculated in real time. Combining this with the missile's position and velocity information from the inertial navigation system enables the determination of the target's motion in inertial space. Short Distance and Low Pulse Repetition Frequency Mode: As the missile approaches the target, the seeker switches to this mode to provide detailed position information of the target. Long Distance and High Pulse Repetition Frequency Doppler Velocity Mode: When the missile is farther from the target, the seeker operates in this mode to measure Doppler velocity (Table 8.4 – 1).

Table 8.4 – 1 Target information acquired by the seeker in different working modes

Working modes of the seeker	Seeker information	Target information acquired in combination with the missile inertial navigation system information
Long distance, high pulse repetition frequency	Velocity measurement, LOS direction measurement	Velocity
Medium distance, medium pulse repetition frequency	Distance measurement, velocity measurement, LOS direction measurement	Position, velocity
Short distance, low repetition frequency	Distance measurement, direction measurement	Position

In the near future, as radar seeker accuracy in measuring velocity and distance continues to improve, it is anticipated that the target's maneuver acceleration can be estimated using the Kalman filter technique across all three operational modes mentioned. This advancement will enable the implementation of advanced proportional navigation laws for precise targeting of maneuvering targets.

The method for estimating target maneuvers using the Kalman filter technique will be briefly outlined in the following section.

First, notice that the preconditions for estimating the system state using the Kalman filter are:

8 Proportional Navigation and Extended Proportional Navigation Guidance Laws

(1) The dynamic model of a certain linear time-varying system is $\dot{X}(t) = A(t)X(t) + B(t)u(t)$, and the output measurement model is $Z(t) = C(t)X(t)$, both are assumed to be known;

(2) A definitive input signal of the system $u(t)$ is given.

When there is a certain amount of disturbance and model parameter variation in the system, as long as the system can measure some state-related output $Z(t)$, the closed-loop state estimation method can be used to estimate the state $X(t)$ of the system. Define the estimated value of state $X(t)$ as $\hat{X}(t)$. The general estimator model is given as

$$\dot{\hat{X}}(t) = A(t)\hat{X}(t) + B(t)u(t) + K(t)[Z(t) - C(t)\hat{X}(t)], \qquad (8.4-1)$$

where $[Z(t) - C(t)\hat{X}(t)] = [Z(t) - \hat{Z}(t)]$ is the measurement estimation error, and $K(t)$ is the estimation feedback gain matrix.

The idea of this closed-loop estimation is that when there is an error between the measurement value $Z(t)$ and the estimated measurement value $\hat{Z}(t) = C(t)\hat{X}(t)$ of the system, these measurement estimation errors will be multiplied by a gain matrix and feedback to the state estimation equation. This feedback will force the state estimation $\hat{X}(t)$ to make proper adjustments to minimize the measurement estimation error. Thus, minimizing the measurement estimation error enables the accurate estimation of the system state (Fig. 8.4-1).

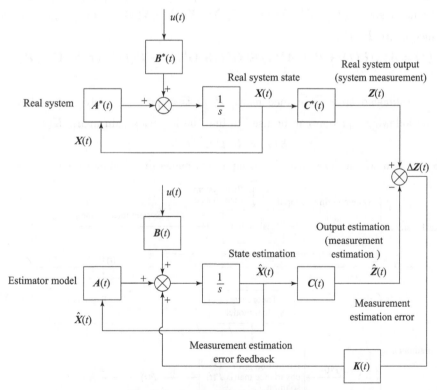

Fig. 8.4-1 Basic estimator model

(System theoretical models $A(t)$, $B(t)$, $C(t)$; system actual models $A^*(t)$, $B^*(t)$, $C^*(t)$)

The difference between different estimator models is that the models for the feedback gain matrix

$K(t)$ are different. The model for the Kalman filter $K(t)$ is given as follows.

Suppose that there are white noise system input w_s and measured white noise input v in the system, and the power spectrum of w_s is S_w and the power spectrum of v is S_v. That is, the system model and measurement model are

$$\dot{X}(t) = A(t)X(t) + B(t)u(t) + G(t)w_s, \qquad (8.4-2)$$

$$Z(t) = C(t)X(t) + v. \qquad (8.4-3)$$

Obviously, to improve the estimation accuracy of the system response to the effect of the system input w_s, it is beneficial to widen the estimator bandwidth. However, to effectively filter out the negative effect of the measurement noise v, it is harmful to enlarge the bandwidth of the estimator. Therefore, to both estimate the system response under the influence of w_s and mitigate the negative effects of measurement noise v, an optimal estimator model must be developed based on the relative values of S_w and S_v.

The design idea of the Kalman filter is that with the given input power spectrum S_w, the measurement noise power spectrum S_v, and the initial state estimation covariance matrix $P(0) = E[\{X(0) - \hat{X}(0)\}\{X(0) - \hat{X}(0)\}^{\mathrm{T}}]$, a feedback matrix $K(t)$ can be derived to make the objective function $J(t) = E[\{X(t) - \hat{X}(t)\}^{\mathrm{T}}\{X(t) - \hat{X}(t)\}]$ minimum at every moment of time t.

The algorithm to calculate $K(t)$ is to first solve the nonlinear time-varying differential equations of the covariance matrix $P(t) = E[\{X(t) - \hat{X}(t)\}\{X(t) - \hat{X}(t)\}^{\mathrm{T}}]$ for the given initial value of the covariance matrix $P(0)$

$$\dot{P}(t) = A(t)P(t) + P(t)A^{\mathrm{T}}(t) + G(t)S_w G^{\mathrm{T}}(t) - P(t)C^{\mathrm{T}}(t)S_v^{-1}C(t)P(t). \qquad (8.4-4)$$

After the estimated error covariance matrix $P(t)$ is acquired by solving the above differential equation, the following equation can be used to find the feedback gain matrix $K(t)$

$$K(t) = P(t)C^{\mathrm{T}}(t)S_v^{-1}. \qquad (8.4-5)$$

Fig. 8.4-2 shows the flow chart for using a continuous-time Kalman filter.

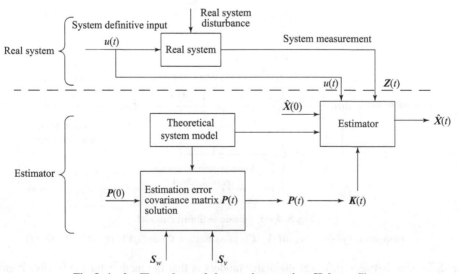

Fig. 8.4-2 Flow chart of the continuous-time Kalman filter

There will always be various external disturbances during the operation of a practical system, which will cause the system to deviate from its ideal operating state. The purpose of introducing the system noise w_s in the Kalman filter model is to simulate the disturbance in the actual operation of a system.

When the system disturbance is relatively large, or the system parameters have a larger deviation from their designed values in applications, the Kalman filter bandwidth should be widened by increasing the value of S_w to improve the estimation accuracy. When the measurement noise of the system is larger, the filter bandwidth should be reduced by increasing the value of S_v to better filter out the noise disturbance. The specific values of S_w and S_v should be determined after iterative adjustments according to the actual operation of the system.

Generally, when selecting the initial value $P(0)$ of the state estimation error covariance matrix, it can be assumed that the estimation errors are statistically independent. That is, $E[(x_i - \hat{x}_i)(x_j - \hat{x}_j)] = 0 (i \neq j)$. Thus, $P(0)$ can be simplified to a diagonal matrix:

$$P(0) = \begin{bmatrix} E[(x_1 - \hat{x}_1)^2] & 0 & 0 \\ 0 & \ddots & 0 \\ 0 & 0 & E[(x_n - \hat{x}_n)^2] \end{bmatrix}. \tag{8.4-6}$$

According to users' experience of the system and the disturbance during the initial operation, each element of $P(0)$ may be taken as the variance value σ^2 of the possible error of the state with respect to its theoretical value.

The Kalman filter described above applies to linear time-varying systems. For nonlinear systems, the extended Kalman filter is used for state estimation.

Consider the nonlinear system model and measurement model as follows

$$\dot{X}(t) = f(X,t) + B(t)u(t) + G(t)w_s, \tag{8.4-7}$$
$$Z(t) = f_z(X,t) + v. \tag{8.4-8}$$

To apply the Kalman filtering method to a nonlinear system, the first step is to perform small disturbance linearization on the nonlinear system model.

$$\begin{cases} A(X,t) = \dfrac{\partial f(X)}{\partial X}, \\ C(X,t) = \dfrac{\partial f_z(X)}{\partial X}. \end{cases} \tag{8.4-9}$$

Therefore, the linearization model of the system will be given as

$$\begin{cases} \dot{X}(t) = A(X,t)X + B(t)u(t) + G(t)w_s, \\ Z(t) = C(X,t)X + v. \end{cases} \tag{8.4-10}$$

The difference from the standard Kalman filter system model is that, the A and C matrices are both not only a function of t but also X.

The extended Kalman filter model corresponding to the above-linearized system model can be given as

$$\begin{cases} \dot{\hat{X}}(t) = f(\hat{X},t) + B(t)u(t) + K(t)[Z(t) - f_z(\hat{X},t)], \\ \dot{P}(t) = A(\hat{X},t)P(t) + P(t)A^{\mathrm{T}}(\hat{X},t) + G(t)S_w G^{\mathrm{T}}(t) - P(t)C^{\mathrm{T}}(\hat{X},t)S_v^{-1}C(\hat{X},t)P(t), \\ K(t) = P(t)C^{\mathrm{T}}(\hat{X},t)S_v^{-1}. \end{cases}$$

$$(8.4-11)$$

It should be noted that at this time, the linearized model is not used in the estimator $\dot{\hat{X}}(t)$ equations, and the nonlinear model is still used. In this way, a higher estimation accuracy can be achieved.

In engineering applications, the Kalman filter often uses its discrete form to facilitate computer programs' realization.

1) Discrete Kalman filter of the linear time-varying system

(1) System model:

$$\begin{cases} \hat{X}_k = \Phi_k X_{k-1} + B_k u_{k-1}, \\ Z_k = C_k X_k. \end{cases} \qquad (8.4-12)$$

(2) Kalman filter model:

$$\begin{cases} \hat{X}_k = \Phi_k \hat{X}_{k-1} + B_k u_{k-1} + K_k(Z_k - C_k \Phi_k \hat{X}_{k-1} - C_k B_k u_{k-1}), \\ M_k = \Phi_k P_{k-1} \Phi_k^{\mathrm{T}} + Q_k, \\ K_k = M_k C_k^{\mathrm{T}} [C_k M_k C_k^{\mathrm{T}} + R_k]^{-1}, \\ P_k = (I - K_k C_k) M_k. \end{cases} \qquad (8.4-13)$$

where Q_k is the white noise variance of the system input, and R_k is the measurement noise variance.

The calculation flow can be referred to in Fig. 8.4 – 3.

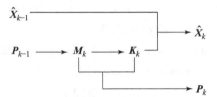

Fig. 8.4 – 3　Kalman filter calculation flow chart of the discrete linear time-varying system

2) Discrete algorithm of the extended Kalman filter

The discrete model of the system should be derived from the linearized model of the nonlinear system.

$$\begin{cases} \hat{X}_k = \Phi_k(X)X_{k-1} + B_k u_{k-1}, \\ Z_k = C_k(X)X_k. \end{cases} \qquad (8.4-14)$$

Its Kalman filter algorithm is

8 Proportional Navigation and Extended Proportional Navigation Guidance Laws

$$\begin{cases} \hat{\boldsymbol{X}}_k^* = \int_{(k-1)T_s}^{kT_s} [f(\hat{\boldsymbol{X}}_{k-1}) + \boldsymbol{B}_k(\hat{\boldsymbol{X}}_{k-1})u_k] \mathrm{d}t + \hat{\boldsymbol{X}}_{k-1}, \\ \hat{\boldsymbol{X}}_k = \hat{\boldsymbol{X}}_k^* + \boldsymbol{K}_k[\boldsymbol{Z}_k - f_z(\hat{\boldsymbol{X}}_k^*)], \\ \boldsymbol{M}_k = \boldsymbol{\Phi}_k(\hat{\boldsymbol{X}}_{k-1})\boldsymbol{P}_{k-1}\boldsymbol{\Phi}_k^\mathrm{T}(\hat{\boldsymbol{X}}_{k-1}) + \boldsymbol{Q}_k(\hat{\boldsymbol{X}}_{k-1}), \\ \boldsymbol{K}_k = \boldsymbol{M}_k\boldsymbol{C}_k^\mathrm{T}(\hat{\boldsymbol{X}}_{k-1}) [\boldsymbol{C}_k(\hat{\boldsymbol{X}}_{k-1})\boldsymbol{M}_k\boldsymbol{C}_k^\mathrm{T}(\hat{\boldsymbol{X}}_{k-1}) + \boldsymbol{R}_k]^{-1}. \end{cases} \quad (8.4-15)$$

where $\boldsymbol{Q}_k$ is the white noise variance of the system input, and $\boldsymbol{R}_k$ is the measurement noise variance.

In the same way, the linearized model here is only used for the solution of $\boldsymbol{M}_k$ and $\boldsymbol{K}_k$, while the estimation model still retains the nonlinear model in the solution of $\hat{\boldsymbol{X}}_k^*$ and $\hat{\boldsymbol{X}}_k$ to improve accuracy.

Refer to its calculation flow chart in Fig. 8.4 – 4.

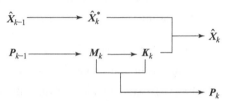

Fig. 8.4 – 4 Calculation flow chart of the Kalman filter of the discrete non-linear time-varying system

As described previously, it is known that three inputs are necessary to calculate the feedback matrix $\boldsymbol{K}(t)$ in addition to the system model. They are the power spectrum $\boldsymbol{S}_w$ of the system noise (or $\boldsymbol{Q}_k$ in the discrete model), the power spectrum $\boldsymbol{S}_v$ of the measurement noise (or $\boldsymbol{R}_k$ in the discrete model), and the initial value $\boldsymbol{P}(0)$ of the state estimation error covariance matrix. These three inputs are discussed in more detail in the following.

(1) Initial value $\boldsymbol{P}(0)$ of the state estimation error covariance matrix.

Since the feedback gain matrix $\boldsymbol{K}(t)$ in the state estimation equation is $\boldsymbol{K}(t) = \boldsymbol{P}(t)\boldsymbol{C}^\mathrm{T}(t)\boldsymbol{S}_v^{-1}$, it can be seen that the higher the covariance matrix $\boldsymbol{P}(t)$ value of the state estimation error, the higher the feedback matrix $\boldsymbol{K}(t)$ value. That is, the wider the frequency bandwidth of the state estimator, the faster the estimation transient process.

For a linear time-varying system, $\boldsymbol{P}(t)$ is obtained by solving the following nonlinear differential equations.

$$\dot{\boldsymbol{P}}(t) = \boldsymbol{A}\boldsymbol{P}(t) + \boldsymbol{P}(t)\boldsymbol{A}^\mathrm{T} + \boldsymbol{G}\boldsymbol{S}_w\boldsymbol{G}^\mathrm{T} - \boldsymbol{P}(t)\boldsymbol{C}^\mathrm{T}\boldsymbol{S}_v^{-1}\boldsymbol{C}\boldsymbol{P}(t)$$

Since the system stimulation w_s and measurement error v in this equation are both white noise, with the increase of time, the optimal solution of the stationary random process should be independent of time after the transient process of $\boldsymbol{P}(t)$ caused by the initial value $\boldsymbol{P}(0)$ is over. Namely, the solution of min $E[\{\boldsymbol{X}(\infty) - \hat{\boldsymbol{X}}(\infty)\}^\mathrm{T}\{\boldsymbol{X}(\infty) - \hat{\boldsymbol{X}}(\infty)\}]$ when $t = \infty$ of the Kalman filter is a constant value $\boldsymbol{P}_\infty$. Because $\dot{\boldsymbol{P}} = 0$, the solution of $\boldsymbol{P}_\infty$ is just the solution of the following Riccati non-linear algebraic equation

$$\boldsymbol{A}\boldsymbol{P}_\infty + \boldsymbol{P}_\infty\boldsymbol{A}^\mathrm{T} + \boldsymbol{G}\boldsymbol{S}_w\boldsymbol{G}^\mathrm{T} - \boldsymbol{P}_\infty\boldsymbol{C}^\mathrm{T}\boldsymbol{S}_v^{-1}\boldsymbol{C}\boldsymbol{P}_\infty = 0. \quad (8.4-16)$$

That is to say that the solution of $\boldsymbol{P}(t)$ has the characteristics shown in Fig. 8.4 – 5.

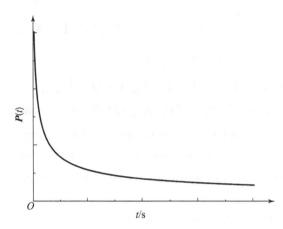

Fig. 8.4 – 5 Typical time characteristics of the covariance matrix solution $P(t)$ of the state estimation error

Therefore, a large value of $P(0)$ makes a large value of $P(t)$ in the initial phase, that is, the wider the Kalman filter bandwidth, the faster the estimation of the initial transient process in the initial phase. However, the value of $P(\infty)$ is independent of the value of $P(0)$, that is to say that, $P(0)$ has no effect on the bandwidth and characteristics of the estimator after the filter enters the steady state.

Since the missile engagement time is relatively short, rapid estimation of target acceleration is crucial. Therefore, for the task of target acceleration estimation, it is very important to ensure that the choice of $P(0)$ value is appropriate to achieve a fast target maneuver acceleration estimation.

At the start of the estimation, the direction and level of the target's maneuver are unknown. Therefore, the initial estimated value $\hat{a}(0)$ of the target acceleration in the estimator's state equation is set to 0. However, the initial estimation error $P(0)$ for the target acceleration should be set to the maximum possible target maneuver acceleration value a_t. That is, the corresponding element of $P(0)$ should be a_t^2. A large initial estimation error necessitates widening the initial filter bandwidth, which helps to quickly reduce the system state estimation error.

Generally, $P(0)$ is taken as a diagonal matrix. The initial value of each state estimation error can be taken as the estimated or known variance of each state.

(2) Power spectrum matrix of the measurement noise S_v (or R_k matrix).

According to users' understanding of the selected measurement sensor, it should not be difficult to choose a reasonable S_v value. It is known from $K(t) = P(t)C^T(t)S_v^{-1}$ that S_v^{-1} and $K(t)$ will decrease with the increase of S_v. This approach will narrow the filter bandwidth, thereby increasing the weighting of the estimator's filtering ability.

(3) Power spectrum matrix S_w of the system noise (or Q_k matrix).

The rational choice of S_w is not so simple and its value will directly affect the bandwidth of the estimator in the initial stage as well as the steady state stage. The Kalman filter theory utilizes system white noise to account for deviations from the ideal system state caused by:

(a) Uncertain external disturbances affecting system operation;

(b) Deviations of system parameters from their designed model;

(c) With a given S_v, the filter bandwidth can be changed by changing the S_w matrix to quickly obtain the best estimate for the system state.

Due to the above reasons, in actual applications, the value of S_w will be adjusted many times to determine its reasonable value with the consideration of the above requirements.

Here's a simple example demonstrating how the Kalman filter can estimate target maneuver acceleration.

Consider a one-dimensional missile-target engagement scenario in which the target suddenly performs a $5g$ constant horizontal maneuver during the engagement to evade the attack. If the missile's active seeker can measure the target maneuver distance y from the moment $t = 0$, the following Kalman filter state dynamics model and measurement model can be used

$$\begin{cases} \dfrac{dy}{dt} = V, \\ \dfrac{dV}{dt} = a, \\ \dfrac{da}{dt} = 0 + w_s, \end{cases} \quad (8.4-17)$$

$$z = y, \quad (8.4-18)$$

where V is the target velocity, a is the target acceleration, z is the measurement, and here z is the target distance y. Suppose that the distance measurement error is $\sigma_y = 20$ m, the initial value of the velocity estimation error is $\sigma_V = 50$ m/s and the initial value of the maneuver acceleration estimation error is $\sigma_a = 5g = 49$ m/s^2. Take the two estimator design parameters, random input, and measurement power spectrum as

$$S_v = 0.04 \ (\text{m}^2\text{s}), \quad S_w = 0.01 \ (\text{m}^2/\text{s}^5),$$

and the initial value $P(0)$ of the state estimation error covariance matrix as

$$P(0) = \begin{bmatrix} 20^2 & 0 & 0 \\ 0 & 50^2 & 0 \\ 0 & 0 & (5 \times 9.81)^2 \end{bmatrix}.$$

The solutions to the nonlinear equation $P(t)$ are shown in Fig. 8.4-6, Fig. 8.4-7, and Fig. 8.4-8.

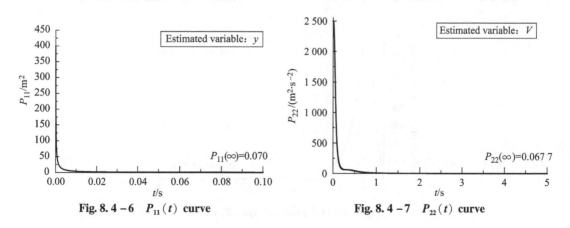

Fig. 8.4-6 $P_{11}(t)$ curve   Fig. 8.4-7 $P_{22}(t)$ curve

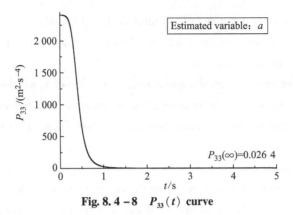

Fig. 8.4-8 $P_{33}(t)$ curve

As can be known from the figures, the initial values of P_{11}, P_{22}, and P_{33} are the initial values of $P(0)$. After the transition process, $P(t)$ enter their respective steady-state value P_∞ independent of $P(0)$.

The measurement estimation error feedback gain K_{11}, K_{21}, and K_{31} can be obtained by using the Equation $K(t) = P(t)C^T(t)S_v^{-1}$ (Fig. 8.4-9, Fig. 8.4-10, and Fig. 8.4-11). The figures show that at the initial stage, the feedback gain is high, which helps quickly reduce the initial estimation error. After the transition phase, the feedback gain stabilizes to a steady-state value. This steady-state gain level determines the system's ability to handle random disturbances and parameter fluctuations during normal operation.

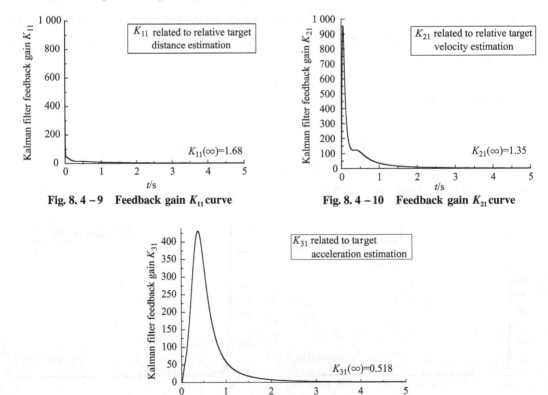

Fig. 8.4-9 Feedback gain K_{11} curve

Fig. 8.4-10 Feedback gain K_{21} curve

Fig. 8.4-11 Feedback gain K_{31} curve

Let us assume the following initial state deviations $y(0)$, $V(0)$, and $a(0)$,

$y(0) = 20$ m,
$V(0) = 50$ m/s,
$a(0) = 5g$ m/s^2.

Since the initial value of the estimator has no prior information, the estimated states $\hat{y}(0)$, $\hat{V}(0)$, and $\hat{a}(0)$ can only be taken as 0, that is

$\hat{y}(0) = 0$ m,
$\hat{V}(0) = 0$ m/s,
$\hat{a}(0) = 0$ m/s^2.

The result of $\boldsymbol{K}(t)$ is substituted into the differential equations of the actual system and the estimator differential equations, and $y(t)$, $V(t)$, $a(t)$ curves of the actual system, and $\hat{y}(t)$, $\hat{y}(t)$, $\hat{a}(t)$ curves of the estimator output are shown in Fig. 8.4 – 12, Fig. 8.4 – 13, Fig. 8.4 – 14, and Fig. 8.4 – 15.

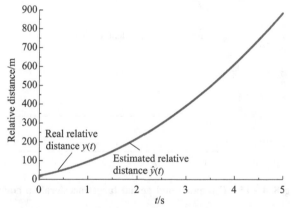

Fig. 8.4 – 12 Estimated and actual relative distance curves

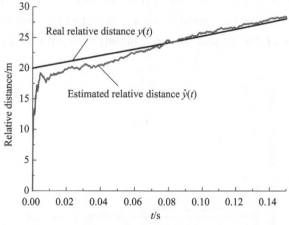

Fig. 8.4 – 13 Estimated and actual relative distance curves of the initial phase

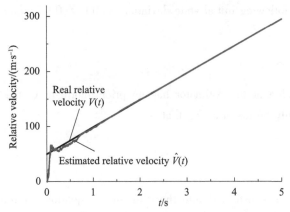

Fig. 8.4 – 14 Estimated and actual relative velocity curves

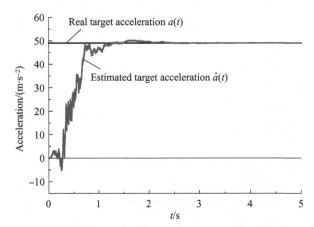

Fig. 8.4 – 15 Estimated and actual target acceleration curves

Fig. 8.4 – 16, Fig. 8.4 – 17, and Fig. 8.4 – 18 show the estimation errors $y - \hat{y}$, $V - \hat{V}$, and $a - \hat{a}$ curves. The figures show that even without prior knowledge of the target maneuver ($\hat{a}(0) = 0$), the Kalman filter technique can quickly estimate the target's maneuver acceleration, provided that some measurements of the target's motion (in this case, the relative distance $y(t)$) are available.

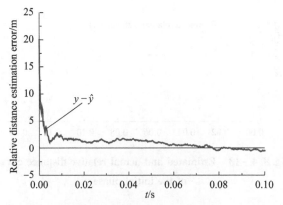

Fig. 8.4 – 16 Relative distance estimation error curve

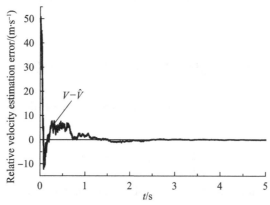

Fig. 8.4 – 17 Relative velocity estimation error curve

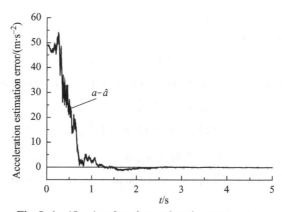

Fig. 8.4 – 18 Acceleration estimation error curve

When the seeker operates in medium pulse repetition frequency mode, and if the velocity measurement is sufficiently accurate, both distance and velocity measurements can be used as system outputs. This approach will enhance the accuracy of state estimation.

In high pulse repetition frequency mode, if the velocity measurement is reliable, it can be used alone as the system measurement output for state estimation.

Currently, the accuracy of distance and velocity measurements from radar seekers is limited. As a result, implementing guidance laws with target maneuver compensation remains challenging. However, with advancements in seeker technology, the application of these techniques holds significant promise.

Finally, it should be noted that the perturbation model of a practical system is not completely consistent with the stochastic model set by the Kalman filter, so it is not necessary to insist on obtaining accurate input values for S_w, S_v and $P(0)$ in practical applications. Havever, the biggest advantage of the Kalman filter is that it has pointed out the specific effects of the filter parameters S_w, S_v and $P(0)$ on the bandwidth of the estimator. When the actual estimation outcomes are found to be unsatisfactory, it is not difficult to know how to adjust the values of S_w, S_v or $P(0)$ to obtain better estimation results.

§8.5 Optimum Trajectory Control Design

The mathematical model for the optimum trajectory control design is given as the system dynamics equations

$$\frac{dx}{dt} = f(x(t), u(t), t). \quad (8.5-1)$$

In the above equation, x is the system state variable and u is the system control variable.

The following design constraints may occur in the trajectory optimization problem.

(1) Initial state constraint:

$$\phi_{0min} < \phi_0(x_0, t_0) < \phi_{0max}. \quad (8.5-2)$$

(2) Terminal state constraint:

$$\phi_{fmin} < \phi_f(x_f, t_f) < \phi_{fmax}. \quad (8.5-3)$$

(3) State initial and terminal state value constraints and state variable constraints:

$$x_{0min} < x_0 < x_{0max},$$
$$x(t)_{min} < x(t) < x(t)_{max}, \quad (8.5-4)$$
$$x_{fmin} < x_f < x_{fmax}.$$

(4) Derived variable $C = C(x, u, t)$ constraints, where C are related to the state x and control u:

$$C(t)_{min} < C(x(t), u(t), t) \leqslant C(t)_{max}. \quad (8.5-5)$$

(5) Control constraints:

$$u(t)_{min} < u(t) < u(t)_{max}. \quad (8.5-6)$$

(6) System parameters constraints:

$$P_{min} < P < P_{max}. \quad (8.5-7)$$

It is required to obtain the optimal control $u(t)$ under the condition that all the above constraints are satisfied and at the same time the following objective function J is minimized

$$J = \phi(x_0, x_f, t_0, t_f) + \int_{t_0}^{t_f} L(x(t), u(t), P, t) dt. \quad (8.5-8)$$

The optimal control problem constrained at both ends of the trajectory is commonly referred to as a two-point boundary value problem (TPBVP). Solving such problems with a continuous system model can be computationally expensive and impractical for engineering applications.

It is noteworthy that the pseudo-spectral software currently available on the market can solve these problems through continuous function parameterization. Currently, pseudo-spectral software available on the market can address these problems by parameterizing continuous functions through polynomial interpolation.

By fitting a continuous function with a polynomial at a finite number of nodes, the function can be approximated. However, to ensure accuracy, a large number of interpolation nodes and a high polynomial order are required. When nodes are equally spaced, and the polynomial order is high, significant interpolation errors can occur. For example, taking the continuous function $f(x) = 1/(1 + 12x^2)$ as an example, when 25 equally spaced nodes are taken in the interval $[-1, 1]$ of the independent variable x, the polynomial fitting result and its error are shown in Fig. 8.5 – 1. It can be seen that the use of equally spaced mode polynomial functions cannot meet the task of approximating a continuous function.

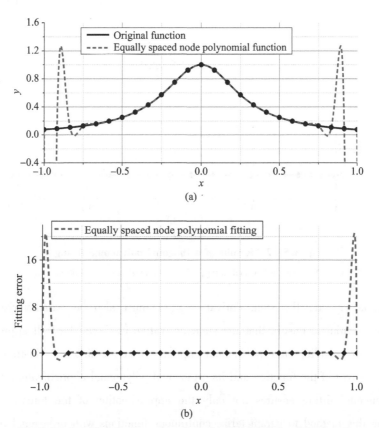

Fig. 8.5 – 1   Equally spaced node polynomial fitting with 25 nodes

(a) Equally spaced node polynomial fitting; (b) Equally spaced node polynomial fitting error

Orthogonal polynomial interpolation fitting employs variable-spaced nodes to create the interpolation function. This approach automatically adjusts the node spacing in areas where large fitting errors might occur, ensuring consistent accuracy across the entire range. Fig. 8.5 – 2 demonstrates the use of orthogonal polynomials with 25 nodes to fit the aforementioned functions. It shows that with an appropriate number of nodes, orthogonal polynomial interpolation can accurately parameterize a continuous function.

Since the system state differential equation is an equality constraint that must be satisfied in the

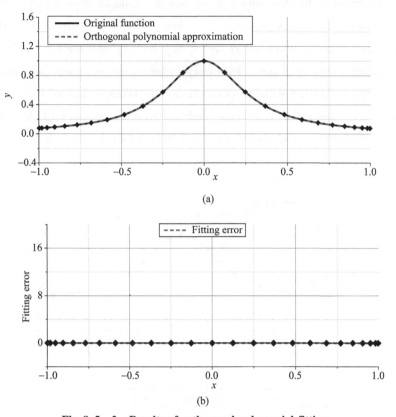

Fig. 8.5 – 2 Results of orthogonal polynomial fitting

(a) Orthogonal polynomial fitting; (b) Orthogonal polynomial fitting error

problems being addressed, the state derivative $\dot{x}(t)$ must also be parameterized accordingly. Therefore, it is essential to ensure that the state derivative $\dot{x}(t)$ can be well approximated by the derivative of its corresponding state polynomial approximation. Fig. 8.5 – 3 illustrates the difference between the derivatives of the function and its orthogonal polynomial approximation. It is evident that orthogonal polynomial fitting ensures not only the approximation of the function but also of its derivative. Using this method to parameterize continuous functions with orthogonal polynomials, the optimal trajectory problem can be converted into a standard-constrained nonlinear programming problem. The standard nonlinear programming problem has the following form.

Objective function is

$$\min J(X),$$

subject to

$$\text{equality constraints} f_E(X) = 0,$$

$$\text{inequality constraints} f_{NE}(X) \leqslant 0,$$

Where X is the design variable vector of the nonlinear programming problem.

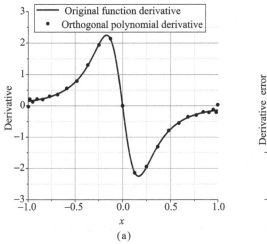 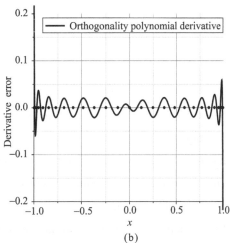

Fig. 8.5-3 Approximation result of the original function derivative and its orthogonal polynomial derivative

(a) Derivative curve; (b) Orthogonal polynomial derivative fitting error

The trajectory optimization problem has the following nonlinear programming design variables after the parameterization of the states and controls.

(1) Corresponding design variables of state functions parameterization, each with n node polynomial parameters as the design variables,

$$x_i(t) \approx L_{xi}(x_{i1},\cdots,x_{ik})\,(i = 1,\cdots,n).$$

In the above eguation, L_{xi} is the orthogonal polynomial corresponding to the i-th state function $x_i(t)$, k is the number of nodes of the $(k-1)$ order orthogonal polynomial. Therefore, the number of the nonlinear programming variables corresponding to the state function x is $n \cdot k$.

(2) Corresponding design variables of m control functions parameterization,

$$u_j(t) \approx L_{uj}(u_{j1},\cdots,u_{jk})\,(j = 1,\cdots,m).$$

Similarly, the number of nonlinear programming design variables corresponding to the parameterization of the control function is $m \cdot k$.

(3) The number of design variables corresponding to the system parameters $P_1,\cdots,P_s$ is s.

(4) The number of design variables corresponding to the initial value of time t_0 is 1.

(5) The number of design variables corresponding to the final time value t_f is 1.

It is known from the above that the nonlinear programming design variable X is composed of

$$X = \begin{bmatrix} x \\ u \\ P \\ t_0 \\ t_f \end{bmatrix}.$$

The total number of design variables is $(m + n)k + s + 2$.

(1) Equality constraint:

The equality constraint at the orthogonal polynomial nodes (note: the original n state equation has been converted into $n \cdot k$ equality constraint at k nodes).

(2) Inequality constraint:

The inequality constraints apply at the initial and final times, as well as at all nodes of the orthogonal polynomial interpolation.

(3) Objective function:

The discrete integral result of the trajectory optimization objective function with J as Equation (8.5 – 8).

The orthogonal polynomial approximate solution of the optimal problem control function $u(t)$ can be obtained by solving the above nonlinear programming problem. It is important to note that the design inputs for this software remain continuous, as specified by Equation (8.5 – 1) ~ Equation (8.5 – 8). The discretization and nonlinear programming processes are handled by the software, so users do not need to manage these details. The software is highly flexible in its applications.

The scope of use of this software is very flexible. For example, the control variables can be taken as missile lateral acceleration $a(t)$, angle of attack $\alpha(t)$, sideslip angle $\beta(t)$, etc. The objective function can be chosen as maximum range, maximum final velocity, minimum control integral, etc. The choices of the constraint are even flexible. For example, there may be dynamic pressure constraints $q = \frac{1}{2}\rho V^2 \leq q_{max}$, normal acceleration constraints $n(t) \leq n_{max}$, heat flow constraints $Q = \frac{c}{\sqrt{R_d}} \rho^{0.5} V^{3.08}$ (of which c is the constant associated with the aircraft characteristics, and R_d is the radius of curvature of the aircraft stagnation point), angle of attack constraints, control surface deflection constraints, and so on. The software can even be utilized to optimize the trajectories of multi-stage rockets.

The following example demonstrates the significance of trajectory optimization. Suppose that there is a guided rocket, the burn-out velocity of the rocket is 1 800 m/s, the control is constrained by maximum lateral acceleration, the terminal inequality constraints are final velocity $V_f \geq 800$ m/s, the impact angle $|\theta_k| \geq 60°$, and the optimum missile lateral acceleration function $a(t)$, which can give the maximum range needs to be solved. The dashed trajectory in Fig. 8.5 – 4 illustrates the optimized missile flight path, achieving a maximum range of 525 km while meeting the final velocity and impact angle constraints. The results show that,

(1) For guided rockets with a range of 400 – 800 km, the maximum altitude for the optimal trajectory typically does not exceed 50 km. This suggests that such rockets can be classified as air-guidance rockets, which can be controlled using air rudders.

(2) To maximize range, guidance rockets should employ low-altitude, high-speed trajectories, allowing for extended flight durations in low-air-density conditions.

The maximum range available for this example with this optimal trajectory is approximately 525 km. If a general ballistic missile scheme is used and the angle of departure θ_0 is taken as an optimization parameter, the range of the missile can only reach 380 km, and the maximum trajectory altitude should reach 104 km (see the solid line trajectory in Fig. 8.5 – 4) under the optimal angle of departure. It is evident that applying a general long-range ballistic missile program in this range is not practical. In summary, even during the concept design stage, the software aids designers in selecting a more effective control scheme.

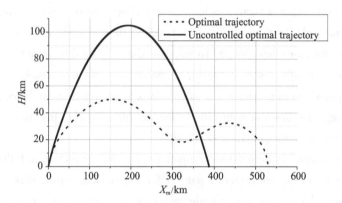

Fig. 8.5 – 4 Curves of the optimal trajectory and the uncontrolled optimal trajectory

9

Optimal Guidance for Trajectory Shaping

§ 9.1 Optimality of Error Dynamics in Missile Guidance

Missile guidance law design fundamentally revolves around finite-time tracking, driven by specific operational objectives. In scenarios where only target interception is considered, the tracking error is typically defined by the Zero-Effort Miss (ZEM) distance. Achieving a ZEM of zero indicates perfect interception with no miss distance. For certain tactical missiles, it is beneficial to constrain the final impact angle or intercept angle to improve the effectiveness of the warhead or to maintain a favorable engagement angle. In these cases, the terminal impact angle error is incorporated as the tracking error in the guidance law design. To improve the survivability of anti-ship missiles against advanced close-in weapon systems on battleships, the concept of salvo attacks[14] is introduced.

A common approach to implementing a salvo attack is impact time guidance, where the missile is required to intercept the target at a specified time. In this scenario, the final impact time error is treated as the tracking error for guidance law design.

The guidance command is formulated to direct the missile's trajectory to follow the desired error dynamics, ensuring a specific convergence pattern for the tracking error. This method represents a general framework for developing new missile guidance laws using nonlinear control techniques. However, most prior research has primarily concentrated on ensuring that the tracking error converges to zero, often overlooking the optimization of error dynamics with respect to meaningful performance criteria.

Building on these observations, this chapter aims to explore the optimal convergence patterns of tracking error and propose error dynamics that are optimized using a meaningful cost function for missile guidance issues. To achieve this, the linear quadratic optimal control problem is addressed for a generalized tracking scenario commonly encountered in missile guidance law design, using Schwarz's inequality approach.

To illustrate, two examples are presented demonstrating the application of the proposed optimal error dynamics in missile guidance law design: impact angle guidance and impact time guidance. By applying these dynamics, the physical implications of these guidance laws can be theoretically analyzed, and the cost functions can guide the selection of appropriate gains, ensuring that terminal guidance commands remain bounded. Theoretical analysis also shows that some existing optimal guidance laws are specific instances of the proposed solutions. Thus, the proposed optimal error dynamics offer a unique and generalized framework for developing optimal guidance laws.

9.1.1 Optimal Error Dynamics

As previously discussed, missile guidance law design is fundamentally a finite-time tracking problem aimed at reducing a mission-specific tracking error to zero within a set timeframe. The general formulation of this tracking problem, frequently encountered in missile guidance law design, is as follows

$$\dot{\varepsilon}(t) = g(t)u(t), \qquad (9.1-1)$$

where $\varepsilon(t)$ represents the tracking error, $g(t)$ stands for a known time-varying function, and $u(t)$ denotes the control input.

Depending on the specific guidance problem, the tracking error may be represented by various metrics such as ZEM, impact angle error, impact time error, or heading error.

In many previous studies, a commonly used error dynamics for guidance law design is given by

$$\dot{\varepsilon}(t) + k\varepsilon(t) = 0, \qquad (9.1-2)$$

where $k > 0$ is the guidance gain to regulate the convergence rate of tracking error.

It is straightforward to verify that the closed-form solution to the differential equation, Equation (9.1-2), is given by

$$\varepsilon(t) = \varepsilon(t_0)e^{-kt}, \qquad (9.1-3)$$

where $\varepsilon(t_0)$ denotes the initial tracking error.

The preceding equation reveals that the tracking error converges to zero asymptotically with an exponential rate governed by the guidance gain k. Accordingly, this desired error dynamics has two significant drawbacks:

(1) The finite-time convergence is not strictly guaranteed;

(2) It focuses solely on driving the tracking error to zero without considering what the optimal error dynamics might be in terms of a meaningful performance index.

Motivated by these observations, this chapter aims to investigate the optimal convergence pattern of the tracking error and propose an error dynamics that achieve this optimal pattern with guaranteed finite-time convergence for missile guidance law design.

The optimal error dynamics for missile guidance law design will be discussed in the following section, with the main results outlined in Theorem 9.1.

Theorem 9.1-1 The desired error dynamics is chosen as

$$\dot{\varepsilon}(t) + \frac{\Gamma(t)}{t_{go}}\varepsilon(t) = 0, \qquad (9.1-4)$$

where

$$\Gamma(t) = \frac{t_{go}R^{-1}(t)g^2(t)}{\int_t^{t_f} R^{-1}(\tau)g^2(\tau)\,d\tau}, \qquad (9.1-5)$$

with $R(t) > 0$ being an arbitrary weighting function.

Then, the resulting control input can be obtained as

$$u(t) = -\frac{\Gamma(t)}{g(t)t_{go}}\varepsilon(t) = -\frac{R^{-1}(t)g(t)}{\int_t^{t_f} R^{-1}(\tau)g^2(\tau)d\tau}\varepsilon(t), \quad (9.1-6)$$

to minimize the performance index

$$\min_u J = \frac{1}{2}\int_t^{t_f} R(\tau)u^2(\tau)d\tau. \quad (9.1-7)$$

Proof The control input that enables the system to follow error dynamics is determined by substituting Equation (9.1–5) into Equation (9.1–4) as Equation (9.1–6).

The goal now is to demonstrate that the control input is the optimal solution to the following optimization problem. The problem is defined by the objective function given in Equation (9.1–7), subject to the following constraints

$$\dot{\varepsilon}(t) = g(t)u(t), \varepsilon(t_f) = 0 \quad (9.1-8)$$

Integrating from t to t_f on both sides of Equation (9.1–8) gives

$$\varepsilon(t_f) - \varepsilon(t) = \int_t^{t_f} g(\tau)u(\tau)d\tau. \quad (9.1-9)$$

Imposing terminal constraint on Equation (9.1–9) gives

$$-\varepsilon(t) = \int_t^{t_f} g(\tau)u(\tau)d\tau. \quad (9.1-10)$$

Introducing a slack variable $R(t)$, Equation (9.1–10) can be transformed as

$$-\varepsilon(t) = \int_t^{t_f} g(\tau)R^{-1/2}(\tau)R^{1/2}(\tau)u(\tau)d\tau. \quad (9.1-11)$$

Applying Schwarz's inequality to the preceding equation yields

$$[-\varepsilon(t)]^2 \leq \left[\int_t^{t_f} R^{-1}(\tau)g^2(\tau)d\tau\right]\left[\int_t^{t_f} R(\tau)u^2(\tau)d\tau\right]. \quad (9.1-12)$$

Inequality Equation (9.1–12) can be rewritten as

$$\frac{1}{2}\int_t^{t_f} R(\tau)u^2(\tau)d\tau \geq \frac{[-\varepsilon(t)]^2}{2\left[\int_t^{t_f} R^{-1}(\tau)g^2(\tau)d\tau\right]}. \quad (9.1-13)$$

As known, the equality of Equation (9.1–13) holds if and only if there exists a constant C such that

$$u(t) = CR^{-1}(t)g(t). \quad (9.1-14)$$

Substituting Equation (9.1–14) into Equation (9.1–10), one has

$$-\varepsilon(t) = C\int_t^{t_f} R^{-1}(t)g^2(\tau)d\tau. \quad (9.1-15)$$

Solving Equation (9.1–15) for C gives

$$C = \frac{-\varepsilon(t)}{\int_t^{t_f} R^{-1}(\tau)g^2(\tau)d\tau}. \quad (9.1-16)$$

Substituting Equation (9.1-16) into Equation (9.1-14) gives the optimal control input as

$$u(t) = -\frac{R^{-1}(t)g(t)}{\int_t^{t_f} R^{-1}(\tau)g^2(\tau)\mathrm{d}\tau}\varepsilon(t), \qquad (9.1-17)$$

which is identical with Equation (9.1-6).

9.1.2 Analysis of Optimal Error Dynamics

A notable feature of the proposed optimal error dynamics is that it guarantees finite-time convergence, as it is derived by directly solving the finite-time optimal tracking problem. The form of the optimal error dynamics is similar to that of existing desired error dynamics, with the primary distinction being in the proportional gain. In conventional methods, the gain is a constant term k, whereas in the proposed method, it is a time-varying term. Unlike the constant gain typically used, the proposed proportional gain starts from a small initial value and increases to infinity as the time-to-go approaches zero. Specifically, the time-varying gain can be expressed as

$$\lim_{t_{go} \to 0} \frac{\Gamma(t)}{t_{go}} = \infty \qquad (9.1-18)$$

According to the value of $\Gamma(t)$, the evolving pattern of the proposed proportional gain further changes.

Let us discuss the characteristics of $\Gamma(t)$. From Equation (9.1-5), for convenience, $\Gamma(t)$ can be rearranged as

$$\Gamma(t) = \frac{\varphi(t)}{\left(\int_t^{t_f} \varphi(\tau)\mathrm{d}\tau / t_{go}\right)} \qquad (9.1-19)$$

where $\varphi(t) = R^{-1}(t)g^2(t)$.

The function of $g(t)$ is given by the missile guidance problem under consideration, and the weighting function $R(t)$ is the design parameter. According to different selections of $R(t)$, the time-varying term $\Gamma(t)$ changes differently. In the following, that $\Gamma(t) > 0, \forall t > 0$ during the homing engagement in Proposition 9.1.

Proposition 9.1-1 For given $g(t)$ and $R(t)$, $\Gamma(t)$ is always greater than zero, i.e., $\Gamma(t) > 0$.

Proof By definition, it is established that $R(t) > 0$ and $g(t) \neq 0$. From Equation (9.1-19), it is obvious that the numerator is positive since $\varphi(t) = R^{-1}(t)g^2(t) > 0$. Additionally, the denominator is positive, as the integral of a positive function results in a positive value. Accordingly, it follows that $\Gamma(t) > 0$.

Proposition 9.1-2 When function $\phi(t)$ keeps constant, the equality $\Gamma(t) = 1$ holds. If $\phi(t)$ decreases as $t \to t_f$, then $\Gamma(t) > 1$. If $\phi(t)$ increases as $t \to t_f$, one has $\Gamma(t) < 1$.

Proof In Equation (9.1-19), the denominator can be considered as the average value of $\phi(t)$, denoted by $\bar{\phi}(t)$, during the remaining time of interception. Accordingly, the time-varying term $\Gamma(t)$ is the ratio of the current value $\phi(t)$ to the average value $\bar{\phi}(t)$. If the term of $\phi(t)$ is given by a constant value, then the $\Gamma(t)$ is unity as $\Gamma(t) = 1$ since $\phi(t) = \bar{\phi}(t)$. If the term $\phi(t)$ *decreases as* $t \to t_f$, $\Gamma(t)$ is greater than unity due to $\phi(t) > \bar{\phi}(t)$. Conversely, if the term of $\phi(t)$ has an increasing pattern, then $\Gamma(t)$ is less than unity because of $\phi(t) < \bar{\phi}(t)$.

Remark 9.1 – 1 It follows from Equation (9.1 – 4) that $\Gamma(t) \geqslant 1$ is desirable to ensure a stable and fast convergence rate at the initial time. Accordingly, Proposition 9.2 reveals that it is desirable to impose a constant value or a decreasing pattern on the function $\phi(t)$.

Remark 9.1 – 2 As shown in Equation (9.1 – 5), the computation of $\Gamma(t)$ contains the integration of $\phi(t)$, which is given as a function of $g(t)$ and $R(t)$. If $\phi(t)$ is given by a closed-form function, $\Gamma(t)$ can be obtained analytically or numerically. Although $g(t)$ is not provided as a closed-form function due to the complexities of the guidance problem, a closed-form solution for $\phi(t)$ can be obtained by selecting an appropriate weighting function $R(t)$. Under this condition, the chosen optimal error dynamics only consider the minimization of a specific performance index since the physical meaning of the performance index is shown in Equation (9.1 – 7) changes according to the choice of $R(t)$.

For a specific mission, the function $g(t)$ is fixed. Thus, by properly choosing the weighting function $R(t)$ for different objectives, the error dynamics are determined by Theorem 9.1. Two specific cases are presented below. In the case of $R(t) = 1$, the term $\Gamma(t)$ is given by

$$\Gamma(t) = \frac{t_{go} g^2(t)}{\int_t^{t_f} g^2(\tau) d\tau}. \tag{9.1-20}$$

The optimal error dynamics with $\Gamma(t)$ shown in Equation (9.1 – 5) minimize the performance index

$$J = \frac{1}{2} \int_t^{t_f} u^2(\tau) d\tau, \tag{9.1-21}$$

which provides an energy-optimal guidance law.

In addition, if with $K \geqslant 1$, the weighting function can be chosen as

$$R(t) = \frac{g^2(t)}{t_{go}^{K-1}}. \tag{9.1-22}$$

Then $\phi(t)$ is given by a function of time-to-go, regardless of $g(t)$, as

$$\varphi(t) = t_{go}^{K-1}. \tag{9.1-23}$$

In this case, $\Gamma(t)$ is a constant as

$$\Gamma(t) = K. \tag{9.1-24}$$

Then, the desired error dynamics are given by

$$\dot{\varepsilon}(t) + \frac{K}{t_{go}} \varepsilon(t) = 0. \tag{9.1-25}$$

Solving Equation (9.1 – 25) gives the closed-form of the tracking error as

$$\varepsilon(t) = \varepsilon(t_0) \left(\frac{t_{go}}{t_f}\right)^K. \tag{9.1-26}$$

In practice, these optimal error dynamics described by Equation (9.1-25) are valuable because they provide a predictable pattern for the tracking error as a function of time-to-go and present the desired error dynamics in a simplified form. Consequently, the next section will apply this desired error dynamics to various missile guidance problems to enhance both simplicity and practicality. According to Theorem 9.1, the desired error dynamics aim to minimize the performance index given by

$$J = \frac{1}{2}\int_{t}^{t_f} \frac{g^2(\tau)}{(t_f - \tau)^{K-1}} u^2(\tau)\,d\tau. \qquad (9.1-27)$$

§9.2 Optimal Predictor-Corrector Guidance

9.2.1 General Approach for Guidance Law Design

OPTIMAL PREDICTOR - CORRECTOR GUIDANCE

This subsection presents the details of a general approach to how to utilize the proposed optimal error dynamics in guidance law design. To this end, let q be the variable of interest for which control is sought to achieve a specific mission objective. Suppose that the dynamics of q is given by a general nonlinear differential equation as

$$\dot{q}(t) = f(t) + b(t)u(t), \qquad (9.2-1)$$

where $f(t)$, $b(t) \neq 0$ are time-varying functions, and $u(t)$ denotes the command input.

By formulating the guidance command as a composite form as

$$u(t) = u_0(t) + u_b(t), \qquad (9.2-2)$$

where $u_0(t)$ denotes the nominal control part and $u_b(t)$ stands for a biased term that can be designed by using the proposed error dynamics.

Based on the general command, the systematic approach to applying the proposed optimal error dynamics in guidance law design involves a prediction correction method.

(1) Predict the terminal error of the variable of interest for a specific mission objective using a nominal trajectory, i.e., only controlled by u_0, as

$$q(t_f) = q(t) + \int_{t}^{t_f} f(\tau) + b(\tau)u_0(\tau)\,d\tau. \qquad (9.2-3)$$

(2) Define $\varepsilon(t) = q(t_f) - q_d$ as the tracking error with q_d being the desired state. Since the nominal control part u_0 imposes no effect on $q(t_f)$, the dynamics of tracking error $\varepsilon(t)$ under composite command can then be readily determined as

$$\dot{\varepsilon}(t) = -b(t)u_b(t). \qquad (9.2-4)$$

(3) Correct the tracking error by using a biased term u_b based on the proposed error dynamics.

9.2.2 Impact Angle Control

Constraining the impact angle is crucial for enhancing warhead effectiveness and maximizing damage, particularly for anti-ship and anti-tank missiles, as it targets the structural weaknesses of the target. This subsection details the application of the proposed optimal error dynamics for trajectory shaping aimed at impact angle control. For simplicity, a stationary target interception scenario is considered. Using the general prediction-correction approach, the guidance command is formulated in a composite form as

$$a_M = a_{Nor} + a_{IA}, \quad (9.2-5)$$

where a_{Nor} denotes the nominal command to guarantee zero ZEM for target interception, and a_{IA} represents the biased command that is utilized to control the impact angle.

For illustration purposes, PNG is used as the nominal guidance command, i.e.,

$$a_{Nor} = NV_M \dot{\sigma} \quad (9.2-6)$$

According to the literature [16], the final flight path angle governed by PNG is given by

$$\gamma_{M_f} = \frac{N}{N-1}\sigma - \frac{1}{N-1}\gamma_M. \quad (9.2-7)$$

Let γ_f be the desired final flight path angle. Then, the impact angle error can be defined as $\varepsilon_\gamma = \gamma_f - \gamma_{Mf}$. To nullify the impact angle error, consider ε_γ as the tracking error, which gives the error dynamics as

$$\dot{\varepsilon}_\gamma = -\dot{\gamma}_{M_f} = -\frac{N}{N-1}\dot{\sigma} + \frac{1}{N-1}\frac{a_M}{V_M}. \quad (9.2-8)$$

Substituting $a_M = a_{Nor} + a_{IA}$ into Equation (9.2-8), one has

$$\dot{\varepsilon}_\gamma = \frac{a_{IA}}{(N-1)V_M}. \quad (9.2-9)$$

For the impact angle error ε_γ, the optimal error dynamics is selected in a similar way as

$$\dot{\varepsilon}_\gamma + \frac{K}{t_{go}}\varepsilon_\gamma = 0, \quad (9.2-10)$$

where $K \geqslant 1$ is the guidance gain to be designed.

The corresponding performance index is obtained as

$$J = \frac{1}{2}\int_t^{t_f} \frac{1}{(N-1)^2 V_M^2 (t_f - \tau)^{K-1}} u^2(\tau) d\tau. \quad (9.2-11)$$

Since the constant terms in the performance index have no effect on the optimal pattern, the performance index is identical to

$$J = \frac{1}{2}\int_t^{t_f} \frac{1}{(t_f - \tau)^{K-1}} u^2(\tau) d\tau. \quad (9.2-12)$$

It can be clearly observed from the preceding performance index that the weighting function with $K > 1$ becomes infinite as $t_{go} \to 0$. Therefore, one can imply that the optimal error dynamics with $K > 1$ guarantees zero impact angle guidance command at the final time. Additionally, if $K = 1$, the above performance index coincides with the energy optimal case.

From Equation (9.2-9) and Equation (9.2-10), the optimal solution of a_{IA} is easily obtained as

$$a_{IA} = -\frac{K(N-1)V_M}{t_{go}}\varepsilon_\gamma. \quad (9.2-13)$$

Combining Equation (9.2-5) with Equation (9.2-13), one can derive the final guidance command for impact angle control as

$$a_M = NV_M\dot{\sigma} - \frac{K(N-1)V_M}{t_{go}}\varepsilon_\gamma. \quad (9.2-14)$$

With $N=3$ and $K=1$, the guidance law becomes the optimal guidance law (OGL) for impact angle control[15]. If the guidance gains satisfy $N \geqslant 3$ and $K=1$, guidance law reduces to interception angle control guidance (IACG) law [16]. If one enforces $N = K+2$ and $K \geqslant 1$, the guidance law is identical to the time-to-go weighted optimal impact angle guidance (TWOIAG) law [17]. Finally, if one selects $N > K+1$ and $K \geqslant 1$, the guidance law turns out to be the time-to-go polynomial guidance (TPG) law[18].

Consequently, the proposed approach yields a generalized optimal impact angle control guidance law that encompasses several previous results. The guidance command can be interpreted as a predictor-corrector guidance law: it first predicts the terminal flight path angle using the optimal PNG as the control input, and then corrects the terminal flight path angle error using the desired error dynamics.

Without loss of generality, the desired impact angle is set as $\gamma_f = 90°$. Fig. 9.2-1 presents the simulation results obtained from guidance law with various guidance gains. From Fig. 9.2-1 (a) and (b), it is evident that the proposed optimal guidance law effectively guides the missile to intercept the target with the desired impact angle. Increasing the guidance gains N and K results in a more curved trajectory and hence requires more control energy. The terminal guidance command converges to zero once the guidance gains satisfy condition $N > 3$, $K > 1$, as confirmed by Fig. 9.2-1 (c). The control effort consumption demonstrated in Fig. 9.21 (d) indicates that guidance law with $N = 3$ and $K = 1$ provides energy optimal interception. These results clearly conform with our analytical analysis.

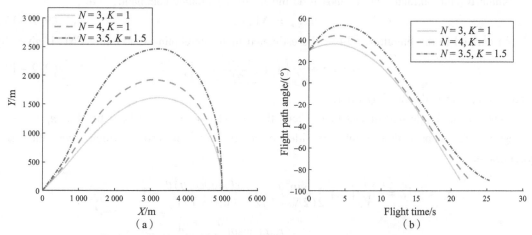

Fig. 9.2-1 Simulation results of impact angle control guidance law

(a) Interception trajectory; (b) Flight path angle

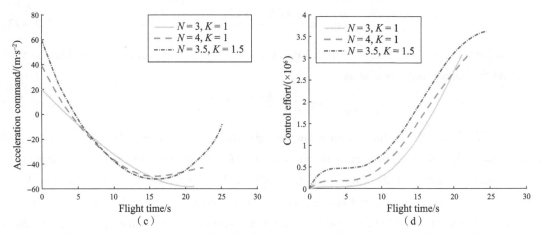

Fig. 9.2-1　Simulation results of impact angle control guidance law（Continued）
(c) Acceleration command; (d) Control effort

9.2.3　Impact Time Control

Constrained final impact time is crucial for increasing the survivability of anti-ship missiles, as it enables multiple missiles to simultaneously intercept a target, effectively penetrating ship-board self-defense systems [14]. A common solution for achieving simultaneous interception is impact time control guidance, where the missile must intercept the target at a specified time. This subsection demonstrates how to use the proposed optimal error dynamics to design an impact time control guidance law. To simplify the analysis, stationary target interception is considered.

Following the systematic prediction-correction approach, the guidance command is formulated as

$$a_M = a_{Nor} + a_{IT}, \quad (9.2-15)$$

where a_{Nor} denotes the nominal command to guarantee zero ZEM for target interception, and a_{IT} stands for the biased correction part to nullify the impact time error.

Similarly, the optimal PNG is chosen as the nominal guidance command, i.e.,

$$a_{Nor} = NV_M \dot{\sigma}. \quad (9.2-16)$$

As derived in the literature [14], the estimated final interception time under PNG is given by

$$t_f = t + \frac{r}{V_M}\left[1 + \frac{\sin^2\theta}{2(2N-1)}\right], \quad (9.2-17)$$

where $\theta = \gamma_M - \sigma$ is the heading error.

Denote t_d as the desired impact time. Then, the impact time error can be defined as $\varepsilon_t = t_d - t_f$. To regulate the impact time error, consider ε_t as the tracking error, which gives the error dynamics as

$$\dot{\varepsilon}_t = -\dot{t}_f = -\frac{\dot{r}}{V_M}\left(1 + \frac{\sin^2\theta}{2(2N-1)}\right) - \frac{r(\sin\theta\cos\theta)\dot{\theta}}{(2N-1)V_M} - 1$$

$$= \cos\theta\left(1 + \frac{\sin^2\theta}{2(2N-1)}\right) - \frac{r\sin\theta\cos\theta\left(\dfrac{a_{M_0} + a_{IT}}{V_M} - \dot{\sigma}\right)}{(2N-1)V_M} - 1 \quad (9.2-18)$$

$$= \cos\theta\left(1 + \frac{\sin^2\theta}{2(2N-1)}\right) + \frac{(N-1)\sin^2\theta\cos\theta}{2N-1} - \frac{r\sin\theta\cos\theta}{(2N-1)V^2}a_{IT} - 1.$$

In practical flight, the lead angle $\theta = \gamma_M - \sigma$ is small, and thus $\sin\theta \approx \theta$, $\cos\theta \approx 1 - \theta^2/2$. Using these approximations and neglecting the higher order term of θ, Equation (9.2-18) can be reduced to

$$\dot{\varepsilon}_t = -\frac{r\sin\theta}{(2N-1)V_M^2}a_{IT}. \tag{9.2-19}$$

In this example, where $u(t) = a_{IT}$, then

$$g(t) = -\frac{r\sin\theta}{(2N-1)V_M^2}. \tag{9.2-20}$$

The optimal error dynamics with respect to the impact time error εt is selected as

$$\dot{\varepsilon}_t + \frac{K}{t_{go}}\varepsilon_t = 0 \tag{9.2-21}$$

Substituting Equation (9.2-21) into Equation (9.2-19) gives the guidance command to nullify the impact time error as

$$a_{IT} = \frac{K(2N-1)V_M^2}{r\sin\theta t_{go}}\varepsilon_t. \tag{9.2-22}$$

Combining Equation (9.2-16) with Equation (9.2-22) leads to the impact time control guidance law as

$$a_M = NV_M\dot{\sigma} + \frac{K(2N-1)V_M^2}{r\sin\theta t_{go}}\varepsilon_t. \tag{9.2-23}$$

Note that with $K=4$ and $N=3$, the guidance command is identical to the impact time guidance law[14]. Also, in the case of $K = N+1$, the obtained impact time guidance law becomes the guidance law. Therefore, the obtained result is a generalized form of previous impact time guidance laws[14].

In addition, it follows from Theorem 9.1 that the corresponding performance index of dynamics is given by

$$J = \frac{1}{2}\int_t^{t_f} \frac{r^2\sin\theta^2}{(2N-1)^2 V_M^4 (t_f - \tau)^{K-1}} u^2(\tau)\,d\tau. \tag{9.2-24}$$

By neglecting the constant terms in the performance index, Equation (9.2-25) further reduces to

$$J = \frac{1}{2}\int_t^{t_f} \frac{r^2\sin\theta^2}{(t_f - \tau)^{K-1}} u^2(\tau)\,d\tau. \tag{9.2-25}$$

The performance index, as specified in Equation (9.2-25), provides valuable insights into the command pattern of the proposed guidance law and elucidates the behavior of guidance laws in the context of impact time control. From Equation (9.2-25), it is apparent that the magnitude of the weighting function decreases as r and θ decrease for a given time-to-go. Therefore, the resultant guidance command of impact time control tends to increase as r and θ decrease. This is a general phenomenon of impact time-control guidance laws. For a stationary target, the parameter θ can be considered as a heading error. Therefore, decreasing r and θ means that the missile approaches a target, converging on a collision course. For the impact time control, the missile must deviate from the intended collision path to adjust the flight time as needed. According to geometric principles, modifying the flight trajectory becomes easier as the missile moves further from the collision course. In other words, once the missile is on a collision trajectory with the target, greater control effort is required to achieve the desired flight time adjustment. This characteristic is inherent to impact time control guidance laws, and the performance index reflects this fact.

Based on the performance index, the command pattern of the proposed guidance law can also be determined. It follows from Equation (9.2-25) that the weighting function $r^2\sin^2\theta/t_{go}^{K-1}$ gradually decreases with the decrease of r and θ. For this reason, it is recommended to choose a relatively large guidance gain K such that t_{go}^{K-1} has a faster-decreasing rate than $r^2\sin4\theta$ to compensate for the decreasing of the weighting function. Notice that the decreasing rates of both relative range and velocity lead angle are governed by the PNG term. Since $r \approx V_M t_{go}$, when the interceptor approaches the target, the decrease rate of r is proportional to t_{go}. In the case of PNG with constant navigation gain N, the closed-form solution of the velocity lead angle is $\theta = Ct_{go}^{K-1}$ with C being a constant determined by the initial condition. Considering the fact that θ is small, the decreasing rate of $r^2\sin^2\theta$ is, therefore, proportional to t_{go}^{2N}. With this in mind, a suitable choice of K that guarantees a zero final guidance command is $K-1 > 2N$.

The performance of generalized optimal impact time guidance law with $t_d = 20$ s is numerically analyzed in this subsection. The simulation results, including interception trajectory, impact time error, acceleration command, and control effort for the proposed guidance law with various guidance gains are presented in Fig. 9.2-2. These results demonstrate that the proposed guidance law effectively guides the missile to intercept the target at the specified impact time. Increasing the value of the biased gain K, the convergence rate of impact time error becomes faster at the sacrifice of more energy consumption, as shown in Fig. 9.2-2 (b) and (d). It can also be observed from Fig. 9.2-2 (c) that when the guidance gain $K > 2N + 1$, the theoretical zero terminal guidance command is achieved. This observation supports the analytical findings.

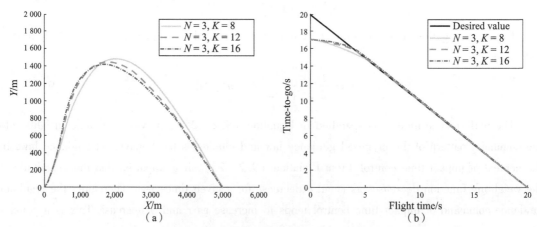

Fig. 9.2-2 Simulation results of impact time control guidance law

(a) Interception trajectory; (b) Time-to-go error

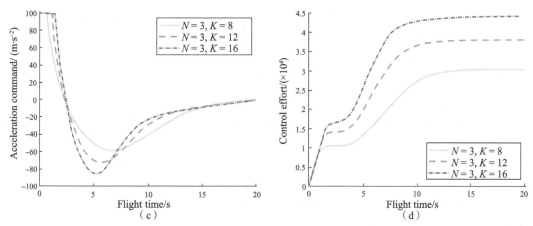

Fig. 9.2-2 Simulation results of impact time control guidance law (Continued)

(c) Acceleration command; (d) Control effort

§ 9.3 Graity-Turn-Assisted Optimal Guidance Law

GRAVITY – TURN – ASSIST

Classical PNG is based on the assumption that the vehicle travels at a constant speed without accounting for gravitational acceleration. In endo-atmospheric scenarios, where aerodynamic forces primarily influence speed, this assumption is generally valid because the speed can be treated as slowly varying or piecewise constant. However, for exo-atmospheric interceptors, such as kinetic kill vehicles, which require substantial impact energy and experience significant accelerations, this assumption is less applicable. In such cases, PNG may fail to guide the missile along the desired straight-line collision course and may not be energy efficient.

To address this issue, the concept of guidance-to-collision (G2C) is introduced for designing guidance laws for accelerating missiles. The fundamental idea behind G2C is to incorporate the effects of axial acceleration into the derivation of the collision triangle. This approach has been implemented through both sliding mode control (SMC)[19] and optimal control techniques[20], enabling the interceptor to follow a straight-line collision course toward the predicted interception point (PIP).

The challenge with the G2C law is that it assumes a gravity-free environment, necessitating an additional compensation term to counteract gravitational effects during implementation. This added term can prevent achieving a zero terminal guidance command, thereby reducing operational margins to address unwanted disturbances. Moreover, direct gravity compensation may demand extra energy, potentially lowering impact energy. Motivated by these issues, this chapter proposes a new gravity-turn-assisted optimal guidance law that leverages gravity rather than compensating for it in accelerating missiles. The approach aims to regulate the ZEM, accounting for both axial acceleration and gravity, to zero within a finite time, guiding the missile along a desired curved trajectory to intercept the target. The proposed algorithm has two key features.

(1) Unlike PNG and G2C, which define interception paths as straight lines, the desired trajectory in the proposed approach is curved, accounting for both axial acceleration and gravity. This curvature complicates the analytical derivation of guidance commands. To overcome this challenge, a new concept known as instantaneous ZEM is introduced. The optimal guidance law is then derived using the optimal error dynamics outlined in Section 9.1.

(2) Detailed analysis shows that both PNG and G2C are specific instances of the proposed guidance law. The advantages of this new approach are evident: it ensures a zero final guidance command and conserves energy by eliminating the need for additional control effort to counteract gravity.

Due to the influence of axial acceleration and gravity, the flight path angle and speed evolve according to

$$\dot{\gamma}_M = \frac{a_x \sin\alpha_M - g\cos\gamma_M}{V_M}, \qquad (9.3-1)$$

$$\dot{V}_M = a_x \cos\alpha_M - g\sin\gamma_M, \qquad (9.3-2)$$

where g stands for the gravitational acceleration. It is assumed to be constant in guidance law design since the duration of the terminal guidance phase is typically very short.

For exo-atmospheric interceptions, the primary control objective is to design a guidance law that eliminates the initial heading error, ensuring that the interceptor's acceleration vector remains aligned with its velocity vector. If gravity is neglected, the missile will follow a straight path to the expected impact point without additional control effort. This feature is particularly valuable for kinematic kill vehicles, as it minimizes the interceptor's angle of attack, thereby enhancing the probability of a successful kill[21]. To realize the concept of a direct hit, the key part is to find the closed-form solution of an ideal collision triangle for the missile that requires no extra control effort, i.e., $\alpha_M = 0$. With the closed-form solution established, a guidance law can be designed to ensure the missile trajectory converges to the optimal collision triangle in finite time, utilizing the optimal error dynamics outlined in Section 9.1. However, because gravity affects the trajectory, the ideal path is no longer a straight line. Consequently, many previous guidance laws were developed under the simplifying assumption of gravity-free conditions. In practical applications, an additional gravity-compensation term $g\cos\gamma_M$ is then leveraged to reject the effect of gravity in implementation. This straightforward approach obviously has two main drawbacks:

(1) Compensating gravity using extra term $g\cos\gamma_M$ cannot guarantee zero terminal guidance command, leading to the sacrifice of operational margins;

(2) The additional term for gravity compensation can lead to increased energy consumption, potentially reducing the effectiveness of a direct hit due to the energy loss.

To address these issues, this chapter proposes a new gravity-turn-assisted optimal guidance law that leverages gravity, rather than compensating for it, to enhance the performance of accelerating missiles.

9.3.1 Zero-Control-Effort Trajectory Considering Gravity

Once the ideal collision triangle is determined, optimal control theory can be applied to design a guidance law that directs the missile to follow this trajectory, achieving the design objectives.

The ideal motion of the interceptor is characterized by the missile's kinematics with zero control input. The instantaneous ZEM is defined as the distance by which the missile would miss the target if the target continues on its current path and the missile maintains a straight-line trajectory based on its current flight path angle, without further corrective maneuvers.

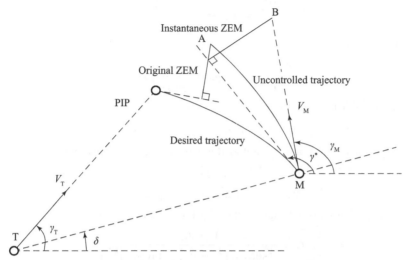

Fig 9.3-1 Definition of instantaneous ZEM

In the derivation of the desired collision triangle, it is natural to enforce the condition of the ideal motion of the interceptor. That is, $\alpha_M = 0$ for the exo-atmospheric case, as our goal is to make the terminal guidance command converge to zero. Under this condition, the missile's kinematics is formulated as

$$\dot{x}_M = V_M \cos\gamma_M, \qquad (9.3-3)$$

$$\dot{y}_M = V_M \sin\gamma_M, \qquad (9.3-4)$$

$$\dot{\gamma}_M = -\frac{g\cos\gamma_M}{V_M}, \qquad (9.3-5)$$

$$\dot{V}_M = a_x - g\sin\gamma_M, \qquad (9.3-6)$$

where (x_M, y_M) represents the inertial position of the interceptor. It follows from Equation (9.3-3) ~ Equation (9.3-6) that these four differential equations are dependent on the flight path angle γ_M and direct integration seems to be intractable. To simplify the problem, the argument is changed from time t to flight path angle γ_M. Then, Equation (9.3-3) ~ Equation (9.3-6) can be reformulated as

$$\frac{dt}{d\gamma_M} = -\frac{V_M}{g\cos\gamma_M}, \qquad (9.3-7)$$

$$\frac{dx_M}{d\gamma_M} = \frac{dx_M}{dt} \cdot \frac{dt}{d\gamma_M} = -\frac{V_M^2}{g}, \qquad (9.3-8)$$

$$\frac{dy_M}{d\gamma_M} = \frac{dy_M}{dt} \cdot \frac{dt}{d\gamma_M} = -\frac{V_M^2}{g}\tan\gamma_M, \qquad (9.3-9)$$

$$\frac{dV_M}{d\gamma_M} = \frac{dV_M}{dt} \cdot \frac{dt}{d\gamma_M} = V_M \tan\gamma_M - \frac{a_x V_M}{g\cos\gamma_M}. \qquad (9.3-10)$$

By changing the argument, it becomes necessary to solve only three independent differential equations to obtain the analytical solution. Let $\kappa = -a_x/g$, Equation (9.3-10) can be rewritten as

$$\frac{dV_M}{d\gamma_M} = V_M \tan\gamma_M + \frac{\kappa V_M}{\cos\gamma_M}. \qquad (9.3-11)$$

Rearranging Equation (9.3-11) as

$$\frac{dV_M}{V_M} = \left(\tan\gamma_M + \frac{\kappa}{\cos\gamma_M}\right) d\gamma_M. \qquad (9.3-12)$$

Integrating both sides of Equation (9.3-12) gives

$$\ln V_M(\gamma_M)\Big|_{\gamma_{M0}}^{\gamma_{Mf}} = -\ln|\cos\gamma_M|\Big|_{\gamma_{M0}}^{\gamma_{Mf}} + \kappa \ln|\sec\gamma_M + \tan\gamma_M|\Big|_{\gamma_{M0}}^{\gamma_{Mf}}, \qquad (9.3-13)$$

where γ_{Mf} and γ_{M0} denote the final and initial flight path angles, respectively.

Solving Equation (9.3-13) for $V_M(\gamma_{Mf})$ yields

$$V_M(\gamma_{Mf}) = C|\sec\gamma_{Mf}||\sec\gamma_{Mf} + \tan\gamma_{Mf}|^\kappa, \qquad (9.3-14)$$

where C is the integration constant, determined by the initial conditions as

$$C = \frac{V_M(\gamma_{M0})}{|\sec\gamma_{M0}||\sec\gamma_{M0} + \tan\gamma_{M0}|^\kappa}. \qquad (9.3-15)$$

Setting γ_{Mf} as γ_M and substituting Equation (9.3-14) into Equation (9.3-7) - Equation (9.3-9) results in

$$\frac{dt}{d\gamma_M} = -\text{sgn}(\cos\gamma_M)\frac{C}{g}\sec\gamma_M|\sec\gamma_M + \tan\gamma_M|^\kappa, \qquad (9.3-16)$$

$$\frac{dx_M}{d\gamma_M} = -\frac{C^2}{g}\sec^2\gamma_M|\sec\gamma_M + \tan\gamma_M|^{2\kappa}, \qquad (9.3-17)$$

$$\frac{dy_M}{d\gamma_M} = -\frac{C^2}{g}\sec^2\gamma_M\tan\gamma_M|\sec\gamma_M + \tan\gamma_M|^{2\kappa}. \qquad (9.3-18)$$

After derivation, the closed-form solution is given by

$$t(\gamma_M) = t(\gamma_{M0}) - \text{sgn}(\cos\gamma_M)\frac{C}{g}[f_t(\gamma_M) - f_t(\gamma_{M0})], \qquad (9.3-19)$$

$$x_M(\gamma_M) = x_M(\gamma_{M0}) - \frac{C^2}{g}[f_x(\gamma_M) - f_x(\gamma_{M0})], \qquad (9.3-20)$$

$$y_M(\gamma_M) = y_M(\gamma_{M0}) - \frac{C^2}{g}[f_y(\gamma_M) - f_y(\gamma_{M0})], \qquad (9.3-21)$$

where

$$f_t(\gamma_M) = \frac{1}{\kappa^2 - 1}(\kappa\sec\gamma_M - \tan\gamma_M)|\sec\gamma_M + \tan\gamma_M|^\kappa, \qquad (9.3-22)$$

$$f_x(\gamma_M) = \frac{1}{4\kappa^2 - 1}(2\kappa\sec\gamma_M - \tan\gamma_M)|\sec\gamma_M + \tan\gamma_M|^{2\kappa}, \qquad (9.3-23)$$

$$f_y(\gamma_M) = \frac{1}{4\kappa^2 - 4}(2\kappa\sec\gamma_M\tan\gamma_M - \tan^2\gamma_M - \sec^2\gamma_M)|\sec\gamma_M + \tan\gamma_M|^{2\kappa}. \qquad (9.3-24)$$

Setting $\gamma_{M0} = \gamma_M$ and $\gamma_M = \gamma_{Mf}$ in Equation (9.3-19) ~ Equation (9.3-21) provides the trajectory of the interceptor with zero control effort as

$$t(\gamma_{Mf}) = t(\gamma_M) - \frac{C}{g}[f_t(\gamma_{Mf}) - f_t(\gamma_M)], \tag{9.3-25}$$

$$x_M(\gamma_{Mf}) = x_M(\gamma_M) - \frac{C^2}{g}[f_x(\gamma_{Mf}) - f_x(\gamma_M)], \tag{9.3-26}$$

$$y_M(\gamma_{Mf}) = y_M(\gamma_M) - \frac{C^2}{g}[f_y(\gamma_{Mf}) - f_y(\gamma_M)]. \tag{9.3-27}$$

Since the target's acceleration is typically not available to the missile in practice, it is assumed that the target uses a gravity compensation scheme while maintaining a constant speed when deriving the collision triangle. With this in mind, the terminal position of the target after t_{go} is given by

$$x_T(t_f) = x_T + V_T\cos\gamma_T t_{go} = x_M + r\cos\sigma + V_T\cos\gamma_T t_{go}, \tag{9.3-28}$$

$$y_T(t_f) = y_T + V_T\sin\gamma_T t_{go} = y_M + r\sin\sigma + V_T\sin\gamma_T t_{go}. \tag{9.3-29}$$

From Equation (9.3-25), the time-to-go t_{go} can be formulated as

$$t_{go} = t(\gamma_{Mf}) - t(\gamma_M) = -\frac{C}{g}[f_t(\gamma_{Mf}) - f_t(\gamma_M)]. \tag{9.3-30}$$

A perfect interception requires

$$x_M(t_f) = x_T(t_f), y_M(t_f) = y_T(t_f). \tag{9.3-31}$$

Equation (9.3-26) ~ Equation (9.3-31) define the ideal instantaneous collision triangle considering gravity, which requires no additional control effort for the missile to intercept the target.

Remark 9.3-1 In the previous derivations, it was assumed that the target utilizes a gravity compensation scheme and maintains a constant flying velocity. For ballistic targets that do not use gravity compensation, incorporating gravitational effects into the target's position prediction can yield a more accurate PIP. This can be easily achieved in a similar way as Equation (9.3-20)- Equation (9.3-21) by setting $\kappa = 0$.

9.3.2 Optimal Guidance Law Design and Analysis

One can directly derive the instantaneous ZEM as

$$z = -\sin(e_\gamma)\left(V_M t_{go} + \frac{1}{2}a_x t_{go}^2\right). \tag{9.3-32}$$

The first-order time derivative of the instantaneous ZEM can be approximated as

$$\dot{z} = -\dot{e}_\gamma\left(V_M t_{go} + \frac{1}{2}a_x t_{go}^2\right) = -a_M\left(t_{go} + \frac{1}{2V_M}a_x t_{go}^2\right). \tag{9.3-33}$$

9.3.2.1 Optimal guidance law design

To ensure that the instantaneous ZEM converges to zero within a finite time, various nonlinear control methods can be applied to the ZEM dynamics as outlined in Equation (9.3-33)

Following the general prediction-correction framework established in Section 9.2, a guidance correction command can be formulated to drive the designed error dynamics towards nullifying the instantaneous ZEM. To achieve this, the optimal error dynamics proposed in Section 9.2 are adopted and are expressed as

$$\dot{z} + \frac{Nz}{t_{go}} = 0, \qquad (9.3-34)$$

where $N > 0$ is a constant guidance gain, which can be tuned to regulate the convergence rate of the instantaneous ZEM. Substituting Equation (9.3 – 33) into Equation (9.3 – 34) gives the guidance command as

$$a_M = \frac{Nz}{\left(1 + \frac{a_x}{2V_M}\right)t_{go}^2}. \qquad (9.3-35)$$

The angle-of-attack command α_M is then obtained as

$$\alpha_M = \arcsin\left(\frac{a_M}{a_x}\right). \qquad (9.3-36)$$

According to Theorem 9.1, guidance command is optimal in the sense of minimizing performance index

$$J = \int_t^{t_f} \frac{\left[1 + \frac{1}{2V_M}a_x(t_f - \tau)\right]^2}{(t_f - \tau)^{N-3}} a_M^2(\tau)\,d\tau. \qquad (9.3-37)$$

For interceptors with fast-moving velocity or short homing phase duration, the performance index can be approximated as

$$J \approx \int_t^{t_f} \frac{1}{(t_f - \tau)^{N-3}} a_M^2(\tau)\,d\tau, \qquad (9.3-38)$$

which indicates that the proposed guidance law with $N = 3$ provides energy-optimal interception, and it is wise to choose $N \geqslant 3$ to guarantee a zero final guidance command. When the interceptor converges to the ideal collision course, the terminal velocity can be obtained as

$$V_{Mf} = V_M + a_x t_{go}. \qquad (9.3-39)$$

Following Equation (9.3 – 39), the average velocity during the homing phase can be readily determined as

$$\bar{V}_M = V_M + \frac{1}{2}a_x t_{go}. \qquad (9.3-40)$$

By using the concept of average velocity, guidance command can also be rewritten as

$$a_M = \frac{\bar{N}z}{t_{go}^2}, \qquad (9.3-41)$$

where the new guidance gain $\bar{N}$ is given by

$$\bar{N} = \frac{NV_M}{\bar{V}_M}. \qquad (9.3-42)$$

It implies that the proposed guidance law is a PNG-type guidance law with a time-varying navigation gain $\bar{N}$. The initial and final value of the guidance gains are given by

$$\bar{N}_0 = \lim_{t \to t_f} N(t) = \frac{N}{1 + \frac{1}{2V_{M0}} a_x t_f},$$
$$\bar{N}_f = \lim_{t \to t_f} \bar{N}(t) = N,$$
(9.3-43)

where V_{M0} denotes the initial vehicle speed.

9.3.2.2 Relationships with PNG

The fundamental concept behind PNG is to generate a lateral acceleration that nullifies the ZEM, thereby ensuring a straight-line interception course. To maintain the collision triangle, it is crucial to equalize the distances traveled by both the interceptor and the target perpendicular to the LOS. In the PNG formulation, assumptions of no axial acceleration and no gravitational effects are made. Under these assumptions, it follows that

$$L_M \sin(\gamma_{PNG}^* - \sigma) - L_T \sin(\gamma_T - \sigma) = 0, \qquad (9.3-44)$$

where γ_{PNG}^* denotes the desired current flight path angle of PNG. L_M and L_T are the vehicles' travelled distances from the current point to PIP.

Solving γ_{PNG}^* using Equation (9.3-45) gives

$$\begin{aligned}\gamma_{PNG}^* &= \pi - \arcsin\left(\frac{L_T \sin(\gamma_T - \sigma)}{L_M}\right) + \sigma \\ &= \pi - \arcsin\left(\frac{V_T \sin(\gamma_T - \sigma)}{V_M}\right) + \sigma.\end{aligned} \qquad (9.3-45)$$

Then, the ZEM dynamics of PNG can be obtained as

$$\begin{aligned} z_{PNG} &= -\sin(e_{\gamma,PNG}) V_M t_{go}, \\ \dot{z}_{PNG} &= -\dot{e}_{\gamma,PNG} V_M t_{go} = -a_M t_{go}, \end{aligned} \qquad (9.3-46)$$

where $e_{\gamma,PNG} = \gamma_M - \gamma_{PNG}^*$ denotes the flight path angle tracking error. Note that the distinction between the PNG ZEM and the proposed ZEM lies in their derivation. For PNG, the desired flight path angle is calculated under the assumption of constant flight velocity in a gravity-free environment. In contrast, the proposed ZEM accounts for gravity and varying missile speeds. It is evident from the dynamics of the PNG ZEM that if gravity and axial acceleration are excluded, the proposed instantaneous ZEM simplifies to the PNG ZEM. The guidance command for PNG is given by

$$a_M = \frac{N z_{PNG}}{t_{go}^2}. \qquad (9.3-47)$$

Since classical PNG leverages the ZEM that is derived based on the gravity-free assumption, it requires extra term $g\cos\gamma_M$ to compensate for the effect of gravity in practical implementation as

$$a_M = \frac{N z_{PNG}}{t_{go}^2} + g\cos\gamma_M, \qquad (9.3-48)$$

which indicates that the terminal guidance command of PNG cannot converge to zero due to the extra term $g\cos\gamma_M$. In comparison, the proposed instantaneous ZEM incorporates gravity, ensuring that the guidance command converges to zero when the interceptor remains on the collision triangle.

9.3.2.3 Relationships with G2C

Like PNG, the objective of G2C is to maintain a straight-line interception course toward the target. This implies that the distances traveled by both the interceptor and the target perpendicular to the LOS are equal at any given moment during the homing phase, expressed as

$$L_M \sin(\gamma_{G2C}^* - \sigma) - L_T \sin(\gamma_T - \sigma) = 0, \quad (9.3-49)$$

where γ_{G2C}^* denotes the desired current flight path angle of G2C. With constant axial acceleration, Equation (9.3-49) can be reformulated as

$$\left(V_M + \frac{1}{2}a_M t_{go}\right)\sin(\gamma_{G2C}^* - \sigma) - V_T \sin(\gamma_T - \sigma) = 0. \quad (9.3-50)$$

Once a straight-line collision course is reached and maintained after the heading error has been nulled, the time-to-go for G2C law can then be computed by the literature [19]

$$r = V_T t_{go} \cos(\gamma_T - \sigma) + V_M t_{go} \cos[\pi - (\gamma_{G2C}^* - \sigma)] + \frac{1}{2}a_M t_{go}^2 \cos[\pi - (\gamma_{G2C}^* - \sigma)]. \quad (9.3-51)$$

Solving Equation (9.3-50) for γ_{G2C}^* gives

$$\gamma_{G2C}^* = \pi - \arcsin\left(\frac{V_T \sin(\gamma_T - \sigma)}{V_M + 0.5 a_M t_{go}}\right) + \sigma. \quad (9.3-52)$$

Then, the G2C ZEM dynamics can be readily obtained as

$$z_{G2C} = -\sin(e_{\gamma, G2C})\left(V_M t_{go} + \frac{1}{2}a_x t_{go}^2\right), \quad (9.3-53)$$

$$\dot{z}_{G2C} = -\dot{e}_{\gamma, G2C}\left(V_M t_{go} + \frac{1}{2}a_x t_{go}^2\right) = -a_M\left(t_{go} + \frac{a_x}{2V_M}t_{go}^2\right), \quad (9.3-54)$$

where $e_{\gamma, PNG} = \gamma_M - \gamma_{G2C}^*$ denotes the flight path angle tracking error. Following the same line shown in the previous subsection, one can easily derive the optimal G2C command given by

$$a_M = \frac{\bar{N} z_{G2C}}{t_{go}^2}. \quad (9.3-55)$$

Similarly, since the ZEM for G2C is derived under the assumption of no gravitational effects, practical implementations of G2C require an additional compensation term as follows

$$a_M = \frac{\bar{N} z_{G2C}}{t_{go}^2} + g\cos\gamma_M. \quad (9.3-56)$$

Since the proposed guidance law utilizes gravity instead of compensating it, one can safely predict that our approach can save energy in the case of instantaneous ZEM ≤ G2C ZEM.

Table 9.3-1 summarizes the relationships between the proposed guidance law and the traditional PNG and G2C methods. Our approach offers a generalized guidance law derived from the collision triangle, incorporating the effects of gravity into the instantaneous ZEM. Notably, both PNG and G2C are special cases of the proposed guidance law. As previously discussed, the key advantage of integrating gravity into the guidance law is that it ensures zero guidance command at impact, thereby providing additional operational margins to address undesired disturbances. Moreover, incorporating gravitational effects directly into the guidance law eliminates the need for additional energy expenditure on gravity compensation.

Table 9.3 – 1 Relationships between the proposed formulation and previous guidance laws

Guidance laws	Guidance commands	Terminal guidance commands	Considerations
PNG	$a_M = \dfrac{\bar{N} z_{PNG}}{t_{go}^2} + g\cos\gamma_M$	Bounded	None
G2C	$a_M = \dfrac{\bar{N} z_{G2C}}{t_{go}^2} + g\cos\gamma_M$	Bounded	Speed variation
Proposed	$a_M = \dfrac{\bar{N} z_{proposed}}{t_{go}^2}$	Zero	Speed variation and gravity

9.3.3 Characteristics Analysis by Simulations

This subsection conducts nonlinear simulations to validate the proposed guidance law in an exo-atmospheric interception scenario. The interceptor is assumed to be equipped with a rocket motor that delivers constant axial acceleration. Additionally, the interceptor's attitude can be adjusted to provide the desired lateral acceleration for changing the heading direction. The initial conditions for a typical exo-atmospheric engagement, referenced from the literatures [14] ~ [16], are summarized in Table 9.3 – 2.

Table 9.3 – 2 Initial conditions for homing engagement

Parameters	Relative range /km	Initial LOS angle /(°)	Missile initial velocity /(m·s^{-1})	Target velocity /(m·s^{-1})	Axial acceleration /g
Values	50	0	1 500	3 000	20

9.3.3.1 Characteristics of the proposed guidance law

The effect of guidance gain α on guidance performance is analyzed first. In the simulations, the initial flight path angle is chosen as $\gamma_M(0) = 150°$ and $\gamma_T(0) = 20°$. The simulation results, including interception trajectory, angle-of-attack command, flight path angle error, control effort, time-to-go estimation and missile velocity, with various guidance gains $N = 2, 3, 4, 5, 6$ are presented in Fig. 9.3 – 2, where the control effort is defined as

$$E = \int_0^{t_f} a_M^2 d\tau. \quad (9.3-57)$$

The results clearly reveal that guidance law with larger guidance gain N provides a faster convergence speed of the flight path angle error. However, a higher guidance gain necessitates a greater angle-of-attack command during the initial flight period, which results in increased control effort. As the flight path angle tracking error approaches zero, the control effort stabilizes, and the angle-of-attack commands remain close to zero across all guidance gain scenarios, indicating that the interceptor remains on the collision trajectory. Since the collision triangle derivation accounts for gravity, the guidance command converges to zero at the time of impact as the guidance gain increases. This characteristic is totally different from previous guidance laws that used an additional term $g\cos\gamma_M$ to compensate for the gravity.

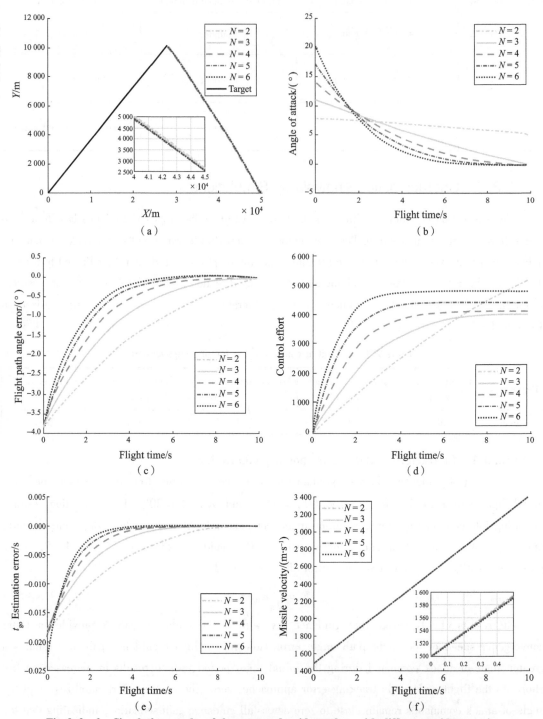

Fig. 9.3-2 Simulation results of the proposed guidance law with different guidance gains
(a) Interception trajectory; (b) Angle-of-attack command; (c) Flight path angle error;
(d) Control effort; (e) Time-to-go estimation; (f) Missile velocity

Fig. 9.3-2 (b) and (d) clearly show that the proposed guidance law with $N \geqslant 3$ ensures zero final guidance command, while $N = 3$ achieves energy-optimal interception. These results support the analytic findings of previous sections.

The results in Fig. 9.3-2 (e) show that our closed-form solution provides an accurate estimation of the time-to-go, with the estimation error converging to zero once the interceptor follows the desired trajectory. The zoomed-in view in Fig. 9.3-2 (f) reveals that, during the initial flight phase, the missile's velocity with a smaller guidance gain N increases slightly faster compared to the velocity with a larger N. This is because a larger guidance gain requires greater control effort, resulting in a larger angle of attack. Since the angle of attack remains very small for most of the flight, the missile's velocity increases almost linearly, as depicted in Fig. 9.3-2 (f).

Now, let us investigate the performance of the proposed guidance law with various interceptors' initial flight path angles. The same scenario is simulated with four different initial flight path angles: $\gamma_M(0) = 90°, 120°, 150°, 180°$, and $\gamma_T(0) = 20°$. The simulation results with guidance gain $N = 9$ are depicted in Fig. 9.3-3, which clearly shows that the proposed guidance law can guide the missile to successfully intercept the target in all tested scenarios.

The interceptor with a larger initial heading error experiences a more curved trajectory to approach the target. For this reason, the duration of acceleration saturation of $\gamma_M(0) = 180°$ is longer than that of other cases. Since the proposed time-to-go is derived from the ideal collision triangle, it tends to underestimate when the missile deviates from the zero-control-effort collision course. Nevertheless, the simulation results show that the estimation accuracy remains acceptable in most cases. As observed in Fig. 9.3-3 (f), the missile's velocity slightly decreases due to the gravitational effect when the guidance command is saturated, meaning that all available control effort is used to nullify the instantaneous ZEM.

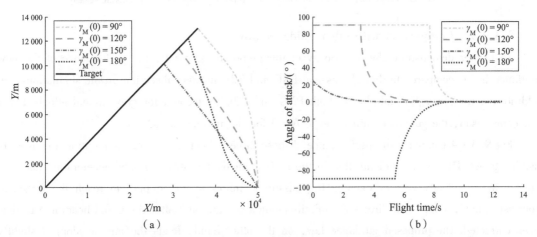

Fig. 9.3-3 Simulation results of the proposed guidance law with different initial flight path angles

(a) Interception trajectory; (b) Angle of attack command

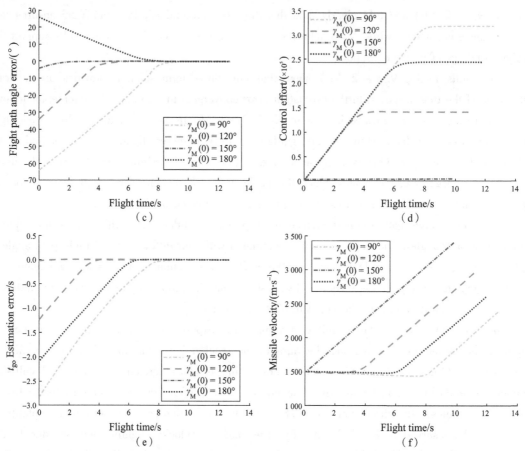

Fig. 9.3-3 Simulation results of the proposed guidance law with different initial flight path angles (Continued)

(c) Flight path angle error; (d) Control effort; (e) Time-to-go estimation; (f) Missile velocity

9.3.3.2 Comparison with other guidance laws

To further demonstrate the superiority of the proposed approach, the performance of the new guidance law is compared to that of classical PNG and G2C in this subsection. In the simulations, an additional term $g\cos\gamma_M$ is included in both PNG and G2C to account for gravitational effects. For a fair comparison, the guidance gain is set to $N = 9$ for all guidance laws.

Fig. 9.3-4 compares the performance of these three guidance laws for intercepting a constantly moving target. The results indicate that in the PNG method, the missile's axial acceleration does not align with its velocity vector, causing the missile to follow a curved path to reach the target. In contrast, the G2C approach directs the missile along a straight path after the initial heading error has been corrected. The proposed guidance law, on the other hand, leads the missile along a slightly curved trajectory by leveraging the gravity-turn concept. Fig. 9.3-4 (b) clearly shows that the angle of attack for all guidance laws remains within bounds over time. The proposed guidance law, in particular, ensures that the guidance command converges to zero at the moment of impact. Consequently, this approach offers greater operational margins compared to the G2C law as the missile nears the target. Since the proposed guidance law does not require an additional term to counteract gravity, as illustrated in Fig. 9.3 – 4 (b), it facilitates energy savings during the terminal guidance phase. The time history of ZEM shown in Fig. 9.3 – 4 (c) demonstrates that all guidance laws effectively nullify their respective ZEM values to ensure target interception. Furthermore, Fig. 9.3 – 4 (d) indicates that the terminal missile velocity is higher under both the proposed guidance law and the G2C law compared to the PNG law, thereby increasing the kill probability.

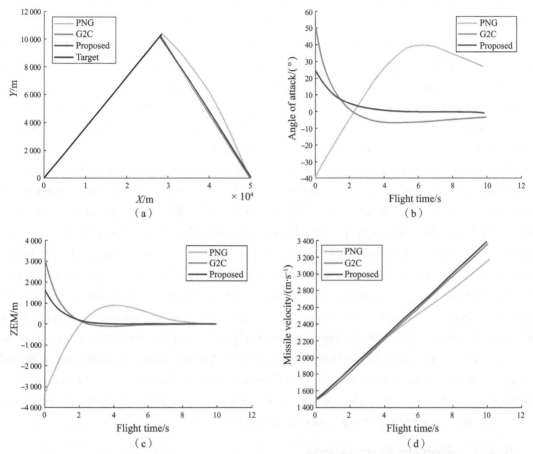

Fig. 9.3 – 4 Comparison results of exo-atmospheric interception with a non-maneuvering target
(a) Interception Trajectory; (b) Angle of attack command; (c) ZEM; (d) Missile velocity

§9.4 3D Optimal Impact Time Guidance for Anti-ship Missiles

THREE DIMENSIONAL OPTIMAL IMPACT TIME GUIDANCE

The primary goal of missile guidance systems is to ensure that a missile intercepts its target with no miss distance. PNG is a well-established and effective guidance algorithm, widely used in missile systems over recent decades[22]. The optimality of PNG was discussed in the literature [23], and its extension to 3D scenarios is covered in the literature [24]. In modern warfare, high-value battleships, such as destroyers and aircraft carriers, are equipped with advanced self-defense systems to counter anti-ship missiles[25]. To overcome these robust defenses, the concept of salvo attack, which involves multiple missiles targeting a battleship simultaneously from different initial positions, has been introduced. A common approach to achieve simultaneous engagement is through impact time control guidance.

Most previous impact time control guidance laws have been designed for two-dimensional (2D) engagement scenarios, thereby neglecting the cross-couplings between horizontal and vertical channels. While 2D guidance laws are often applicable to roll-stabilized interceptors in realistic 3D engagements, developing a dedicated 3D guidance law is advantageous because it can fully leverage the synergistic effects of both horizontal and vertical planes. This approach becomes more beneficial when cross-couplings are significant. Although the Lyapunov-based guidance law[28] and the consensus-based guidance law[31] are formulated using 3D kinematics, they fall short in addressing optimality issues. These laws were derived through nonlinear control methods rather than an optimal control framework and only guaranteed asymptotic convergence.

Motivated by these observations, this chapter introduces a generalized optimal 3D guidance law for anti-ship missiles that meet impact time constraints. A composite guidance command is employed, similar to the methods in the literatures [25], [26], and [27]. This approach integrates an optimal 3D PNG component with a feedback loop to control the impact time error. For the error feedback term, the 2D PNG-based time-to-go estimation method[26] is extended to the 3D homing scenario. To ensure that the predicted time-to-go conforms to its desired trajectory, the optimal error dynamics method from the literature [27] is utilized to design the feedback command. This approach guarantees finite-time convergence of the impact time error and evaluates the optimality of the proposed guidance law. The developed guidance algorithm can be seen as an extension of the literature [27] to a realistic 3D scenario, offering a more thorough analysis of guidance commands and gain selection. Compared to the 2D optimal impact time guidance law[27], the proposed 3D guidance law delivers more stable acceleration commands and reduced energy consumption, particularly in the presence of strong cross-coupling effects.

9.4.1 Problem Formulation

This section outlines the problem formulation for the proposed guidance law, based on the following three key assumptions:

Assumption 1: The target is stationary;

Assumption 2: The missile is assumed to be an ideal point-mass model;

Assumption 3: The missile is flying with constant velocity.

These assumptions are commonly accepted in the design of impact time guidance laws for anti-ship missiles. In Assumption 1, the target's movement is considered negligible compared to the missile's capabilities. Assumption 2 separates the guidance and control processes, where the guidance kinematics operate in an outer loop to generate commands that are then executed by an inner dynamic control loop, or autopilot. Assumption 3 considers the missile's velocity as nearly constant over short intervals, given that it changes slowly relative to the guidance dynamics.

Under these assumptions, the 3D homing engagement geometry is shown in Fig. 9.4 – 1, in which (X_I, Y_I, Z_I) denotes the inertial reference coordinate system and V is the missile velocity. The missile-target relative range is denoted as R. The notations θ_M and ϕ_M stand for two velocity lead angles with respect to the LOS line in the pitch and yaw planes, respectively. Note that both θ_M and ϕ_M can be obtained indirectly from the onboard seeker's gimbal angles[32]. The variables θ_L and ϕ_L represent the LOS angles in the azimuth and elevation directions, respectively. The angle σ is the missile velocity lead angle in the engagement plane, e.g., total velocity lead angle, also known as heading error. The differential equations describing the 3D kinematics can be formulated as[32]

$$\dot{R} = -V\cos\theta_M\cos\phi_M, \quad (9.4-1)$$

$$\dot{\theta}_L = -\frac{V}{R}\sin\theta_M, \quad (9.4-2)$$

$$\dot{\phi}_L = -\frac{V}{R\cos\theta_L}\cos\theta_M\sin\phi_M, \quad (9.4-3)$$

$$\dot{\theta}_M = \frac{a_z}{V} + \frac{V}{R}\cos\theta_M\sin^2\phi_M\tan\theta_L + \frac{V}{R}\sin\theta_M\cos\phi_M, \quad (9.4-4)$$

$$\dot{\phi}_M = \frac{a_y}{V\cos\theta_M} - \frac{V}{R}\sin\theta_M\sin\phi_M\cos\phi_M\tan\theta_L + \frac{V}{R\cos\theta_M}\sin^2\theta_M\sin\phi_M + \frac{V}{R}\cos\theta_M\sin\phi_M, \quad (9.4-5)$$

where a_y and a_z are missile accelerations in the yaw and pitch directions, respectively.

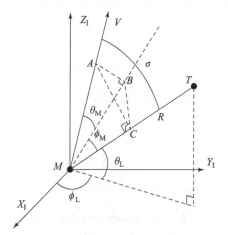

Fig. 9.4 – 1 3D homing engagement geometry

The complementary equation defining the relationship between the heading error and the projected velocity lead angles can be obtained from Fig. 9.4 – 1 as

$$\cos\sigma = \cos\theta_M\cos\phi_M. \quad (9.4-6)$$

The aim of this note is to design a 3D optimal guidance law such that the missile can intercept a stationary target with a specific impact time t_d. Our solution to this problem is given by a composite guidance command consisting of an optimal baseline 3D PNG and an optimal impact time error feedback term.

9.4.2 3D Optimal Impact Time Guidance Law Design

This section details the design of the proposed impact time guidance law. The impact time is first predicted using a 3D PNG approach. Subsequently, an error feedback term is designed based on optimal error dynamics to refine the guidance law.

9.4.2.1 Impact time prediction in 3D engagement

Accurate impact time prediction is crucial for effective impact time control. Therefore, this subsection generalizes the 2D PNG-based time-to-go estimation methods[26],[30] to practical 3D scenarios. In classical PNG, the commanded acceleration of the interceptor is proportional to the rate of change of the LOS angle. For a 3D engagement, PNG is expressed in vector form as[32]

$$\boldsymbol{a}^{PNG} = N\boldsymbol{\Omega}_L \times \boldsymbol{V}, \tag{9.4-7}$$

where $N > 0$ denotes the navigation gain. The notations $\boldsymbol{\Omega}_L$ and $\boldsymbol{V}$ represent the LOS angular rate and missile velocity vectors, respectively. These two vectors define the engagement plane in the 3D interception geometry[33]. Since the relative range is typically fixed during the terminal guidance phase, 3D PNG is commonly implemented in two velocity planes as[32]

$$a_y^{PNG} = -NV\dot{\lambda}_y \sin\theta_M \sin\phi_M + NV\dot{\lambda}_z \cos\theta_M, \tag{9.4-8}$$

$$a_z^{PNG} = -NV\dot{\lambda}_y \cos\phi_M, \tag{9.4-9}$$

where $\dot{\lambda}_y$ and $\dot{\lambda}_z$ are LOS angular velocity vector components in the LOS coordinate, which can be directly measured using onboard seekers, and are determined as

$$\dot{\lambda}_y = \frac{V}{R}\sin\theta_M, \tag{9.4-10}$$

$$\dot{\lambda}_z = -\frac{V}{R}\cos\theta_M \sin\phi_M, \tag{9.4-11}$$

Substituting Equation (9.4-10) and Equation (9.4-11) into Equation (9.4-8) and Equation (9.4-9) and yields

$$a_y^{PNG} = -\frac{NV^2}{R}\sin\phi_M, \tag{9.4-12}$$

$$a_z^{PNG} = -\frac{NV^2}{R}\sin\theta_M \cos\phi_M. \tag{9.4-13}$$

Differentiating Equation (9.4-6) and substituting Equation (9.4-4), Equation (9.4-5), Equation (9.4-12) and Equation (9.4-13) into it results in

$$\begin{aligned}\dot{\sigma} &= \frac{1}{\sin\sigma}(\sin\theta_M \cos\phi_M \dot{\theta}_M + \cos\theta_M \sin\phi_M \dot{\phi}_M) \\ &= \frac{1}{\sin\sigma}\left[-\frac{(N-1)V}{R}\sin^2\theta_M \cos^2\phi_M - \frac{(N-1)V}{R}\sin^2\phi_M\right] \\ &= -\frac{(N-1)V}{R\sin\sigma}(\cos^2\phi_M - \cos^2\theta_M \cos^2\phi_M + \sin^2\phi_M) \\ &= -\frac{(N-1)V}{R}\sin\sigma. \end{aligned} \tag{9.4-14}$$

Dividing Equation (9.4-1) by Equation (9.4-14) yields

$$\frac{dR}{d\sigma} = \frac{R\cot\sigma}{N-1}. \quad (9.4-15)$$

Solving differential Equation (9.4-15) in terms of σ gives

$$R = \frac{R_0}{(\sin\sigma_0)^{1/(N-1)}}(\sin\sigma)^{1/(N-1)}, \quad (9.4-16)$$

where R_0 and σ_0 stand for the initial relative range and velocity lead angle.

Assume that the velocity leading angle satisfies $|\sigma| < \pi/2$, which implies that R strictly decreases from Equation (9.4-1). Define an auxiliary variable $\eta = \sin\sigma$; then, Equation (9.4-1) can be reformulated as

$$\frac{dt}{dR} = -\frac{1}{V\sqrt{1-\eta^2}}. \quad (9.4-17)$$

Integrating the preceding expression using binomial series gives the predicted impact time t_f as

$$t_f = \frac{1}{V}\int_0^{R_0}\frac{1}{\sqrt{1-\eta^2}}dR = \frac{1}{V}\int_0^{R_0}\left(1 + \frac{1}{2}\eta^2 + \frac{3}{8}\eta^4 + \frac{5}{16}\eta^6 + \cdots\right)dR. \quad (9.4-18)$$

Substituting Equation (9.4-16) into Equation (9.4-18) and after integration, the following expression can be written

$$t_f = \frac{R_0}{V}\left[1 + \frac{\eta_0^2}{2(2N-1)} + \frac{3\eta_0^4}{8(4N-3)} + \frac{5\eta_0^6}{16(6N-5)} + \cdots\right], \quad (9.4-19)$$

where η_0 denotes the initial value of η.

Replacing R_0 and η_0 with R and η, respectively, yields the predicted time-to-go under 3D PNG as

$$t_{go} = \frac{R}{V}\left[1 + \frac{\eta^2}{2(2N-1)} + \frac{3\eta^4}{8(4N-3)} + \frac{5\eta^6}{16(6N-5)} + \cdots\right]. \quad (9.4-20)$$

By neglecting the higher-order terms of η^2, the time-to-go estimation can be approximated as follows

$$t_{go} = \frac{R}{V}\left[1 + \frac{\sin^2\sigma}{2(2N-1)}\right]. \quad (9.4-21)$$

Remark 9.4-1 For practical interceptors equipped with roll stabilization, the 3D guidance problem can be effectively addressed through two separate channels. Specifically, 3D homing guidance can be achieved by implementing two distinct 2D PNGs in the pitch and yaw planes for roll-stabilized airframes. The commanded accelerations in these planes are defined as[34]

$$a_y^{PNG} = -\frac{NV^2}{R}\sin\phi_M, \quad (9.4-22)$$

$$a_z^{PNG} = -\frac{NV^2}{R}\sin\theta_M. \quad (9.4-23)$$

It follows from Equation (9.4-12) and Equation (9.4-13) that separate 2D PNG is identical to 3D PNG if the cross-coupling between the pitch and the yaw planes is ignorable.

However, if the relative motions in the two planes cannot be decoupled, performance degradation of separately implementing 2D PNG is inevitable due to the cross-coupling effect.

Remark 9.4-2 In a 2D engagement scenario, such as the pitch plane, with $\sigma = \theta_M$ and $\phi_M = 0$, the predicted time-to-go equation, Equation (9.4-20), simplifies to

$$t_{go} = \frac{R}{V}\left[1 + \frac{\sin^2\theta_M}{2(2N-1)}\right] \quad (9.4-24)$$

which coincides with the results derived in the literature [30] when the velocity leading angle is small, e.g., $\sin\theta_M \approx \theta_M$. Comparing Equation (9.4-20) and Equation (9.4-24), it can be concluded that the proposed time-to-go estimation extends the 2D algorithm to a projected plan containing the LOS vector and missile velocity vector in the 3D scenario.

Remark 9.4-3 In impact time guidance law design, the desired impact time t_d should be set to be achievable, e.g., the problem is well-posed. From a practical standpoint, the desired impact time t_d is required to be larger than the predicted impact time t_f. For this reason, a suitable choice of t_d is

$$t_d > \frac{R_0}{V}\left[1 + \frac{\sin^2\sigma_0}{2(2N-1)}\right] \quad (9.4-25)$$

9.4.2.2 Impact time guidance law design

To achieve precise impact time control, it is crucial to address both target interception and zero impact time error. A composite 3D guidance law is proposed, which combines an optimal baseline 3D PNG with a feedback term designed for optimal impact time error control. Unlike traditional methods that use separate biased terms, our approach employs a single unified feedback command. This command is dynamically allocated to both the vertical and horizontal planes and is defined as follows

$$a_y = a_y^{PNG} + a_b \sin\phi_M = \left(-\frac{NV^2}{R} + a_b\right)\sin\phi_M \quad (9.4-26)$$

$$a_z = a_z^{PNG} + a_b \sin\theta_M \cos\phi_M = \left(-\frac{NV^2}{R} + a_b\right)\sin\theta_M \cos\phi_M \quad (9.4-27)$$

where a_b denotes the error feedback term to be determined.

Define $e_t = t_d - t_{go} - t$ as the impact time error. Substituting Equation (9.4-21) into e_t and taking the time derivative using Equation (9.4-26) and Equation (9.4-27) gives

$$\dot{e}_t = -\dot{t}_{go} - 1$$

$$= -\frac{\dot{R}}{V} - \frac{\dot{R}}{V}\frac{\sin^2\sigma}{2(2N-1)} - \frac{R}{V}\cdot\frac{\sin\sigma\cos\sigma\dot{\sigma}}{2N-1} - 1$$

$$= \cos\sigma\left[1 + \frac{\sin^2\sigma}{2(2N-1)}\right] - \frac{R\cos\sigma}{(2N-1)V^2}\left[-\frac{(N-1)V^2}{R}\sin^2\sigma + a_b\sin^2\sigma\right] - 1.$$

$$(9.4-28)$$

$$= \cos\sigma\left[1 + \frac{\sin^2\sigma}{2(2N-1)}\right] - \frac{R\cos\sigma\sin^2\sigma}{(2N-1)V^2}\left[-\frac{(N-1)V^2}{R} + a_b\right] - 1$$

$$= \cos\sigma\left[1 + \frac{\sin^2\sigma}{2(2N-1)}\right] + \frac{(N-1)\cos\sigma\sin^2\sigma}{2N-1} - \frac{R\cos\sigma\sin^2\sigma}{(2N-1)V^2}a_b - 1$$

Assume that the missile velocity lead angle σ is small. Then, $\sin\sigma \approx \sigma$ and $\cos\sigma \approx 1 - \sigma^2/2$ are obtained. Using these two approximations and neglecting higher-order terms of σ, Equation (9.4 – 28) reduces to

$$\dot{e}_t = -\frac{R\sin^2\sigma}{(2N-1)V^2}a_b. \tag{9.4-29}$$

Consider the optimal error dynamics[21]

$$\dot{e}_t + \frac{K}{t_{go}}e_t = 0, \tag{9.4-30}$$

where $K > 0$ is the guidance gain to be designed.

Combining Equation (9.4 – 29) with Equation (9.4 – 30) gives the guidance command to nullify the impact time error as

$$a_b = \frac{K(2N-1)V^2}{R\sin^2\sigma t_{go}}e_t. \tag{9.4-31}$$

Substituting Equation (9.4 – 31) into Equation (9.4 – 26) and Equation (9.4 – 27) yields the explicit guidance command as

$$a_y = \left[-\frac{NV^2}{R} + \frac{K(2N-1)V^2}{R\sin^2\sigma t_{go}}e_t\right]\sin\phi_M, \tag{9.4-32}$$

$$a_z = \left[-\frac{NV^2}{R} + \frac{K(2N-1)V^2}{R\sin^2\sigma t_{go}}e_t\right]\sin\theta_M\cos\phi_M. \tag{9.4-33}$$

Similar to 3D PNG, the proposed guidance law can be formulated as a vector located in the engagement plane as

$$\boldsymbol{a} = \left[-\frac{NV^2}{R}\sin\sigma + \frac{K(2N-1)V^2}{R\sin\sigma t_{go}}e_t\right]\boldsymbol{e}_a \tag{9.4-34}$$

where $\boldsymbol{e}_a = [0, \sin\phi_M/\sin\sigma, \sin\theta_M\cos\phi_M/\sin\sigma]^T$ denotes the unit vector that specifies the direction of the commanded acceleration in the velocity coordinate.

Remark 9.4 – 4 Although the proposed 3D guidance law is developed with the assumption of stationary targets, it can be readily adapted for non-maneuvering target scenarios by employing the widely used concept of PIP[29].

9.4.3 Analysis of Proposed Guidance Law

This section examines the properties of the proposed 3D optimal impact time guidance law, focusing on several key aspects.

9.4.3.1 Singularity issue

From Equation (9.4 – 31), it can be observed that $\sigma = 0$ is a singular point, which will result in an infinite guidance command. However, it is easy to verify that this singular point is trivial since the velocity lead angle $\sigma \neq 0$ except for the final impact point. To see this, taking the time derivative of σ and substituting Equation (9.4 – 32) and Equation (9.4 – 33) into it yields

$$\dot{\sigma} = -\frac{(N-1)V}{R}\sin\sigma + \frac{K(2N-1)V}{R\sin\sigma t_{go}}e_t. \tag{9.4-35}$$

By choosing the desired impact time t_d that satisfies the condition, it can be readily concluded that $e_{t,0} > 0$, where $e_{t,0}$ denotes the initial impact time error. With this in mind, one can imply that the term $K(2N-1)Ve_t/(R\sin\sigma t_{go})$ initially tries to increase the magnitude of the velocity lead angle to reduce the impact time error. Also, note that the PNG term $-(N-1)V\sin\sigma/R$ is used to regulate the velocity lead angle to zero to guarantee target interception. It is well known that the velocity lead angle under PNG converges to zero only at the time of impact[14]. Therefore, if the guidance gain K satisfies

$$\frac{K(2N-1)V}{R_0|\sin\sigma_0|t_f}e_{t,0} > \frac{(N-1)V}{R_0}|\sin\sigma_0|, \qquad (9.4-36)$$

or the equivalent form

$$K > \frac{(N-1)\sin^2\sigma_0 t_f}{(2N-1)e_{t,0}} \qquad (9.4-37)$$

The error feedback term a_b will play a dominant role initially in the guidance command, thus forcing the magnitude of the velocity lead angle to increase until a certain time instant t_1. When $t \geq t_1$, the PNG term will dominate over a_b, hence regulating the magnitude of the velocity-lead angle to zero at the time of impact. It is important to note that the condition for ensuring the proposed guidance law is nonsingular is easily met in practical scenarios. This ensures that the guidance law remains effective and stable under realistic conditions. An example of the velocity-leading angle response under the proposed guidance law is illustrated in Fig. 9.4-2.

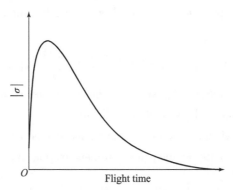

Fig. 9.4-2　An example of the velocity-leading angle response

9.4.3.2　Finite-time convergence of impact time error

Under the optimal error dynamics, it can be readily verified that the closed-form solution for the impact time error is given by

$$e_t = e_{t,0}\left(\frac{t_{go}}{t_f}\right)^K, \qquad (9.4-38)$$

which clearly reveals that the impact time error e_t will converge to zero at the time of impact if $K > 0$, thus satisfying the impact time control requirement. Furthermore, the convergence rate of the impact time error is determined by the guidance gain K: larger K results in faster convergence since $t_{go}/t_f \leq 1$.

9.4.3.3　Optimality of error feedback term

According to Theorem 1 in the literature [27], error dynamics is optimal in terms of performance index

$$J = \frac{1}{2}\int_{t}^{t_f} \frac{R^2\sin^4\sigma}{(2N-1)^2 V^4 (t_f - \tau)^{K-1}} a_b(\tau)\,\mathrm{d}\tau. \quad (9.4-39)$$

Since the constant terms in the performance index do not influence the optimal pattern, the performance index simplifies to

$$J = \frac{1}{2}\int_{t}^{t_f} \frac{R^2\sin^4\sigma}{(t_f - \tau)^{K-1}} a_b(\tau)\,\mathrm{d}\tau. \quad (9.4-40)$$

It follows from Eq. that the weighting function $R^2\sin^4\sigma/(t_f - \tau)^{K-1}$ gradually decreases with the decrease of R and σ. This means that the magnitude of the error feedback term a_b tends to increase when the missile approaches the target. This property is obviously not desirable to guarantee the finite guidance command. For this reason, it is recommended to choose a relatively large guidance gain K such that t_{go}^{K-1} is larger than $R^2\sin^4\sigma$ to compensate for the decreasing of the weighting function. Notice that the decreasing rates of both the relative range and velocity lead angle are governed by the PNG term; a suitable choice of K is $K > N$.

9.4.3.4 Relationship with 2D optimal impact time guidance law

When considering only the 2D homing engagement, such as the pitch plane, the equations simplify to $\sigma = \theta_M$ and $\phi_M = 0$. Then, the proposed 3D impact time guidance law, shown in Equation (9.4-32) and Equation (9.4-33) reduces to

$$a_y = 0, \quad (9.4-41)$$

$$a_z = -\frac{NV^2}{R}\sin\theta_M + \frac{K(2N-1)V^2}{R\sin\theta_M t_{go}} e_t, \quad (9.4-42)$$

which coincides with the generalized 2D optimal impact time guidance law proposed in the literature [27].

It is well known that the 2D guidance law can be directly applied to 3D scenarios for roll-stabilized airframes by ignoring the crosscoupling effect between the horizontal and the vertical channels, e.g., assuming θ_M and ϕ_M are small. Under the condition that the relative motions in the two planes are decoupled, achieving impact time control in a 3D scenario can be accomplished by employing separate 2D guidance laws for each plane. One effective strategy is to use the 2D optimal impact time guidance law[27] in the vertical plane to manage impact time control while applying 2D PNG in the horizontal plane to satisfy the homing constraint. Thus, the individual 2D guidance commands can be expressed as follows

$$a_y^{2D} = -\frac{NV^2}{R}\sin\phi_M, \quad (9.4-43)$$

$$a_z^{2D} = -\frac{NV^2}{R}\sin\theta_M + \frac{K(2N-1)V^2}{R\sin\theta_M t_{go}} e_t. \quad (9.4-44)$$

Comparing Equation (9.4-42) and Equation (9.4-43) with Equation (9.4-43) and Equation (9.4-44), it is evident that the proposed 3D guidance law automatically allocates the error feedback command to both the horizontal and vertical planes. In contrast, the 2D guidance law utilizes only one channel for impact time control. This means that the proposed 3D guidance law fully exploits the synergism between these two channels and thus is beneficial when $\theta_M \neq 0$ and $\phi_M \neq 0$, especially when the effect of cross-coupling is strong. For example, if $\sin\theta_M \cos\phi_M > \sin\phi_M$, the proposed guidance law will mainly use the vertical plane for impact time control. Similarly, if $\sin\phi_M$ is dominant over $\sin\theta_M \cos\phi_M$, the horizontal plane will play an important role in impact time control. This property will be empirically evaluated in simulations. It is worth pointing out that the performance of 3D impact time guidance law is close to its 2D counterpart only when ϕ_M is small. As separately implementing the 2D guidance law ignores the cross-coupling effect, performance degradation is inevitable if this small angle approximation is used. For example, if θ_M approaches to near zero before interception, the pitch guidance command will suffer from a singular issue, as can be observed from the simulation studies.

Notice that the proposed 3D guidance leverages a kind of automatic command allocation; it is therefore helpful in saving energy consumption, compared to the separate 2D guidance law. To see this, define $E = a_y^2 + a_z^2$ as the quadratic energy consumption at each time instant; then, the required energy of the proposed 3D guidance law can be obtained as

$$E_{3D} = \left[-\frac{NV^2}{R} + \frac{K(2N-1)V^2}{R\sin^2\sigma t_{go}}e_t\right]^2 (\sin^2\phi_M + \sin^2\theta_M\cos^2\phi_M)$$

$$= \left[-\frac{NV^2}{R} + \frac{K(2N-1)V^2}{R\sin^2\sigma t_{go}}e_t\right]^2 \sin^2\sigma \quad (9.4-45)$$

$$= \frac{N^2 V^4}{R^2}\sin^2\sigma - \frac{2KN(2N-1)V^4 e_t}{R^2 t_{go}} + \frac{K^2(2N-1)^2 V^4}{R^2 \sin^2\sigma t_{go}^2}e_t^2.$$

The required energy of using the separate 2D guidance law is given by

$$E_{2D} = \left(-\frac{NV^2}{R}\sin\phi_M\right)^2 + \left(-\frac{NV^2}{R}\sin\theta_M + \frac{K(2N-1)V^2}{R\sin\theta_M t_{go}}e_t\right)^2$$

$$= \frac{N^2 V^4}{R^2}(\sin^2\phi_M + \sin^2\theta_M) - \frac{2KN(2N-1)V^4 e_t}{R^2 t_{go}} + \frac{K^2(2N-1)^2 V^4}{R^2 \sin^2\theta_M t_{go}^2}e_t^2.$$

$$(9.4-46)$$

Since $\cos^2\sigma = \cos^2\theta_M \cos^2\phi_M \leq \cos^2\theta_M$, it follows that $\sin^2\sigma \geq \sin^2\theta_M$. Then, it follows from Equation (9.4-47) that

$$E_{2D} \geq \frac{N^2 V^4}{R^2}(\sin^2\phi_M + \sin^2\theta_M\cos^2\phi_M) - \frac{2KN(2N-1)V^4 e_t}{R^2 t_{go}} + \frac{K^2(2N-1)^2 V^4}{R^2 \sin^2\sigma t_{go}^2}e_t^2$$

$$= \frac{N^2 V^4}{R^2}\sin^2\sigma - \frac{2KN(2N-1)V^4 e_t}{R^2 t_{go}} + \frac{K^2(2N-1)^2 V^4}{R^2 \sin^2\sigma t_{go}^2}e_t^2 \quad (9.4-47)$$

$$= E_{3D},$$

where the equality holds if and only if $\phi_M = 0$. This expression clearly indicates that the proposed 3D guidance law is more energy-efficient compared to implementing the 2D guidance law separately.

9.4.4 Numerical Simulations

This section demonstrates the effectiveness of the proposed 3D optimal impact time guidance law using numerical simulations. These simulations involve an anti-ship missile intercepting a stationary target. In the considered scenario, the target is located at (0 m, 0 m, 0 m). The interceptor is initially located at (6 000 m, 6 000 m, 0 m) with initial velocity lead angles $\theta_{M,0} = 10°$ and $\phi_{M,0} = 10°$. The missile flies with constant velocity $V = 250$ m/s. For implementing the proposed guidance law, the navigation gain of the baseline PNG is set as $N = 3$. In practice, the achieved acceleration of the missile is always bounded due to physical limits. For this reason, the magnitudes of both a_y and a_z are constrained by 100 m/s² in simulations.

9.4.4.1 Characteristics of proposed 3D impact time guidance law

This subsection will empirically analyze the properties of the proposed 3D optimal impact time guidance law under various conditions. It is clear that the guidance gain K plays an important role in the proposed guidance law since it governs the convergence rate of the impact time error. Therefore, simulations are first performed with various guidance gains $K = 4, 8, 12$. In these simulations, the desired impact time is set as $t_d = 45$ s, which satisfies the condition. The simulation results, including the interception trajectories, history of the relative range, and acceleration command, are presented in Fig. 9.4–3. From this figure, it is clear that the proposed 3D guidance law successfully drives the missile to intercept the stationary target with the desired impact time. With higher guidance gain K, the response pattern of the relative range becomes more curved and thus is closer to the

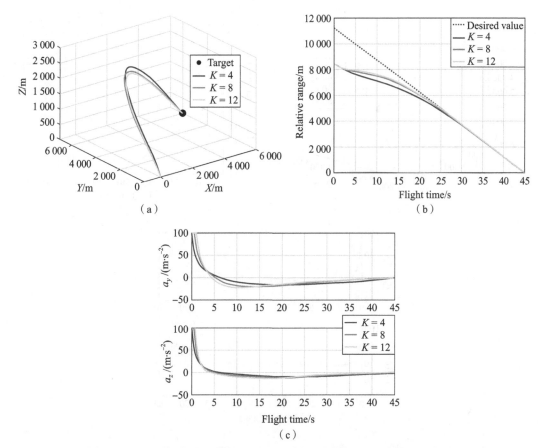

Fig. 9.4–3 Simulation results with respect to different guidance gains K
(a) Interception trajectories; (b) Relative range; (c) Acceleration command

desired pattern $(t_d - t)V$. That is, higher guidance gain K helps to increase the convergence speed of the impact time error. However, Fig. 9.4 - 3 (c) reveals that the proposed guidance law with higher guidance gain K requires more control energy during the initial flight phase. Furthermore, by selecting $K > N$, it is evident from Fig. 9.4 - 3 (d) that the guidance commands in both the vertical and horizontal planes converge to zero at the time of impact. This demonstrates that the proposed guidance law provides sufficient operational margins to handle undesired disturbances as the missile approaches the target.

Now, let us investigate the performance of the proposed 3D impact time guidance law with respect to different desired impact time $t_d = 40$, 60, and 80 s. For implementing the proposed guidance law, the guidance gain of the impact time error feedback term is chosen as $K = 12$. Fig. 9.4 - 4 (a) compares the interceptor trajectory for these three different impact time constraints. This figure clearly shows that the interceptor takes a longer flight path with a larger desired impact time. The history of the relative range with the three cases of desired impact time t_d is presented in Fig. 9.4 - 4 (b), which reveals that the proposed guidance law satisfies the impact time constraint precisely. The impact time error of the proposed guidance law turns out to be less than

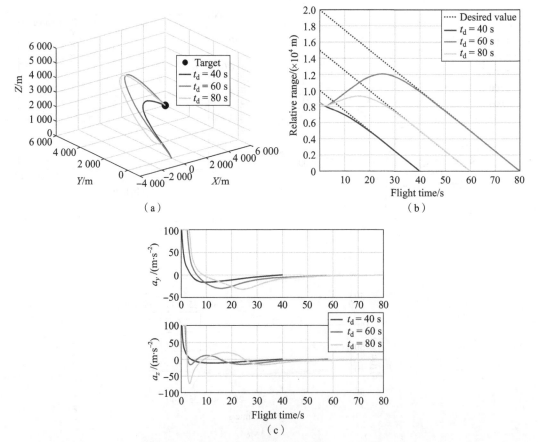

Fig. 9.4 - 4 Simulation results with respect to different desired impact time t_d

(a) Interception trajectories; (b) Relative range; (c) Acceleration command

0.01 s with different impact time constraints in our simulations. From this figure, it is obvious that the longer convergence phase is required to regulate the impact time error with a larger desired impact time t_d under the same initial conditions. The missile acceleration command produced by the proposed guidance law with different t_d is provided in Fig. 9.4 – 4 (c). Clearly, more energy consumption is required during the initial phase for a larger t_d. For this reason, the duration of the initial acceleration saturation of $t_d = 80$ s is longer than that of $t_d = 40$ s and $t_d = 60$ s.

9.4.4.2 Comparison with 2D impact time guidance law

To further show the advantages of the proposed 3D impact time guidance law, comparisons with generalized 2D optimal impact time guidance law[21] are conducted in this subsection. For the purpose of comparison, three different initial conditions are considered as ① $\theta_{M,0} = 40°$, $\phi_{M,0} = 0°$; ② $\theta_{M,0} = 40°$, $\phi_{M,0} = 40°$; and ③ $\theta_{M,0} = 40°$, $\phi_{M,0} = 80°$. It is clear that Case 1 represents a 2D homing scenario within the vertical plane, Case 2 involves a moderate level of coupling between the two planes, and Case 3 illustrates a scenario with strong cross-coupling effects. To ensure a fair comparison, the guidance gain for the impact time error feedback term is set to $K = 12$ for both guidance laws.

The simulation results, including the interception trajectories, relative range, and acceleration command, are shown in Fig. 9.4 – 5 with the desired impact time $t_d = 45$ s. From the first row of Fig. 9.4 – 5, it is evident that both guidance laws produce identical results for case 1, as anticipated. This is because, as previously discussed, the proposed 3D impact time guidance law simplifies to its 2D counterpart when only the vertical plane is considered. For case 2, since $\sin\theta_M\cos\phi_M < \sin\phi_M$, the proposed 3D guidance law will mainly use the horizontal plane to regulate the impact time error, as shown in the second row of Fig. 9.4 – 5. Since the same guidance gain K is applied to both guidance laws, the dynamics of the impact time error are identical under each law. As a result, the convergence patterns of the relative range under both guidance laws exhibit similar characteristics, as shown in Fig. 9.4 – 5 (e). Fig. 9.4 – 5 (f) reveals that the proposed 3D guidance law utilizes both the vertical and horizontal channels for impact time control, automatically distributing the error feedback command across these channels. In contrast, the 2D impact time guidance law relies solely on the vertical plane for impact time control. In the strong cross-coupling scenario depicted in Case 3, both guidance laws successfully direct the missile to intercept the target while adhering to the desired impact time constraint, as demonstrated in the third row of Fig. 9.4 – 5. However, the 2D impact time guidance law shows oscillating patterns when the interceptor is close to the target, which is not desirable for onboard control systems. The control effort $\int_{t_0}^{t_f} [a_y^2(t) + a_z^2(t)] dt$ obtained from both the 2D and 3D impact time guidance laws for these three different cases are summarized in Table 9.4 – 1. This table clearly shows that, except for case 1, the proposed 3D impact guidance law effectively reduces energy consumption. This confirms the theoretical findings presented in the previous section.

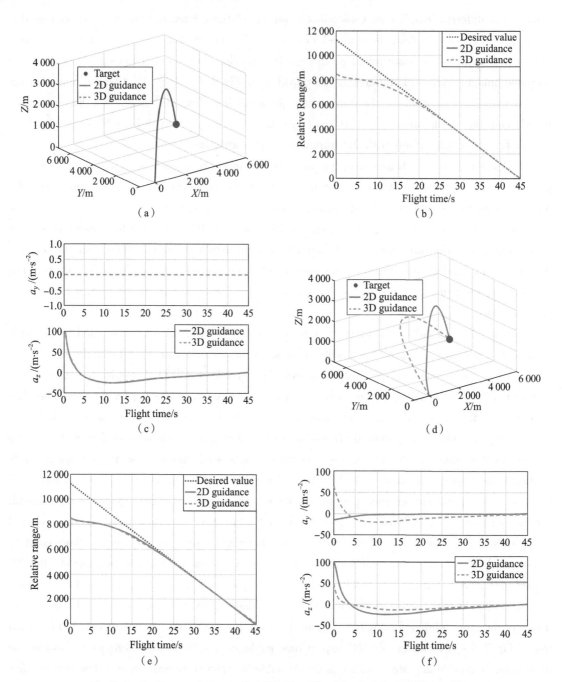

Fig. 9.4−5 Comparison results with 2D optimal impact time guidance law

(a) Interception trajectories for case ①; (b) Relative range for case ①; (c) Acceleration command for case ①;
(d) Interception trajectories for case ②; (e) Relative range for case ②;
(f) Acceleration command for case ②

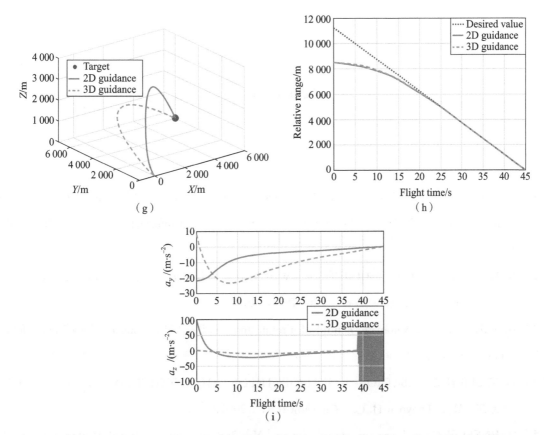

Fig. 9.4−5 Comparison results with 2D optimal impact time guidance law (Continued)

(g) Interception trajectories for case ③; (h) Relative range for case ③;

(i) Acceleration command for case ③

Table 9.4−1 Control effort comparisons

Guidance	Case ①	Case ②	Case ③
2D	18 733	18 951	62 803
3D	18 733	13 080	9 926

References

[1] BAI W, XUE W, HUANG Y, et al. On extended state based Kalman filter design for a class of nonlinear time-varying uncertain systems [J]. Science China Information Sciences, 2018, 61(4).

[2] BALAKRISHNAN S N, TSOURDOS A, WHITE B A. Advances in missile guidance, control, and estimation [M]. New York: CRC Press, 2012.

[3] BLAKELOCK J H. Automatic control of aircraft and missiles [M]. 2nd ed. New York: John Wiley & Sons, 1991.

[4] ГОЛУБЕВ И С, СВЕТЛОВ В Г. ПРОЕКТИРОВАНИЕ ЗЕНИТНЫХ УПРАВЛЯЕМЫХ РАКЕТ [M]. Голубев И. С., Светлов В. Г., 2001.

[5] GARNELL P. Guided weapon control system [M]. 2nd ed. London: Royal Military College of Science, 1980.

[6] LIN C F. Modern navigation, guidance, and control processing, volume 2 [M]. Englewood Cliffs, NJ: Prentice Hall, 1991.

[7] MUSOFF H, ZARCHAN P. Fundamentals of kalman filtering: a practical approach [M]. Virginia: American Institute of Aeronautics and Astronautics, 2009.

[8] SHNEYDOR N A. Missiles guidance and pursuit: kinematics, dynamics and control [M]. New York: Elsevier, 1998.

[9] SIOURIS G M. Missile guidance and control systems [M]. Springer Science & Business New York: Media, 2004.

[10] TEWARI A. Advanced control of aircraft, spacecraft and rockets [M]. New York: John Wiley & Sons, 2011.

[11] YANUSHEVSKY R. Modern missile guidance [M]. New York: CRC Press, 2007.

[12] ZARCHAN P. Tactical and strategic missile guidance, 6th edition [M]. Virginia: American Institute of Aeronautics and Astronautics, Inc, 2012.

[13] LIN Defu, WANG Hui, WANG Jiang, et al. Design and Guidance Law Analysis of Tactical Missile Autopilot [M]. Beijing: Beijing Institute of Technology Press, 2012.

[14] JEON I S, LEE J I, TAHK M J. Impact-time-control guidance law for anti-ship missiles [J]. IEEE Trans Control Systems Technology, 2006, 14 (2): 260-266.

[15] RYOO C K, CHO H, TAHK M J. Optimal guidance laws with terminal impact angle constraint [J]. Journal of Guidance Control and Dynamics, 2005, 28 (4): 724-732.

[16] LEE C H, KIM T H, TAHK M J. Interception angle control guidance using proportional navigation with error feedback [J]. Journal of Guidance Control and Dynamics, 2013, 36 (5): 1556-1561.

[17] OHLMEYER E J, PHILLIPS C A. Generalized vector explicit guidance [J]. Journal of Guidance Control and Dynamics, 2006, 29 (2): 261-268.

[18] LEE C H, KIM T H, TAHK M J, et al. Polynomial guidance laws considering terminal impact angle and acceleration constraints [J]. IEEE Transactions on Aerospace and Electronic Systems, 2013, 49 (1): 74-92.

[19] DWIVEDI P, BHALE P, BHATTACHARYYA A, et al. Generalized estimation and predictive guidance for evasive targets [J]. IEEE Transactions on Aerospace and Electronic Systems, 2016, 52 (5): 2111-2122.

[20] HE S, LEE C H. Gravity-turn-assisted optimal guidance law [J]. Journal of Guidance Control and Dynamics, 2018, 41 (1): 171-183.

[21] SHIMA T, GOLAN OM. Exo-atmospheric guidance of an accelerating interceptor missile [J]. Journal of the Franklin Institue, 2012, 349 (2): 622-637.

[22] ZARCHAN P. Tactical and strategic missile guidance [M]. Reston, VA: AIAA, 2012: 163-184.

[23] JEON I S, LEE J I. Optimality of Proportional Navigation Based on Nonlinear Formulation [J]. IEEE Transactions on Aerospace and Electronic Systems, 2010, 46 (4): 2051-2055. DOI: 10.1109/TAES.2010.5595614.

[24] CHO N, KIM Y. Optimality of augmented ideal proportional navigation for maneuvering target interception [J]. IEEE Transactions on Aerospace and Electronic Systems, 2016, 52 (2): 948-954. DOI: 10.1109/TAES.2015.140432.

[25] JEON I S, LEE J I, TAHK M J. Impact-time-control guidance law for anti-ship missiles [J]. IEEE Transactions on Control Systems Technology, 2006, 14 (2): 260-266. DOI:

10.1109/TCST.2005.863655.

[26] JEON I S, LEE J I, TAHK M J. Impact-Time-Control Guidance with Generalized Proportional Navigation Based on Nonlinear Formulation [J]. Journal of Guidance, Control, and Dynamics, 2016, 39 (8): 1887-1892. DOI: 10.2514/1.G001681.

[27] HE S, LEE C H. Optimality of Error Dynamics in Missile Guidance Problems [J]. Journal of Guidance, Control, and Dynamics, 2018, 41 (7): 1620-1629. DOI: 10.2514/1.G003343.

[28] KIM M, JUNG B, HAN B, et al. Lyapunov-based impact time control guidance laws against stationary targets [J]. IEEE Transactions on Aerospace and Electronic Systems, 2015, 51 (2): 1111-1122. DOI: 10.1109/TAES.2014.130717.

[29] CHO D, KIM H J, TAHK M J. Nonsingular sliding mode guidance for impact time control [J]. Journal of Guidance, Control, and Dynamics, 2016, 39 (1): 61-68. DOI: 10.2514/1.G001167.

[30] JEON I S, LEE J I, TAHK M J. Homing guidance law for cooperative attack of multiple missiles [J]. Journal of Guidance, Control, and Dynamics, 2010, 33 (1): 275-280. DOI: 10.2514/1.40136.

[31] HE S, WANG W, LIN D, et al. Consensus-based two-stage salvo attack guidance [J]. IEEE Transactions on Aerospace and Electronic Systems, 2018, 53 (3): 1555-1566. DOI: 10.1109/TAES.2017.2773272.

[32] SONG S H, HA I J. A Lyapunov-like approach to performance analysis of 3-dimensional pure PNG laws [J]. IEEE Transactions on Aerospace and Electronic Systems, 1994, 30 (1): 238-248. DOI: 10.1109/7.250424.

[33] SHIN H S, TSOURDOS A, LI K B. A new three-dimensional sliding mode guidance law variation with finite time convergence [J]. IEEE Transactions on Aerospace and Electronic Systems, 2017, 53 (5): 2221-2232. DOI: 10.1109/TAES.2017.2689938.

[34] ZHOU D, SUN S, TEO K L. Guidance laws with finite time convergence [J]. Journal of Guidance, Control, and Dynamics, 2009, 32 (6): 1838-1846. DOI: 10.2514/1.42976.

新能源汽车
电机及电机控制系统
原理与检修

理论+实训一体工单

主编 赵振宁 赵宇

北京理工大学出版社
BEIJING INSTITUTE OF TECHNOLOGY PRESS